METHODS IN MOLECULAR BIOLOGY

Series Editor
John M. Walker
School of Life and Medical Sciences
University of Hertfordshire
Hatfield, Hertfordshire, UK

For further volumes:
http://www.springer.com/series/7651

For over 35 years, biological scientists have come to rely on the research protocols and methodologies in the critically acclaimed *Methods in Molecular Biology* series. The series was the first to introduce the step-by-step protocols approach that has become the standard in all biomedical protocol publishing. Each protocol is provided in readily-reproducible step-by-step fashion, opening with an introductory overview, a list of the materials and reagents needed to complete the experiment, and followed by a detailed procedure that is supported with a helpful notes section offering tips and tricks of the trade as well as troubleshooting advice. These hallmark features were introduced by series editor Dr. John Walker and constitute the key ingredient in each and every volume of the *Methods in Molecular Biology* series. Tested and trusted, comprehensive and reliable, all protocols from the series are indexed in PubMed.

Schistosoma mansoni

Methods and Protocols

Edited by

David J. Timson

School of Pharmacy and Biomolecular Sciences, University of Brighton, Brighton, UK

Humana Press

Editor
David J. Timson
School of Pharmacy and Biomolecular
Sciences
University of Brighton
Brighton, UK

ISSN 1064-3745 ISSN 1940-6029 (electronic)
Methods in Molecular Biology
ISBN 978-1-0716-0634-6 ISBN 978-1-0716-0635-3 (eBook)
https://doi.org/10.1007/978-1-0716-0635-3

This Humana imprint is published by the registered company Springer Science+Business Media, LLC, part of Springer Nature.
The registered company address is: 1 New York Plaza, New York, NY 10004, U.S.A.

Preface

Schistosomiasis, infection with blood flukes from the genus *Schistosoma*, is the second most prevalent parasitic disease of humans. The vast majority of infected people live in poorer countries, and schistosomiasis remains a neglected tropical disease [1]. There is no effective vaccine against this disease, but there is a very effective drug—praziquantel. This drug is cheap and well tolerated by most patients. However, its molecular mechanism of action remains uncertain [2, 3].

Resistance to antimicrobials is well documented and considered to be a global threat to public health. Many antiparasitic drugs are becoming less effective as resistance is selected for. Unusually, there are no definitive reports of resistance to praziquantel. However, resistance has been generated in the laboratory [4]. It is widely accepted that praziquantel resistance is a question of "when, not if"; sooner or later it will appear. There are few other safe and effective drugs to treat *Schistosoma* infections. Thus, the emergence of sustained, widespread praziquantel resistance would be a disaster affecting some of the world's poorest people.

The key to new treatments is to understand the biochemistry of the causative agent—a "know thine enemy" strategy. Sadly, our understanding of invertebrate biochemistry lags behind that of higher mammals and even commercially important microbes such as yeasts. Thus, tools are needed to interrogate the cell biology, gene expression, signaling pathways, and enzymes of helminth pathogens such as *Schistosoma mansoni*. This basic science is the essential foundation on which novel therapies will be built. We need to understand the differences, both subtle and substantial, in the biochemistry of host and parasite.

This volume brings together some protocols for studying *S. mansoni*. It begins with two short reviews on current drugs and drug discovery. There follows a series of methods to interrogate the cell biology of the worm, which are likely to be useful to any laboratory working with this parasite. We then move to some protocols to look at proteome-level changes—essential approaches to understand global changes in response to new drugs. Then we have two bioinformatics chapters which demonstrate how to interpret the increasing amounts of sequence data available. Next, there is a series of chapters describing how to prepare and work with specific proteins from *S. mansoni*. Finally, two chapters describe what, for many, are the end goals of research on this organism: one describing a drug screening method and the other how to test potential vaccine targets.

I thank my fellow authors for their contributions and hope this volume will be useful in the global efforts to understand *S. mansoni* better.

Brighton, UK ***David J. Timson***

References

1. Utzinger J, Becker SL, Knopp S, Blum J, Neumayr AL, Keiser J, Hatz CF (2012) Neglected tropical diseases: diagnosis, clinical management, treatment and control. Swiss Med Wkly 142:w13727. https://doi.org/10.4414/smw.2012.13727
2. Thomas CM, Timson DJ (2018) The mechanism of action of praziquantel: six hypotheses. Curr Top Med Chem 18 (18):1575–1584. https://doi.org/10.2174/1568026618666181029143214

3. Thomas CM, Timson DJ (2020) The mechanism of action of praziquantel: can new drugs exploit similar mechanisms? Curr Med Chem 27 (5):676–696. https://doi.org/10.2174/0929867325666180926145537

4. Mwangi IN, Sanchez MC, Mkoji GM, Agola LE, Runo SM, Cupit PM, Cunningham C (2014) Praziquantel sensitivity of Kenyan *Schistosoma mansoni* isolates and the generation of a laboratory strain with reduced susceptibility to the drug. Int J Parasitol Drugs Drug Resist 4 (3):296–300. https://doi.org/10.1016/j.ijpddr.2014.09.006

Contents

Contributors

ALBERTO BRESCIANI • *Translational Biology Department, IRBM S.p.A., Rome, Italy*
LUIS G. BRIEBA • *Laboratorio Nacional de Genómica para la Biodiversidad, Centro de Investigación y de Estudios Avanzados del IPN, Guanajuato, Mexico; Langebio-Cinvestav Sede Irapuato, Guanajuato, Mexico*
EDUARDO CASTRO-TORRES • *Laboratorio Nacional de Genómica para la Biodiversidad, Centro de Investigación y de Estudios Avanzados del IPN, Guanajuato, Mexico*
CRISTIAN CHAPARRO • *IHPE, Univ. Montpellier, CNRS, Ifremer, Univ. Perpignan Via Domitia, Perpignan, France*
JAMES J. COLLINS • *Department of Pharmacology, University of Texas Southwestern Medical Center, Dallas, TX, USA*
CELINE COSSEAU • *IHPE, Univ. Montpellier, CNRS, Ifremer, Univ. Perpignan Via Domitia, Perpignan, France*
HUMBERTO D'MUNIZ PEREIRA • *Physics Institute of São Carlos, University of São Paulo, São Paulo, SP, Brazil*
RONALDO DE CARVALHO AUGUSTO • *IHPE, Univ. Montpellier, CNRS, Ifremer, Univ. Perpignan Via Domitia, Perpignan, France; LBMC, Laboratoire de Biologie et Modé lisation de la Cellule Univ Lyon, ENS de Lyon, Université Claude Bernard Lyon 1, CNRS, UMR 5239, INSERM, U1210, Lyon, France*
RICARDO DEMARCO • *Departamento de Física e Ciência Interdisciplinar, Instituto de Física de São Carlos, Universidade de São Paulo, São Carlos, SP, Brazil*
JÉSSIKA DE OLIVEIRA VIANA • *Federal University of Paraíba, Health Center, João Pessoa, PB, Brazil*
CORINA DIAZ-QUEZADA • *Laboratorio Nacional de Genómica para la Biodiversidad, Centro de Investigación y de Estudios Avanzados del IPN, Guanajuato, Mexico*
JAN DVOŘÁK • *Institute of Organic Chemistry and Biochemistry, Czech Academy of Sciences, Prague, Czech Republic; Department of Zoology and Fisheries, Faculty of Agrobiology, Food and Natural Resources, Czech University of Life Sciences in Prague, Prague, Czech Republic; Institute of Specific Prophylaxis and Tropical Medicine, Medical University of Vienna, Vienna, Austria*
RASHIKA EL RIDI • *Zoology Department, Faculty of Science, Cairo University, Giza, Egypt*
LUIS G. V. FERNANDES • *Laboratório de Desenvolvimento de Vacinas, Instituto Butantan, São Paulo, SP, Brazil*
RAFAELA S. FERNANDES • *Laboratório de Desenvolvimento de Vacinas, Instituto Butantan, São Paulo, SP, Brazil; Institute of Physics of São Carlos, University of São Paulo, São Carlos, SP, Brazil*
MARINA GABRIEL FONTES • *Instituto de Física de São Carlos, Universidade de São Paulo, São Carlos, SP, Brazil*
ROBERTO GIMMELLI • *National Research Council, Institute of Biochemistry and Cell Biology, Rome, Italy*
MAKAN GOLIZEH • *National Reference Centre for Parasitology, Montreal, QC, Canada; Infectious Diseases and Immunity in Global Health Program, Research Institute of the McGill University Health Centre, Montreal, QC, Canada*

CHRISTOPH GRUNAU • *IHPE, Univ. Montpellier, CNRS, Ifremer, Univ. Perpignan Via Domitia, Perpignan, France*
ALESSANDRA GUIDI • *National Research Council, Institute of Biochemistry and Cell Biology, Rome, Italy*
MARTIN HORN • *Institute of Organic Chemistry and Biochemistry, Czech Academy of Sciences, Prague, Czech Republic*
ADÉLA JÍLKOVÁ • *Institute of Organic Chemistry and Biochemistry, Czech Academy of Sciences, Prague, Czech Republic*
PEDRO JIMENEZ-SANDOVAL • *Laboratorio Nacional de Genómica para la Biodiversidad, Centro de Investigación y de Estudios Avanzados del IPN, Guanajuato, Mexico*
EUNJEE KIM • *Department of Internal Medicine, Texas Tech University Health Sciences Center, Lubbock, TX, USA*
LUCIANA C. C. LEITE • *Laboratório de Desenvolvimento de Vacinas, Instituto Butantan, São Paulo, SP, Brazil*
ADRIAN LEONTOVYČ • *Institute of Organic Chemistry and Biochemistry, Czech Academy of Sciences, Prague, Czech Republic; First Faculty of Medicine, Charles University, Prague, Czech Republic*
ALEX LOUKAS • *Centro Nacional de Microbiologia, Instituto de Salud Carlos III, Madrid, Spain*
TIMOTHY LYNAGH • *Sars International Centre for Marine Molecular Biology, University of Bergen, Bergen, Norway*
LUCAS F. MACIEL • *Instituto Butantan, São Paulo, SP, Brazil; Instituto de Química, Universidade de São Paulo, São Paulo, SP, Brazil*
MICHAEL MAREŠ • *Institute of Organic Chemistry and Biochemistry, Czech Academy of Sciences, Prague, Czech Republic*
ADEBAYO MOLEHIN • *Center for Tropical Medicine and Infectious Diseases, Texas Tech University Health Sciences Center, Lubbock, TX, USA*
ANA L. T. O. NASCIMENTO • *Laboratório de Desenvolvimento de Vacinas, Instituto Butantan, São Paulo, SP, Brazil*
MOMAR NDAO • *National Reference Centre for Parasitology, Montreal, QC, Canada; Infectious Diseases and Immunity in Global Health Program, Research Institute of the McGill University Health Centre, Montreal, QC, Canada; Department of Experimental Medicine, Faculty of Medicine, McGill University, Montreal, QC, Canada*
PARTH PATEL • *Department of Internal Medicine, Texas Tech University Health Sciences Center, Lubbock, TX, USA*
MARK S. PEARSON • *Centro Nacional de Microbiologia, Instituto de Salud Carlos III, Madrid, Spain*
ADAM PEÑA • *Department of Internal Medicine, Texas Tech University Health Sciences Center, Lubbock, TX, USA*
DILHAN PERERA • *National Reference Centre for Parasitology, Montreal, QC, Canada; Infectious Diseases and Immunity in Global Health Program, Research Institute of the McGill University Health Centre, Montreal, QC, Canada; Department of Experimental Medicine, Faculty of Medicine, McGill University, Montreal, QC, Canada*
GISELE S. PHILIPPSEN • *Universidade Federal do Paraná, Jandaia do Sul, Paraná, Brazil*
MARION A. L. PICARD • *Centre National de la Recherche Scientifique (CNRS), Laboratory MIVEGEC (CNRS IRD Uni Montpellier), Montpellier, France*
DAVID ROQUIS • *Amélioration des Grandes Cultures et Ressources Génétiques, Agroscope, Changins, Switzerland*

GIOVINA RUBERTI • *National Research Council, Institute of Biochemistry and Cell Biology, Rome, Italy*
JÉSSICA F. SCORTECCI • *Physics Institute of São Carlos, University of São Paulo, São Paulo, SP, Brazil; Department of Biochemistry and Molecular Biology, University of British Columbia, Vancouver, BC, Canada*
LUCIANA SCOTTI • *Federal University of Paraíba, Health Center, João Pessoa, PB, Brazil*
MARCUS T. SCOTTI • *Federal University of Paraíba, Health Center, João Pessoa, PB, Brazil*
VITOR HUGO B. SERRÃO • *Physics Institute of São Carlos, University of São Paulo, São Paulo, SP, Brazil; Laboratory Medicine and Pathobiology, University of Toronto, Toronto, ON, Canada*
AFZAL A. SIDDIQUI • *Center for Tropical Medicine and Infectious Diseases, Texas Tech University Health Sciences Center, Lubbock, TX, USA*
JAVIER SOTILLO • *Centro Nacional de Microbiologia, Instituto de Salud Carlos III, Madrid, Spain*
HATEM TALLIMA • *Zoology Department, Faculty of Science, Cairo University, Giza, Egypt; Department of Chemistry, School of Science and Engineering, American University in Cairo, New Cairo, Egypt*
DAVID J. TIMSON • *School of Pharmacy and Biomolecular Sciences, University of Brighton, Brighton, UK*
LENKA ULRYCHOVÁ • *Institute of Organic Chemistry and Biochemistry, Czech Academy of Sciences, Prague, Czech Republic; Department of Parasitology, Faculty of Science, Charles University, Prague, Czech Republic*
SERGIO VERJOVSKI-ALMEIDA • *Instituto Butantan, São Paulo, SP, Brazil; Instituto de Química, Universidade de São Paulo, São Paulo, SP, Brazil*
GEORGE WENDT • *Department of Pharmacology, University of Texas Southwestern Medical Center, Dallas, TX, USA*
ANA ELIZA ZERAIK • *Instituto de Física de São Carlos, Universidade de São Paulo, São Carlos, SP, Brazil*
WEIDONG ZHANG • *Center for Tropical Medicine and Infectious Diseases, Texas Tech University Health Sciences Center, Lubbock, TX, USA*

Chapter 1

Praziquantel: An Enigmatic, Yet Effective, Drug

David J. Timson

Abstract

Praziquantel is a remarkably effective drug for the treatment of schistosomiasis. It has few side effects, some of which have been attributed to its inactive enantiomer. Few, if any, verified cases of drug resistance have been reported in a clinical setting. The preponderance of scientific evidence suggests that the drug works by dysregulating calcium homeostasis in the worm. Voltage-gated calcium channels have been proposed as the main pharmacological target of praziquantel, although no direct evidence of interaction with this protein is available. Here, the biochemical pharmacology of praziquantel is briefly reviewed and a hypothesis for its mechanism proposed. This hypothesis suggests that the drug works, in part, by disrupting an interaction between a voltage-gated calcium channel (SmCav1B) and an accessory protein, SmTAL1.

Key words Schistosomiasis, Voltage-gated calcium channels, SmTAL1, Calcium signaling, Neglected tropical disease

1 Introduction: Schistosomiasis Chemotherapy

Schistosomiasis (bilharzia) is a neglected tropical disease which affects hundreds of millions of people in poorer, tropical regions of the world [1–3]. It is caused by infection by trematodes from the genus *Schistosoma*. These worms have a complex life cycle involving several morphologically distinct forms and an intermediate snail host. Since the infectious life cycle stage requires water as a habitat, communities which live close to lakes and rivers are particularly at risk, especially if these are relied upon for drinking and washing [4].

In the mammalian host, worms live primarily in the blood, hence their common name "blood flukes." While in the mammalian host, they mate and produce many eggs. Some of the pathology resulting from infection is caused by eggs accumulating and immune reactions to egg surface proteins. Eggs are passed in the urine or feces (depending on the species of schistosome). These eggs hatch, and the miracidia locate and penetrate the snail intermediate host where they live as sporocysts, sometimes for several generations. They exit the snail as cecariae, which are the infectious

David J. Timson (ed.), *Schistosoma mansoni: Methods and Protocols*, Methods in Molecular Biology, vol. 2151,
https://doi.org/10.1007/978-1-0716-0635-3_1,

form. These cecariae can penetrate the skin of humans, shed their tails, and enter the circulatory system. On migration to the portal blood of the liver, they mature into adult flukes. These migrate again to the mesenteric venules of the bladder or rectum where they mate, produce eggs, and complete the life cycle.

Symptoms of infection include itching shortly after infection. Within 1–2 months, fevers, chills, and muscle aches are often observed. The liver and intestinal stages of infection can result in scarring and inflammation of these tissues. Repeated, or long-term, infection can cause permanent damage to the liver, intestine, lungs, and bladder. Anemia, malnutrition, and learning difficulties can occur in children who are repeatedly infected. Although the disease is rarely immediately fatal, it causes significant distress and loss of function for patients [2]. It also affects their communities due to reduced productivity of workers and disrupted education for children.

Historically, schistosomiasis was treated with trivalent antimony compounds. These kill the worms by targeting the glycolytic enzyme phosphofructokinase [5–7]. However, they have a low therapeutic index and cause significant and debilitating side effects in patients. Oxamniquine (CAS: 21738-42-1; Fig. 1) and metrifonate (CAS: 52-68-6; Fig. 1) can be used to treat the disease [8–

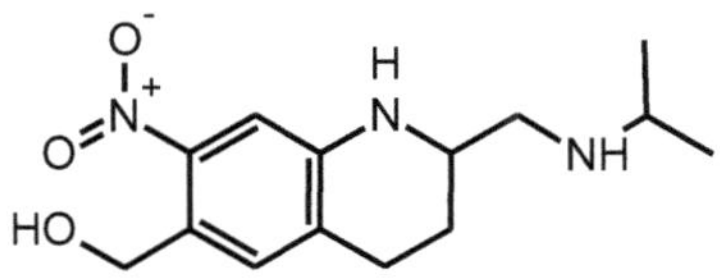

Oxamniquine

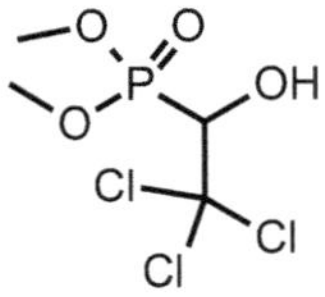

Metrifonate

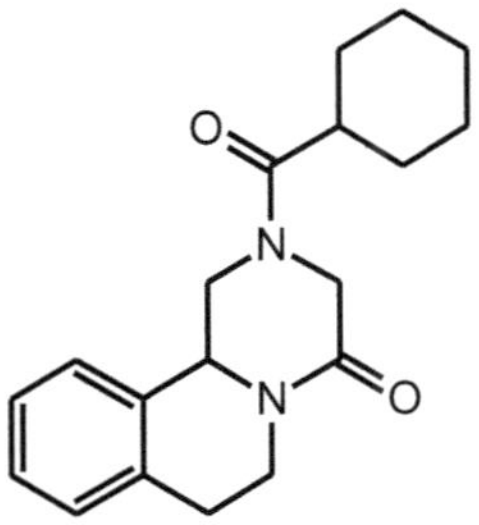

Praziquantel

Fig. 1 Drugs used to treat schistosomiasis

11]. These have better therapeutic indices than antimony compounds. The current treatment of choice is praziquantel (CAS: 55268-74-1; Fig. 1), which is generally highly effective and well tolerated by patients [12]. WHO includes it on their list of essential medicines [13]. In the absence of an effective vaccine to protect against schistosome infections, some countries have initiated mass drug administration (MDA) with praziquantel. In MDA, large numbers of people are given praziquantel prophylactically in order to eliminate the worm in the entire populations and regions, breaking the cycle of infection. This approach is effective and greatly reduces the number of new cases of schistosomiasis, sometimes for several years [14].

2 Praziquantel: Chemistry and Pharmacology

Praziquantel was originally developed as a putative novel tranquilizer [15]. However, it proved much more effective as an antiparasitic agent with particular success in the treatment of schistosomiasis. The patent for the drug has expired and its cost is low. It is manufactured as a racemic mixture of (*R*)- and (*S*)-praziquantel. Only the (*R*)-enantiomer is pharmacologically active, and the (*S*)-enantiomer is responsible for the bitter taste of the drug and some side effects [16–18]. Typical treatment involves the administration of two or three doses of 20 mg/kg approximately 4–6 h apart, or a single dose of 40 mg/kg [19, 20]. The current cost of this treatment, excluding salaries of the healthcare workers, is approximately US$0.20–0.30 [21]. Although resistance to other antiparasitic drugs is known, widespread, and increasing, few if any cases of clinically relevant resistance to praziquantel have been reported. However, resistance can be induced in the laboratory [22]. Thus, the almost exclusive reliance on one drug is a cause for concern. When widespread resistance to praziquantel does emerge, there will be few alternatives.

The biochemical mechanism of action of praziquantel is not known [23]. Early studies demonstrated that administration of the drug results in a rapid release of calcium ions into the cytoplasm of the parasite's cells [24]. This causes contraction of the worm's muscles and paralysis. The paralysis prevents mating and egg production and enables the host's immune system to clear the worms from the body. In the process, some worms lyse, releasing their cellular contents into the host. The immune reactions to this cause some of the adverse effects of praziquantel treatment.

This influx of calcium ions suggests a dysregulation of calcium homeostasis. In eukaryotes, calcium ion concentrations in the cytoplasm are normally maintained at low (micromolar) levels. Calcium ions are actively transported out of the cell and into organelles such as the endoplasmic reticulum. In response to signaling pathways, controlled calcium release into the cytoplasm can occur. This stimulates processes such as muscle contraction and apoptosis. The release and reuptake of calcium ions is normally tightly controlled, and most cells express a variety of calcium-specific membrane pumps and channels to facilitate this.

The preponderance of evidence suggests that voltage-gated calcium channels are involved in the pharmacological mechanism of praziquantel [25, 26]. However, this has yet to be conclusively demonstrated. Other targets have been suggested including a myosin regulatory light chain and adenosine signaling [27, 28]. In addition, recent work suggests that host proteins may also be targeted by the drug. Specifically, the human serotoninergic 5HT2B receptor is partially agonized by praziquantel. This may result in contractions of the blood vessels which help clear the parasites from the host [29].

3 Mechanism of Action: A Hypothesis

Mammalian voltage-gated calcium channels are regulated by the calcium receptor protein calmodulin. This acts as a control on the channel's activity, and to the calcium ion potential across the membrane. As calcium ions enter the cell, their concentration increases to a level which enables them to bind calmodulin. This alters the interaction between calmodulin and the calcium channel and results in the inhibition of calcium influx [30]. It could be hypothesized that praziquantel interferes with this process. If it antagonized the interaction, it would dysregulate the influx of calcium ions, potentially leading to uncontrolled influx. An IQ-motif from the α-subunit of a *Schistosoma mansoni*voltage-gated calcium channel (SmCa$_v$1B) has been identified as a binding partner of calmodulin. However, this interaction is not disrupted by praziquantel suggesting that the drug does not act on this interaction [31].

Another calcium regulatory pro
tein, SmTAL1 (Sm22.6), also interacts with SmCa$_v$1B [32]. This protein belongs to a family of calcium-binding proteins which appears to be unique to helminths [33–36]. These proteins combine a calcium-binding, N-terminal EF-hand domain with a C-terminal, dynein light chain-like domain. Their functions are largely unknown [33]. Intriguingly, SmTAL1 (along with SmTAL4, SmTAL5, and SmTAL8) also interacts with praziquantel [35, 36]. The functional consequences of this interaction are unknown, but preliminary data suggest that it affects the interaction with SmCa$_v$1B [32]. Thus, a hypothesis arises in which praziquantel acts, at least in part, through the disruption of this interaction which leads to the dysregulation of calcium pumping and the uncontrolled influx of calcium ions into the cytoplasm (Fig. 2).

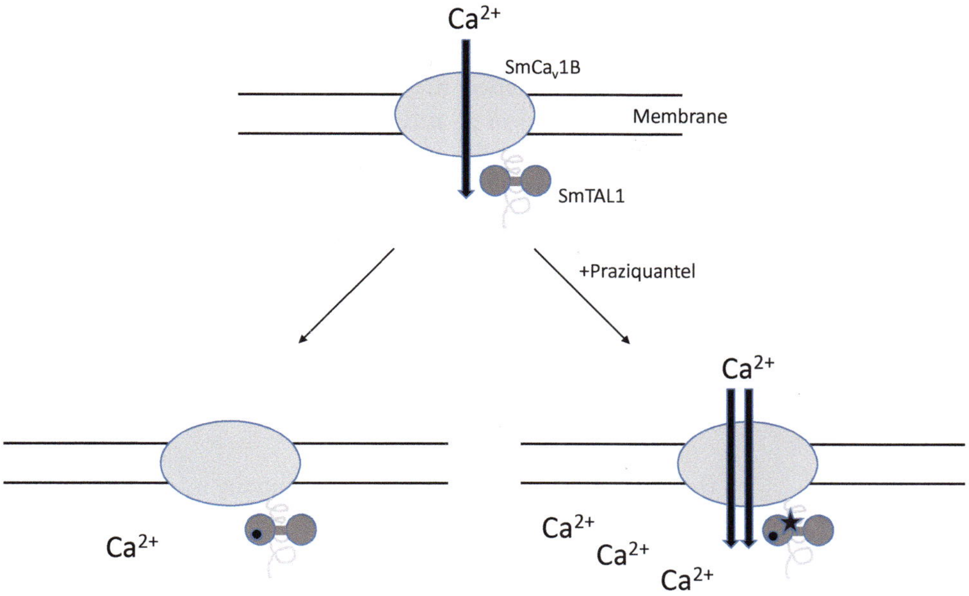

Fig. 2 Proposed biochemical mechanism for the action of praziquantel. Voltage-gated calcium channels (e.g., $SmCa_V1B$) permit the regulated influx of calcium ions. These channels can be regulated by calmodulin- and other calcium-binding proteins. These bind to IQ-motifs in a cytoplasmic region of the channel protein (top). It is hypothesized that SmTAL1 functions in this way. As calcium ion concentrations rise, SmTAL1 binds to the ions (black circles) and undergoes a conformational change which induces further changes in the calcium channel such that is closes regardless of the membrane potential. This prevents the further influx of calcium ions (bottom left). Praziquantel (black star) binds to SmTAL1, disrupting the interaction and preventing SmTAL from inducing the changes in the channel protein which result in it closing. Thus, the channel remains open, and calcium ions continue to enter the cell (bottom right). This excess influx of calcium ions results in uncontrolled muscle contraction and/or apoptosis, depending on the cell type

While there is no evidence to refute this hypothesis, evidence in favor is currently limited. Proof of interaction in vitro is not proof of pharmacologically relevant action in vivo. The demonstration of interaction was performed with full length SmTAL1 and a peptide from $SmCa_V1B$ which includes the IQ-motif. To understand the functional consequences of this interaction, and the consequences of praziquantel binding, experiments with full-length, membrane-bound $SmCa_V1B$ are required. Furthermore, many drugs work through multiple targets; therefore, some of the other potential targets may also be important pharmacologically [37].

4 Conclusions

Praziquantel is a remarkably successful drug despite lack of knowledge of its precise mechanism of action. Knowledge of the mechanism might enable the discovery of alternative drugs or help predict how resistance might arise. The possibility that praziquantel may act through disrupting an interaction between SmTAL1 and $SmCa_v1B$ is intriguing. Even if this hypothesis is subsequently refuted, this interaction may be worth exploring as a possible target for a new generation of antischistosomal drugs.

Note added in proof: During the production of this chapter, strong pharmacological evidence was published implicating a transient receptor potential calcium channel (SmTRP-PZQ) as the main target for the drug [38, 39]. The mechanism described in this chapter is, therefore, likely to represent a minor or secondary mechanism.

Acknowledgments

I thank my colleagues and students for many interesting discussions on molecular parasitology.

References

1. Utzinger J, Becker SL, Knopp S, Blum J, Neumayr AL, Keiser J, Hatz CF (2012) Neglected tropical diseases: diagnosis, clinical management, treatment and control. Swiss Med Wkly 142:w13727. https://doi.org/10.4414/smw.2012.13727
2. Colley DG, Bustinduy AL, Secor WE, King CH (2014) Human schistosomiasis. Lancet 383(9936):2253–2264. https://doi.org/10.1016/S0140-6736(13)61949-2
3. Houweling TA, Karim-Kos HE, Kulik MC, Stolk WA, Haagsma JA, Lenk EJ, Richardus JH, de Vlas SJ (2016) Socioeconomic inequalities in neglected tropical diseases: a systematic review. PLoS Negl Trop Dis 10(5):e0004546. https://doi.org/10.1371/journal.pntd.0004546
4. Steinmann P, Keiser J, Bos R, Tanner M, Utzinger J (2006) Schistosomiasis and water resources development: systematic review, meta-analysis, and estimates of people at risk. Lancet Infect Dis 6(7):411–425. https://doi.org/10.1016/S1473-3099(06)70521-7
5. Bueding E, Mansour JM (1957) The relationship between inhibition of phosphofructokinase activity and the mode of action of trivalent organic antimonials on *Schistosoma mansoni*. Br J Pharmacol Chemother 12(2):159–165
6. Timson DJ (2016) Metabolic enzymes of Helminth parasites: potential as drug targets. Curr Protein Pept Sci 17(3):280–295
7. Su JG, Mansour JM, Mansour TE (1996) Purification, kinetics and inhibition by antimonials of recombinant phosphofructokinase from *Schistosoma mansoni*. Mol Biochem Parasitol 81(2):171–178. https://doi.org/10.1016/0166685196027028
8. Foster R (1987) A review of clinical experience with oxamniquine. Trans R Soc Trop Med Hyg 81(1):55–59
9. Kramer CV, Zhang F, Sinclair D, Olliaro PL (2014) Drugs for treating urinary schistosomiasis. Cochrane Database Syst Rev 8:Cd000053. https://doi.org/10.1002/14651858.CD000053.pub3
10. Feldmeier H, Doehring E (1987) Clinical experience with metrifonate. Review with emphasis on its use in endemic areas. Acta Trop 44(3):357–368
11. Cioli D, Pica-Mattoccia L, Archer S (1995) Antischistosomal drugs: past, present and future? Pharmacol Ther 68(1):35–85
12. Thetiot-Laurent SA, Boissier J, Robert A, Meunier B (2013) Schistosomiasis chemotherapy. Angew Chem Int Ed Engl 52(31):7936–7956. https://doi.org/10.1002/anie.201208390
13. Anonymous (2015) 19th WHO model list of essential medicines. 2015. World Health Organization. World Health Organization, Geneva
14. Olveda DU, McManus DP, Ross AG (2016) Mass drug administration and the global control of schistosomiasis: successes, limitations and clinical outcomes. Curr Opin Infect Dis 29(6):595–608. https://doi.org/10.1097/qco.0000000000000312

15. Cioli D, Pica-Mattoccia L, Basso A, Guidi A (2014) Schistosomiasis control: praziquantel forever? Mol Biochem Parasitol 195 (1):23–29. S0166-6851(14)00075-9
16. Meyer T, Sekljic H, Fuchs S, Bothe H, Schollmeyer D, Miculka C (2009) Taste, a new incentive to switch to (*R*)-praziquantel in schistosomiasis treatment. PLoS Negl Trop Dis 3(1):e357. https://doi.org/10.1371/journal.pntd.0000357
17. Liu YH, Qian MX, Wang XG, Jia J, Wang QN, Jiang YF, Wang RQ, Yan SH, Chen BY, Li JS et al (1986) Comparative efficacy of praziquantel and its optic isomers in experimental therapy of schistosomiasis japonica in rabbits. Chin Med J 99(12):935–940
18. Xiao SH, Catto BA (1989) Comparative *in vitro* and *in vivo* activity of racemic praziquantel and its levorotated isomer on *Schistosoma mansoni*. J Infect Dis 159(3):589–592
19. Organization WHO (1995) WHO model prescribing information: drugs used in parasitic diseases. World Health Organization, Geneva
20. Manders S (2019) 9.11 tropical infectious diseases. In: Textbook of adult emergency medicine (E-book), 5th edn. Elsevier Health Sciences, Amsterdam
21. Kusel J, Hagan P (1999) Praziquantel--its use, cost and possible development of resistance. Parasitol Today 15(9):352–354
22. Mwangi IN, Sanchez MC, Mkoji GM, Agola LE, Runo SM, Cupit PM, Cunningham C (2014) Praziquantel sensitivity of Kenyan *Schistosoma mansoni* isolates and the generation of a laboratory strain with reduced susceptibility to the drug. Int J Parasitol Drugs Drug Resist 4 (3):296–300. https://doi.org/10.1016/j.ijpddr.2014.09.006
23. Thomas CM, Timson DJ (2020) The mechanism of action of praziquantel: can new drugs exploit similar mechanisms? Curr Med Chem 27(5):676–696
24. Coles GC (1979) The effect of praziquantel on *Schistosoma mansoni*. J Helminthol 53 (1):31–33
25. Greenberg RM (2005) Are Ca^{2+} channels targets of praziquantel action? Int J Parasitol 35 (1):1–9. https://doi.org/10.1016/j.ijpara.2004.09.004
26. Chan JD, Zarowiecki M, Marchant JS (2013) Ca^{2+} channels and praziquantel: a view from the free world. Parasitol Int 62(6):619–628. https://doi.org/10.1016/j.parint.2012.12.001
27. Gnanasekar M, Salunkhe AM, Mallia AK, He YX, Kalyanasundaram R (2009) Praziquantel affects the regulatory myosin light chain of *Schistosoma mansoni*. Antimicrob Agents Chemother 53(3):1054–1060. https://doi.org/10.1128/AAC.01222-08
28. Angelucci F, Basso A, Bellelli A, Brunori M, Pica Mattoccia L, Valle C (2007) The antischistosomal drug praziquantel is an adenosine antagonist. Parasitology 134 (Pt 9):1215–1221. https://doi.org/10.1017/S0031182007002600
29. Chan JD, Cupit PM, Gunaratne GS, McCorvy JD, Yang Y, Stoltz K, Webb TR, Dosa PI, Roth BL, Abagyan R, Cunningham C, Marchant JS (2017) The anthelmintic praziquantel is a human serotoninergic G-protein-coupled receptor ligand. Nat Commun 8(1):1910. https://doi.org/10.1038/s41467-017-02084-0
30. Ben-Johny M, Yue DT (2014) Calmodulin regulation (calmodulation) of voltage-gated calcium channels. J Gen Physiol 143 (6):679–692. https://doi.org/10.1085/jgp.201311153
31. Thomas CM, Timson DJ (2018) Calmodulins from *Schistosoma mansoni*: biochemical analysis and interaction with IQ-motifs from voltage-gated calcium channels. Cell Calcium 74:1–13
32. Thomas CM, Timson DJ (2020) The *Schistosoma mansoni* tegumental allergen protein, SmTAL1: binding to an IQ-motif from a voltage-gated ion channel and effects of praziquantel. Cell Calcium 86:102161
33. Thomas CM, Timson DJ (2016) A mysterious family of calcium-binding proteins from parasitic worms. Biochem Soc Trans 44 (4):1005–1010. https://doi.org/10.1042/BST20150270
34. Fitzsimmons CM, Jones FM, Stearn A, Chalmers IW, Hoffmann KF, Wawrzyniak J, Wilson S, Kabatereine NB, Dunne DW (2012) The *Schistosoma mansoni* tegumental-allergen-like (TAL) protein family: influence of developmental expression on human IgE responses. PLoS Negl Trop Dis 6(4):e1593. https://doi.org/10.1371/journal.pntd.0001593
35. Thomas CM, Fitzsimmons CM, Dunne DW, Timson DJ (2015) Comparative biochemical analysis of three members of the *Schistosoma mansoni* TAL family: differences in ion and drug binding properties. Biochimie 108:40–47. https://doi.org/10.1016/j.biochi.2014.10.015
36. Carson J, Thomas CM, McGinty A, Takata G, Timson DJ (2018) The tegumental allergen-like proteins of *Schistosoma mansoni*: a biochemical study of SmTAL4-TAL13. Mol Biochem Parasitol 221:14–22. https://doi.org/10.1016/j.molbiopara.2018.02.002

37. Thomas CM, Timson DJ (2018) The mechanism of action of Praziquantel: six hypotheses. Curr Top Med Chem 18(18):1575–1584. https://doi.org/10.2174/1568026618666181029143214
38. Park S-K, Gunaratne GS, Chulkov EG, Moehring F, McCusker P, Dosa PI, Chan JD, Stucky CL, Marchant JS (2019) The anthelmintic drug praziquantel activates a schistosome transient receptor potential channel. J Biol Chem 294(49):18873–18880
39. Park S-K, Marchant JS (2020) The journey to discovering a flatworm target of praziquantel: a long TRP. Trends Parasitol 36(2):182–194

Chapter 2

Computer-Aided Drug Design for the Identification of Novel Antischistosomal Compounds

Jéssika de Oliveira Viana, Marcus T. Scotti, and Luciana Scotti

Abstract

Schistosomiasis is a chronic neglected tropical disease, highlighted by the presence of *Schistosoma* worms, which presents in advanced cases in approximately 80 countries, affecting almost 300 million people. The treatment is based on only one drug, praziquantel, a drug discovered in the 1970s that shows moderate efficacy against the adult parasite, but low efficacy against the larval stages of the parasite. Therefore, the use of only one drug has brought concerns and losses on drug-resistance cases, necessitating the development of new effective chemotherapeutic agents against *Schistosoma* species. One of the strategies that have been implemented in drug development is the computer-aided drug design (CADD), investigating the structural characteristics of the compounds and targets in order to understand their actions and biological activities through 3D virtual manipulation, as the QSAR applied to ligands and molecular docking applied to a respective biological target. These studies help to extract information and characteristics relevant to the activity, as well as to predict potential applications and activity. Therefore, this chapter will present the main validated biological targets of the genus *Schistosoma*, as thioredoxin glutathione reductase (TGR), histone deacetylases (HDAC 1, HDAC 8), dihydroorotate dehydrogenase, sirtuin protein and cathepsin L1, as well as reports of CADD in literature applied to the development of drugs against schistosomiasis, providing compounds with high pharmacological potential and high specificity.

Key words Schistosomiasis, In silico studies, Antischistosomal activity, QSAR, Docking studies

1 Introduction

The neglected tropical diseases (NTDs) are a group of 17 health conditions, highly prevalent in tropical areas, with a higher incidence in urban or rural areas with poor sanitary conditions [1], affecting approximately one and a half billion people throughout the countries of the world.

Schistosomiasis, also known as bilharziasis, is one of these highly prevalent and chronic neglected tropical diseases (NTD) caused by flatworm parasites of the *Schistosoma* genus, involving *S. mansoni*, *S. haematobium*, and *S. japonicum* [2] and *S. mekongi*, *S. guineensis*, and *S. intercalatum* [3]. The World Health Organization (WHO) indicated that about 220.8 million people require

David J. Timson (ed.), *Schistosoma mansoni: Methods and Protocols*, Methods in Molecular Biology, vol. 2151, https://doi.org/10.1007/978-1-0716-0635-3_2,

preventive treatment in 2017, which threatens 800 million with risk of infection. Schistosomiasis is prevalent in tropical and subtropical areas, especially in areas of poor communities without access to safe drinking water and adequate sanitation [3, 4].

People infected with schistosomes release the parasite eggs into feces or urine, where the freshwater becomes contaminated with feces or urine containing the eggs. When people come into contact with contaminated water, the eggs hatch and release the larvae, known as miracidia. The larvae come in contact with the snail and thus multiply in several cycles producing numerous new parasites, known as cercariae, being released again into the water. The cercariae, when in contact, infect the human being through penetration of the skin in the water [2–4].

In humans, schistosomiasis can be characterized as intestinal or urogenital, which will depend on the location of the adult worms. The intestinal schistosomiasis is caused by the species *Schistosoma intercalatum*, *S. japonicum*, *S. mansoni*, and *S. mekongi*, where the worms occupy the mesenteric veins and their eggs are deposited in the intestine, arriving at the feces. Thus, the chronic intestinal schistosomiasis progresses from abdominal pain and bladder diarrhea, periportal liver fibrosis, and portal hypertension [5, 6].

On the other hand, urogenital schistosomiasis is caused by adult *S. haematobium*, residing in veins that drain the urinary tract, releasing the eggs through the urine [5, 6]. Thus, it results in hematuria, hydronephrosis, dysuria, and calcification of the bladder and is associated with bladder cancer [7] and HIV infection [8]. This involvement results in malnutrition, anemia, stunted growth, and reduced capacity to work [3, 9, 10].

2 Treatment and Drugs

In a 40-year time frame, only one drug praziquantel (PQZ) has been used to control schistosomiasis and recommended by the World Health Organization (WHO) [11] as a base drug in population chemotherapy. In the early 1970s, the PZQ was synthesized among a series of pyrazino isoquinolines and evaluated as potential tranquilizers. Only after 8 years was the drug evaluated as an antischistosomal drug candidate in animal models against numerous helminth infections and, subsequently, marketed as the drug Biltricide® and, later, by several generic manufacturers under a variety of brand names (Distocide®, Bilharzid, Prazitel, and Piquiton). Therefore, of all helminth infection treatment drugs, only PQZ is still considered the ideal drug for the treatment of schistosomiasis [12].

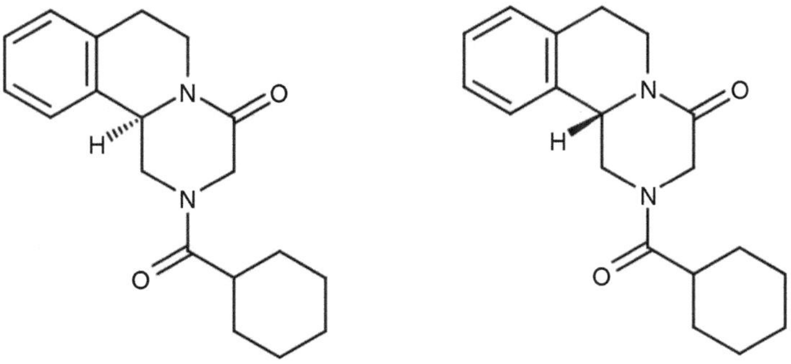

(R)-PZQ (S)-PZQ

Fig. 1 Enantiomers of the medicament PQZ: (*R*)-PZQ and (*S*)-PZQ

Although PQZ is given orally and rapidly absorbed after 15 min [12], its racemic nature raises cases of toxicity and side effects. Praziquantel is normally distributed as a racemate drug, including both the biologically active enantiomers (*R*)-PZQ, which has an antischistosomal effect [13], while (*S*)-PZQ acts as an ineffective enantiomer against worms (Fig. 1). This requires a 600 mg tablet to have the final dose capacity of 40 mg/kg. These disadvantages arise as inefficiency in treatment [14, 15]. PZQ is a class II drug with high tissue permeability and low solubility. However, its metabolism via hydroxylation turns it into inactive metabolites, causing only low concentrations of the drug to reach the blood system.

Currently, PZQ is recommended for humans in 3 doses of 40–60 mg/kg per day. However, there is only a decrease in parasite burden between 60 and 90% [16, 17], and cure is not fully achieved [18]. Trials performed by the WHO and Bayer have demonstrated that a single dose of PZQ of 40 mg/kg is effective against *S. mansoni* and *S. haematobium*, with a cure rate of 61–89% [19–21] and 40–89%, respectively. On the other hand, for *S. japonicum*, an administration of 30 mg/kg is used, with cure rate of 75–100% [22]. However, when using a higher dose of 60 mg/kg PZQ, there was no significant additional efficacy against *S. japonicum* [23].

Despite all these studies applied to the use of the PZQ, its mechanism of action is not yet fully established. What is known is that effects on the schistosome can occur by calcium influx throughout the parasite, muscle contraction, and superficial changes [24, 25].

However, recent studies by Sun et al. [26] evaluated the cytotoxicity of enantiomers in eight cell lines, demonstrating that (*R*)-PZQ presented lower cytotoxicity against L-02, SH-SY5Y, HUVEC, A549, HCT-15, and Raw264.7 cells, but the (*S*)-isomer showed higher cytotoxicity against L-02, prf-plc-5, and HepG2 cells.

In addition to cases of PZQ toxicity, cases of resistance have been evidenced by the low sensitivity to PZQ in areas where the drug has been extensively used. Among the mechanisms of resistance, the following are intensified: (1) the decrease or lack of drug activity at the younger stages of the parasite, which leads to survival under exposure to the drug; (2) the racemic nature of PZQ, where only the levo form has schistosomicidal activity, with half the dose having a useful pharmacological effect; and (3) the rapid metabolism of the drug, transforming it into an inactive metabolite, further decreasing the concentration of the drug in the blood circulation [15]. Almeida and coworkers [27] studied proteinases of PZQ-resistant and PZQ-sensitive strains of *S. mansoni*, of which 60 proteins were identified, some of them being considered protein targets, expressly differently in both strains and involved in PZQ resistance.

However, the resources to provide treatment to patients with schistosomiasis are still insufficient for the large number of patients who need it. To this end, new strategies have led to the implementation of mass drug administration (MDA) programs for control and elimination of the human helminthiasis [28]. The PZQ has been used through "preventive chemotherapy" programs in the distribution of the drug to school-age children and at-risk populations. By the year 2018, it was estimated that about 235 million people received PZQ [29], as a treatment of schistosomiasis and cestode infections [30].

In addition to PZQ, other medications have been used historically in the treatment of schistosomiasis. One of them is Oxamniquine (Fig. 2), an aminoethyltetrahydroquinolone derivative, which proved effective against *S. mansoni* [31, 32]. This drug, like PZQ, is rapidly absorbed, ensuring metabolism by oxidation and excretion through the urine. Its efficacy is achieved in a time of 2–3 days, under the dose of 15–60 mg/kg. Studies have demonstrated its safety and efficacy compared to PZQ, but less effective when measured with quantitative oogram by biopsy of rectal mucosa [31].

Another alternative group of drugs, known as derivatives of artemisinin (Fig. 2), has been used in the treatment of

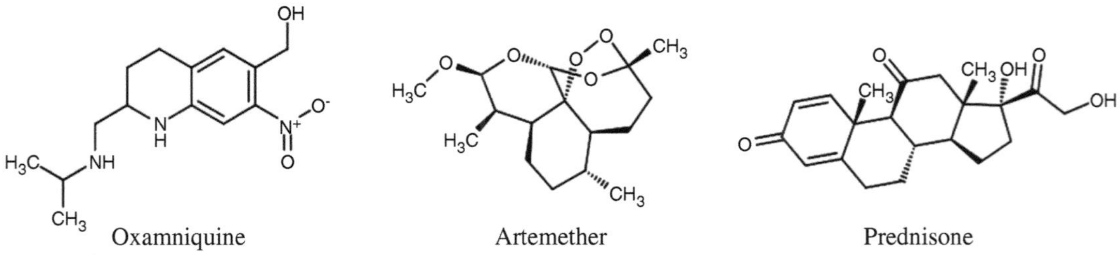

Fig. 2 Compounds used in the treatment of schistosomiasis

schistosomiasis [33]. These derivatives are best known as antimalarial drugs but have shown antischistosomal activity on *S. japonicum* [32], being more effective against younger stages of *Schistosoma* [34]. People's Republic of China was assessed that the efficacy of PZQ was not increased when used in combination with artemether [35], an artemisinin derivative. However, the combination of artemisinin derivatives with sulfadoxine/sulfalene and pyrimethamine showed higher cure rates, from 14% to 100% in *S. haematobium* [34, 36], and may be considered as a treatment alternative to be explored.

The adjuvant treatment in cases of neuroschistosomiasis has been applied with the use of corticosteroids (Fig. 2) [37, 38], reducing the immune response and avoiding tissue inflammation [38, 39]. Despite the need for further clinical trials in the treatment of schistosomiasis, case reports indicated the benefit of the combination of praziquantel and anti-schistosomiasis drugs [40, 41].

3 Computer-Aided Drug Design

Therapeutic innovation has become challenging, with the advent of computational studies, resulting from molecular targets derived from advances in genomics, structural, and molecular biology. Thus, bioinformatics and chemoinformatics have a central purpose in the development process, presenting a synergistic relationship with medicinal chemistry, molecular modeling, and drug development, where pharmaceutical companies have been increasing their ability to find new compounds by means of three-dimensional molecular targets, obtained through genomic data [42].

The molecular information storage of biological systems has evolved, which has generated a concomitant increase in tools to store and analyze these data. As an example, we have the wide variety of databases available, with information on transcriptomes, proteomes, metabolomes, protein sequence and structure, ligand structure, as well as protein-ligand interactions [43–45]. The use of these databases has provided support for use in academia and in the pharmaceutical industry in the design and development of new drugs and new targets.

Among the computational methods applied in medicinal chemistry are two classifications: structure-based drug design (SBDD) and ligand-based drug design (LBDD) [46]. SBDD is applied when 3D structural information on the molecular target from crystallography and nuclear magnetic resonance (NMR) is used [47], simulating intermolecular events between molecules. As examples, we apply structure-based virtual screening (SBVS), molecular docking, to predict binding conformations, and molecular dynamics, to evaluate conformational stability and flexibility [48, 49]. In the SBDD Protein Data Bank (PDB; www.rcsb.org)

has been widely used, containing three-dimensional information of over 100,000 biological structures, of which 92% are proteins [50].

On the other hand, LBDD methods are based on the investigation of small binders with known activity. In this method, we extract its molecular characteristics, known as molecular descriptors, essential for biological activity, producing predictive patterns of pharmacophoric hypotheses, classification models, quantitative structure–activity relationships (QSAR), and quantitative structure–property relationships (QSPR) [51, 52]. These models are generated in order to predict the biological activity of a compound, based on its molecular characteristics, by designing new predicted compounds based on ligand-based virtual screening (LBVS) [53]. Databases, such as ZINC, PubChem, and ChEMBL, are used in LBVS containing information about small compounds, such as their structure, physicochemical properties, and biological activity data [44, 54, 55].

Therefore, the use of these databases requires the constant development of computational tools of chemo- and bioinformatics, converting available information into useful and applicable knowledge.

4 Targets

The genomes of *S. mansoni*, *S. haematobium*, and *S. japonicum* are now available and provide a fundamental basis for understanding the metabolism of these parasite [56–58], the use of computational methods, and the development of schistosomiasis drugs.

The SchistoDB database has been used for drug development, gene sequence search mechanisms, gene similarity analyses, and predicting gene function [59]. The publication of the schistosome genomes has provided much information for the determination of protein structures. The PDB contains more than 200 schistosome protein structures, of which more than 170 are from *S. mansoni*.

The understanding of these *Schistosoma* genomes has also provided clues to predict the validity of proteins as drug targets [60, 61], such as the thioredoxin glutathione reductase (TGR), dihydroorotate dehydrogenase (DHODH), the sirtuin class (Sirtl and Sirt2), cathepsine L1 (CL1), and histone deacetylase (HDAC) enzymes.

4.1 Thioredoxin Glutathione Reductase (TGR)

The protein TGR is a key chimeric flavoenzyme expressed by schistosomes and is involved in two detoxification pathways crucial for the parasite survival in the host's organism [62, 63]. Recent studies have pointed to thioredoxin glutathione reductase as one of the well-characterized alternate targets for drug development for schistosomiasis [64].

The selenium moiety in this protein functions to reduce free radicals and is essential for the survival of the parasite. In addition, the protein has role in protein folding control, regulation of various enzymes, and transcription factors and provides electrons in deoxyribonucleotide synthesis. In schistosomes, this protein has two functions in a single enzyme that protects the body from oxidative stress and host-induced damage [65], providing the contribution that this enzyme in schistosomes was exploited in search of new schistosomidal drugs [66, 67].

Its structure has the fusion of the TR domain with glutaredoxin domain. The active site of SmTGR features three cysteine dimers or thiol centers and a FAD that transfers electrons from Cys154, Cys159 dimer to Cys596, selenoCys 597 dimer upon NADPH binding [68]. In the portion of Cys596 and selenoCys597, the protein has a selenium that can transfer hydrogen or oxidizer substrate.

4.2 Histone Deacetylases (HDAC 1, HDAC 8)

Another protein investigated as a therapeutic target in the treatment of schistosomiasis is the histone modifying enzymes, which have the role of catalyzing posttranslational ubiquitination, methylation, phosphorylation, and acetylation. The insertion of functional groups to the histone structure influences the organization of the chromatin filaments, maintaining the balance of transcriptional activity [69].

Thus, histone deacetylases (HDACs) are a family of enzymes found in numerous organisms, catalyzing the removal of acetyl groups from histone acetylated lysine, altering transcriptional activity. The HDAC family has 18 different members in humans. Two have been well-characterized in *S. mansoni* histone deacetylase 8 (HDAC 8) and histone deacetylase 1 (HDAC 1). SmHDAC has had inhibitors identified, as Trichostatin A (TSA) and valproic acid (VPA), which caused protein acetylation in schistosomes and dose-dependent mortality of adult worms [70, 71], being expressed at all life cycle stages. These studies indicate that the enzymes act as a promising molecular target in the fight against schistosomiasis [61].

4.3 Dihydroorotate Dehydrogenase

The enzyme dihydroorotate dehydrogenase (DHODH) is a flavoenzyme responsible for catalyzing the stereospecific oxidation of (*S*)-dihydroorotate (DHO) to orotate in the biosynthetic pathway of pyrimidine nucleotide [72–75]. The DHODHs are divided into two major classes, according to their cell localization: class 1 are cytosolic enzymes and use soluble substances, fumarate in the final electron acceptor [76], and class 2 are found in conjunction with membranes and use quinones to reoxidize the flavin group [77, 78].

SmDHODH has been indicated as a possible therapeutic target in the development of schistosomidal drugs [72–74] due to depletion of nucleotide pools following the inhibition of DHODH [79]. Structural differences were observed between human DHODH and *Schistosoma* enzymes [80]. smDHODH cannot synthesize purine bases, meaning that the worm depends exclusively on the salvage pathway and that the orotic acid produced control transcription, suggesting that its inhibition interferes with the cellular mechanism [80].

4.4 Sirtuin Protein

The sirtuin proteins are classified into five different classes (I, II, III, IV, and V) which vary according to conserved motifs in their core domain [81]. In parasites, class I has been found and exploited as antiparasitic targets [82], being identified as a vital pathway of parasite survival. The function of sirtuin protein is to catalyze the deacetylation reaction of acetylated lysine residues in nuclear histones and other substrates, using the NAD^+ as a cofactor [82]. One of the compounds tested was Salermide, which induces cell death in *S. mansoni* by binding in Sirt1 and Sirt2 [83], suggesting that these proteins may be useful in the development of new targets for schistosomiasis [84].

4.5 Cathepsin L1

The protein cathepsin L1 (CL1) in *Schistosoma* was primarily described by Smith et al. [85], suggesting that the protein is located in the gastrodermal cells lining the cecum of the parasite [86], with the function of degrading the hemoglobin of the host, therefore, source of nutrient for the adult *Schistosoma* [87]. Thus, the importance of SmCL1 and SmCL2 presents distinct activities, where one of these is expressed in cell lining the parasite's gut and other was present in the tegument. With the supply of the metabolic activity of the schistosome, these enzymes have become a crucial target in the development of new drugs with anti-schistosome activity [88, 89].

5 Compounds

Scientific advances in the field of medicinal chemistry have expanded the area of drug target space available for pharmacological intervention. Chemical diversity, followed by combinatorial chemistry, has been included in the large number of small-molecule screening, increasing the investments in research approaches based on the computational methods [90, 91]. Herein will be described several compounds identified as drug lead in development of new compounds with act as antischistosomal activity, showing their features and interactions with active site.

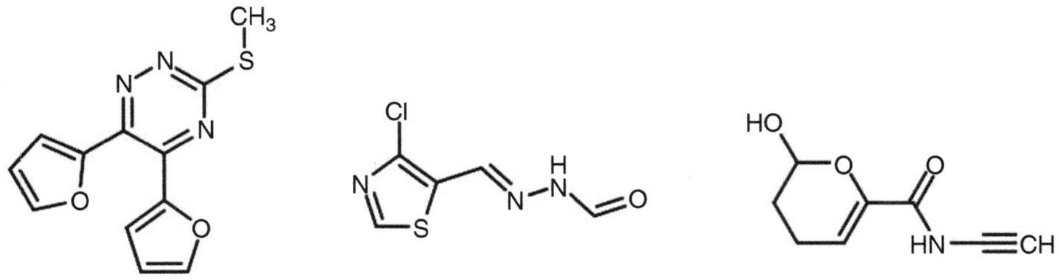

Scaffold 1 Scaffold 3 Scaffold 5

Fig. 3 Three of the ten scaffolds discovered by Gaba [92] for schistosomal activity

Gaba and colleagues [92] performed computer modeling combined with machine learning in the development approach based on the generation of molecular descriptors for the discovery of inhibitors of the enzyme thioredoxin glutathione reductase (TGR) of *Schistosoma mansoni*. The study was conducted in order to perform small molecule screening from PubChem against TGR, where the random forest algorithm was identified as the best classifier among all classifiers (Naive Bayes and J48) with an accuracy of around 80%. In conjunction with the molecular docking in the enzyme, thioredoxin glutathione reductase (PDB ID: 2V6O) showed that the active molecules predicted in the model presented different docking scores, suggesting that the prediction model was viable and can be used in the discovery of potential compounds with antischistosomal activity. In addition, in the study, it was possible to identify 10 scaffolds with the best profile (Fig. 3) for the *S. mansoni* TGR protein.

Kannan et al. [93] performed a target-based virtual screening of hydroxamate and sulfonamide-thiazole derivatives with the enzyme *S. mansoni* histone deacetylase 8 (smHDAC8). Using the ZINC drug-like database, 5000 compounds with zinc-binding moieties (thiazoles, hydroxamates, and anilinobenzamides) were used, all of which were docked with the active site of the SmDHAC8. In all, the researchers identified new hydroxamates derivatives (Fig. 4) as potent inhibitors of smHDAC8 with high flexibility at the active site on residues of Tyr341, Phe151, and Lys20.

Melo-Filho et al. [94] identified new chemical scaffolds using a series of 25 oxadiazoles-2-oxides against TGR from *S. mansoni*. In this study, computational methods were applied in order to combine 2D and 3D combi-QSAR descriptors with a partial least squares (PLS) and virtual screening (VS) of a commercial database. With a good prediction, Q^2_{ext} more than 80%, the consensus model of HQSAR + CoMFA + CoMSIA was considered the best performance and unites the characteristics of the three QSAR approaches explored. From the 35 compounds, 10 oxadiazoles-2-oxide was selected for presenting favorable ADME properties and did not

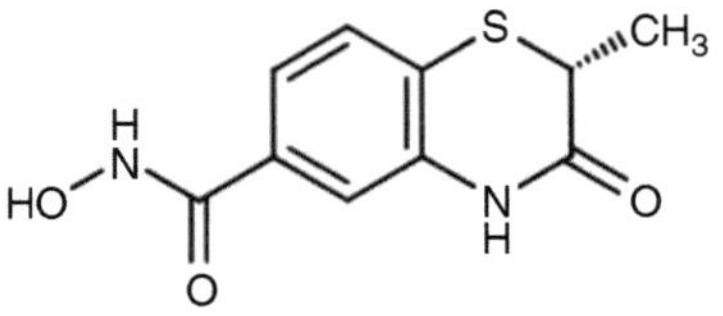

T5979345

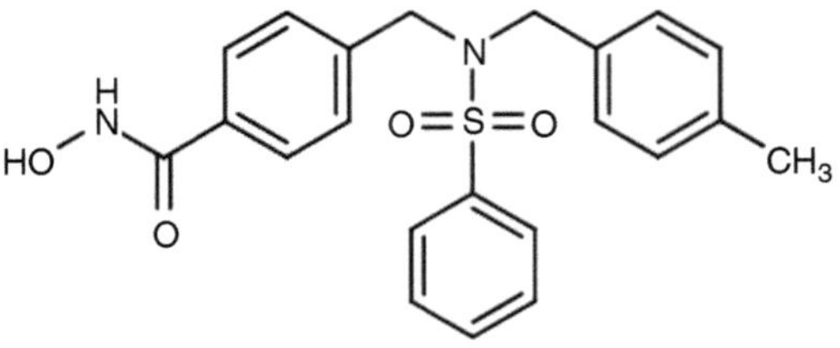

K783-3813

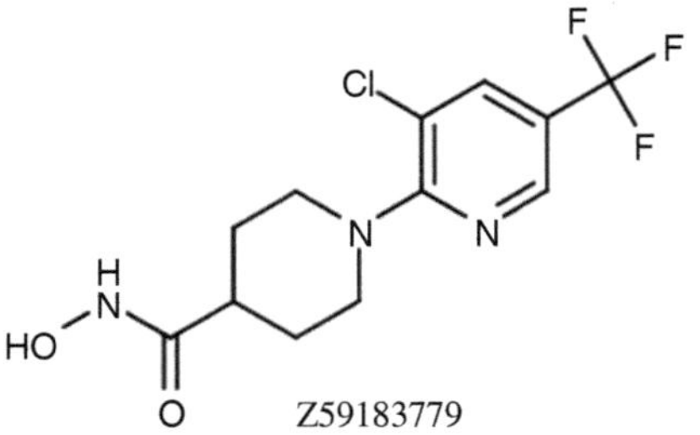

Z59183779

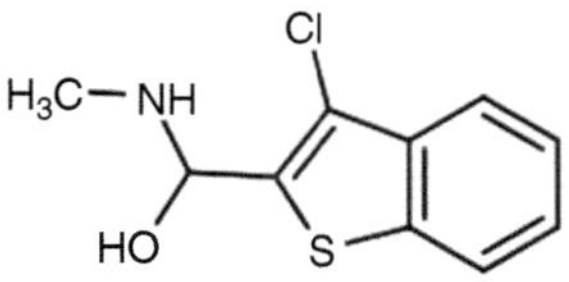

T6072858

Fig. 4 Chemical structure of the hits identified by Kannan et al. [93] by virtual screening

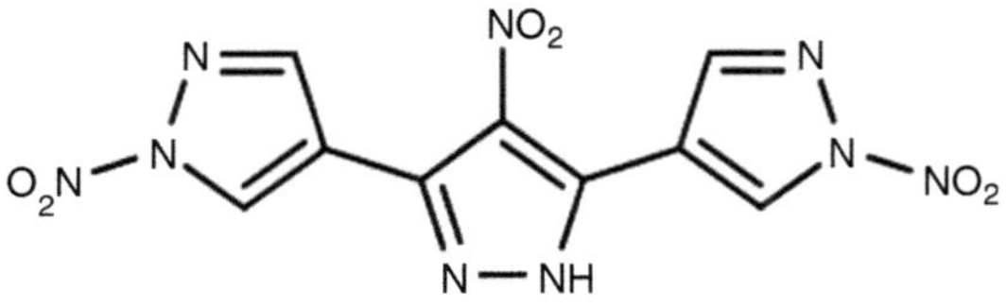

LabMol-17

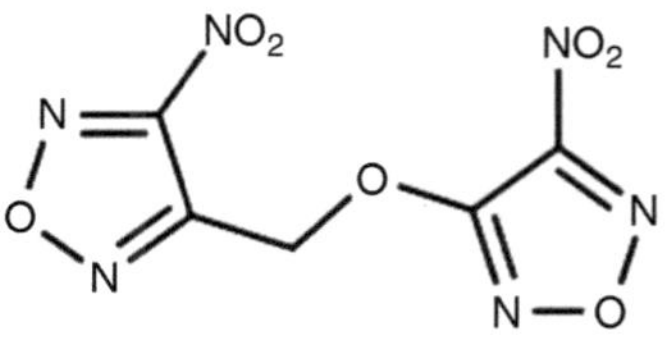

LabMol-19

Fig. 5 Chemical structures of the two lead compounds (LabMol-17 and LabMol-19) related by Melo-filho [94]

show any potential of being hERG blocker or 1 CYP3A4 inhibitors. After virtual screening, two compounds (Fig. 5) representing new chemical scaffolds by in vitro experimental evaluation in larval *S. mansoni* (schistosomula), demonstrating that the compounds showed high activity in schistosomula and adult worms.

Studies conducted by Otarigho [95] employed computational approaches to screen 12,110 compounds from DrugBank against *S. mansoni* dihydroorotate dehydrogenase (DHODH), a protein druggable target for schistosomiasis chemotherapy. For this study, the compounds were docked with DHODH, which showed that the drugs used in the treatment of other diseases have the potential to inhibit the activities of *Sm*DHODH, with binding affinity of less than —6 kcal/mol. This study suggests the strong binding and affinity of the compounds DB07978 and DB07976 (Fig. 6), potential antischistosomal drugs targeting the enzyme *Sm*DHODH.

Simoben et al. [96] investigated an inhibitory activity of 550,000 new small molecules for treatment of *S. mansoni* histone

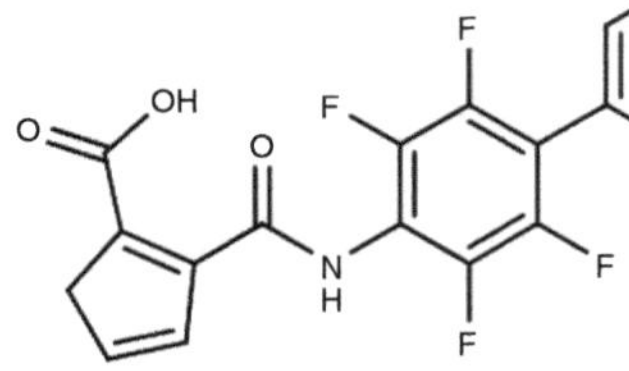

DB07978

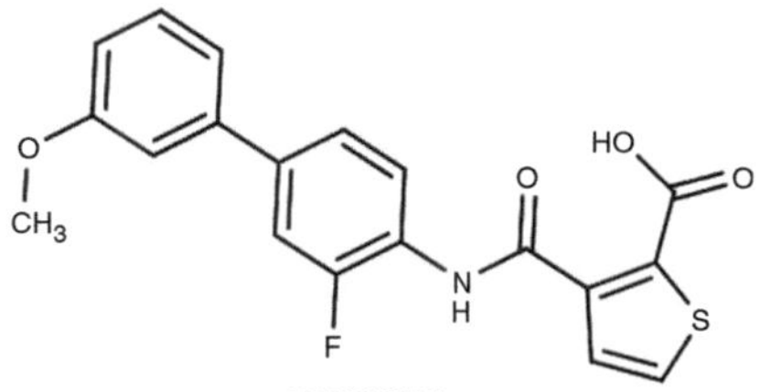

DB07976

Fig. 6 Chemical structures of the compounds DB07978 and DB07976, related by Otarigho [95]

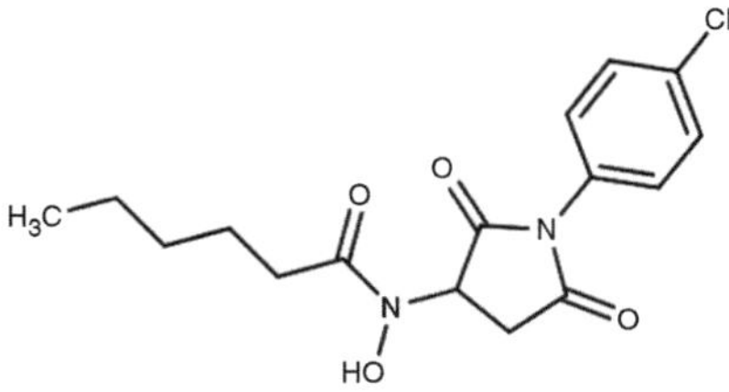

J1036

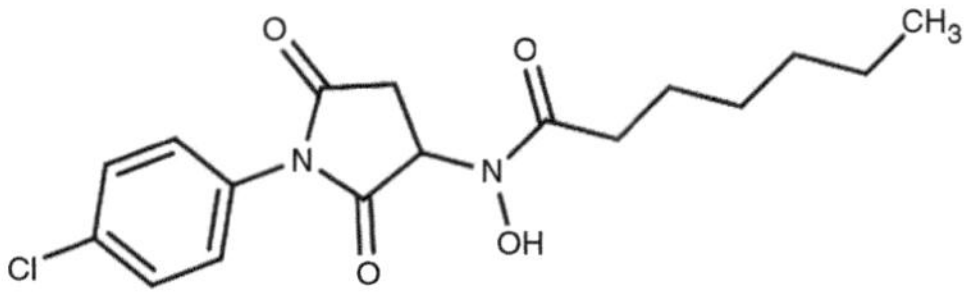

J1061

Fig. 7 Chemical structure of the inhibitors of the enzyme SmHDAC8, reported by Simoben et al. [96]

deacetylase 8 (smHDAC8) using docking-based virtual screening. Eight compounds identified as possible inhibitors were selected by docking scores, pan-assay interference compounds (PAINS) filtering, and molecular weight. From the docking studies, the researchers demonstrated that the *n*-alkyl group binds to the acetate-binding well of *Sm*HDAC8, whereas pyrrolidinedione moiety can undergo hydrogen bond interaction with the His292 and can be considered a major feature for inhibition with the enzyme *Sm*HDAC8. These eight compounds also showed considerable in vitro activity of IC_{50} values ranging from 4.4 to 20.3 μM against *Sm*HDAC8 (Fig. 7), demonstrating that the prediction of activity for these compounds was successful; in addition to that, these new compounds could be considered new smHDAC8 inhibitors.

In order to find new bioactive compounds against *Schistosoma*, Singh and Pandey [97] generated using homology modeling for the three-dimensional structure of the protein SmHDAC1, expressed protein at all stages of the *Schistosoma* life cycle, being considered an important target in inhibitor planning. Therefore, using virtual screening studies, a total of 70 class I HDAC inhibitors were submitted to molecular docking with the enzymes SmHDAC1 and HsHDAC1. Among the binding affinity studies, compound 24 (*N*, 8-dihydroxy-8-(naphthalen-2-yl) octanamide zinc (ID: ZINC13474421)) (Fig. 8) was shown to be the best inhibitor due to its high affinity (−8.1 kcal/mol) and favorable IC_{50} (0.035 μM), demonstrating the reliability of drug development based on receptor structure.

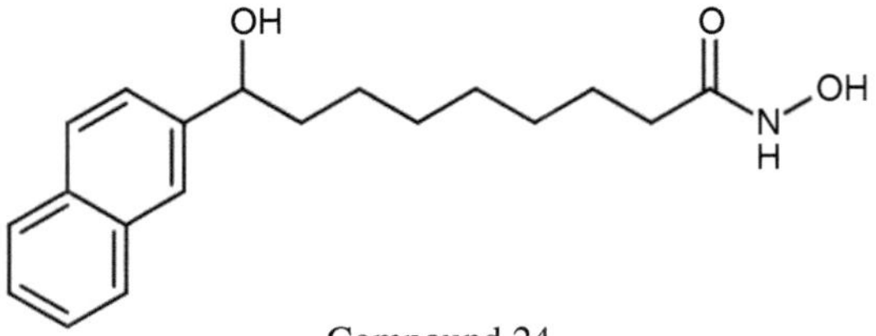

Compound 24

Fig. 8 Chemical structure of the inhibitor 24 with the enzyme SmHDAC1, reported by Singh e Pandey [97]

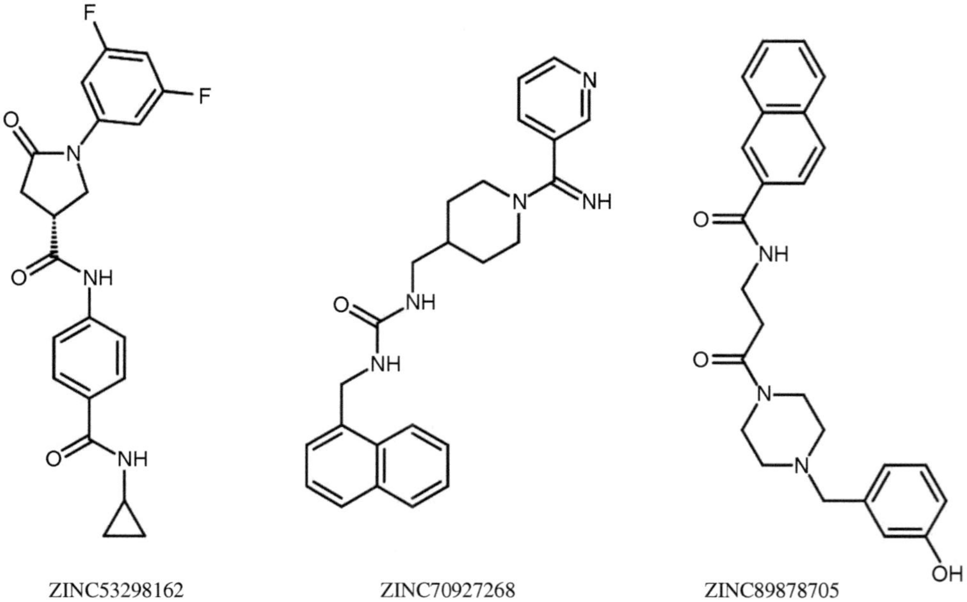

ZINC53298162 ZINC70927268 ZINC89878705

Fig. 9 Chemical structure of the ZINC compounds as inhibitors of SmSIRT2

Singh and his research group [98] developed homology models to determine the characteristic structures of the SmSirt2 protein, in addition to molecular docking and virtual screening to select the lead inhibitor for *S. mansoni*. From the 18,560 compounds selected from the ZINC database, three compounds (Fig. 9) showed the best binding results with the enzyme SmSirt2, ZINC53298162 (−9.4 kcal/mol), ZINC70927268 (−10.0 kcal/mol), and ZINC89878705 (−11.2 kcal/mol), demonstrating high-binding affinity when compared to the selective AGK2 (−8.8 kcal/mol) ligand.

Singh et al. [99] investigated the molecular interaction between Sirtl protein and its inhibitors (sirtinol and Salermide derivatives) in order to identify its target-binding affinities by constructing the protein from sequencing, homology modeling, molecular docking, and ADMET screening. The high affinity binding (−11 and −11.01 kcal/mol) of two sirtinol derivatives (Fig. 10) showed better energy when compared to the standard

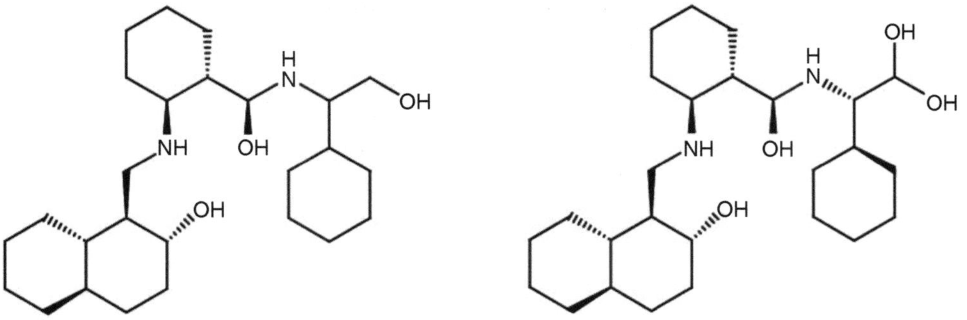

Sirt2 compound Sirt3 compound

Fig. 10 Inhibitor structure of the Sirt1 protein, related from Singh et al. [99]

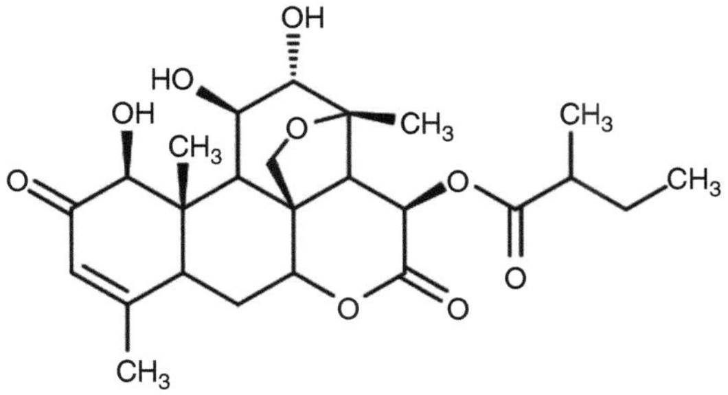

Simalikalactone-D

Fig. 11 Chemical structure of the simalikalactone-D related by Zafar as potential lead drug against SmCL1 enzyme

Salermide (−8 kcal/mol), making them possible candidates for antischistosomal drugs.

Zafar [100] performed in silico studies, using receiver operating characteristic (ROC) curve analysis and molecular dynamics (MD) simulation, to screen non-peptide inhibitors against SmCL1 via structure-based drug discovery. In their study, the researchers identified the phytochemical simalikalactone-D as a potential lead against CL1 of *S. mansoni*, identifying another compound lead as a drug candidate for antischistosomal therapy (Fig. 11).

6 Conclusion

In this chapter, we present studies reported in the literature on numerous techniques applied in medicinal chemistry for the development of drugs. Among them, we highlight the studies using structure-based drug design (SBDD) applying homology modeling tools, molecular docking, and molecular dynamics. Such studies have also brought the ligand-based drug design (LBDD) methods, which have employed several software molecular descriptors, chemometric analysis, and machine learning for model generation. To this end, validated and present PDB molecular targets such as thioredoxin glutathione reductase (TGR), dihydroorotate dehydrogenase (DHODH), the sirtuin class (Sirt1 and Sirt2), cathepsin L1 (CL1), and histone deacetylase (HDAC) enzymes have been reported. In addition, this chapter presented the best profile and scaffold lead compounds selected from in silico tools, which presented similar characteristics in their structure. Therefore, the use of in silico tools demonstrated the action and prediction of these compounds in greater detail and specificity, accelerating and favoring the offer of future studies to be implemented in the development of antischistosomal compounds.

References

1. World Health Organization (2010) First WHO report on neglected tropical diseases: working to overcome the global impact of neglected tropical diseases. https://apps.who.int/iris/bitstream/handle/10665/44440/9789241564090_eng.pdf;jsessionid=FB90F2B5E9FB386995329EAE7523301B?sequence=1. Accessed 25 May 2019
2. Colley DG, Bustinduy AL, Secor WE et al (2014) Human schistosomiasis. Lancet 383:2253–2264
3. World Health Organization (2019) Schistosomiasis. https://www.who.int/en/news-room/fact-sheets/detail/schistosomiasis. Accessed 20 May 2019
4. Gray DJ, Ross AG, Li YS et al (2011) Diagnosis and management of schistosomiasis. BMJ 342:2651
5. Steinmann P, Keiser J, Bos R et al (2006) Schistosomiasis and water resources development: systematic review, meta-analysis, and estimates of people at risk. Lancet Infect Dis 6:411–425
6. Amaral RSD, Tauil PL, Lima DD et al (2006) An analysis of the impact of the Schistosomiasis control programme in Brazil [in English]. Mem Inst Oswaldo Cruz 101:79–85
7. Parkin DM (2008) The global burden of urinary bladder cancer. Scand J Urol Nephrol Suppl 42:12–20
8. Kjetland EF, Ndhlovu PD, Gomo E et al (2006) Association between genital schistosomiasis and HIV in rural Zimbabwean women. AIDS 20:593–600
9. King CH, Dangerfield-Cha M (2008) The unacknowledged impact of chronic schistosomiasis. Chronic Illn 4(1):65–79
10. King CH (2017) The evolving schistosomiasis agenda 2007–2017. Why we are moving beyond morbidity control toward elimination of transmission. PLoS Negl Trop Dis 11(4): e0005517
11. World Health Organization (2002) Prevention and control of schistosomiasis and soil-transmitted helminthiasis: report of a WHO expert committee. https://apps.who.int/iris/bitstream/handle/10665/42588/WHO_TRS_912.pdf?sequence=1&isAllowed=y. Accessed 15 May 2019
12. Valencia CI, Catto BA, Webster LT et al (1994) Concentration time course of praziquantel in Filipinos with mild *Schistosoma japonicum* infection. Southeast Asian J Trop Med Public Health 25:409–414
13. Xiao S, You J, Mei J et al (1998) In vitro and in vivo effect of levopraziquantel, dextropraziquantel versus racemic praziquantel on different developmental stages of *Schistosoma japonicum*. Zhongguo Ji Sheng Chong Xue Yu Ji Sheng Chong Bing Za Zhi 6 (5):335–341
14. Olliaro P, Delgado-Romero P, Keiser J (2014) The little we know about the pharmacokinetics and pharmacodynamics of praziquantel (racemate and R-enantiomer). J Antimicrob Chemother 69:863–870
15. Bühring KU, Diekmann HW, Müller H et al (1978) Metabolism of praziquantel in man. Eur J Drug Metab Pharmacokinet 3:179–190
16. Wegner DHG (1984) The profile of the trematodicidal compound praziquantel. Arzneimittelforschung 34:1132–1136
17. Black CL, Steinauer ML, Mwinzi PN et al (2009) Impact of intense, longitudinal retreatment with praziquantel on cure rates of *Schistosomiasis mansoni* in a cohort of

occupationally exposed adults in western Kenya. Tropical Med Int Health 14:450–457

18. N'Goran EK, Gnaka HN, Tanner M et al (2003) Efficacy and side-effects of two praziquantel treatments against *Schistosoma haematobium* infection, among schoolchildren from cote d'Ivoire. Ann Trop Med Parasitol 97(1):37–51
19. Obonyo CO, Muok EM, Mwinzi PN (2010) Efficacy of artesunate with sulfalene plus pyrimethamine versus praziquantel for treatment of *Schistosoma mansoni* in Kenyan children: an open-label randomized controlled trial. Lancet Infect Dis 10(9):603–611
20. Erko B, Degarege A, Tadesse K et al (2012) Efficacy and side effects of praziquantel treatment of *Schistosomiasis mansoni* in Shesha Kekele elementary school, Wondo genet, southern Ethiopia. Asian Pac J Trop Biomed 2(3):235–239
21. Reta B, Erko B (2013) Efficacy and side effects of praziquantel in the treatment for *Schistosoma mansoni* infection in school children in Senbete town, northeastern Ethiopia. Tropical Med Int Health 18(11):1338–1343
22. World Health Organization (1998) International Strategies for Tropical Disease Treatment: Experiences with Praziquantel. Edited by Reich, MR. http://apps.who.int/medicinedocs/pdf/whozip48e/whozip48e.pdf. Accessed 15 May 2019
23. Olliare PL, Vaillant MT, Belizario VJ et al (2011) A multicentre randomized controlled trial of the efficacy and safety of single dose praziquantel at 40 mg/kg vs 60 mg/kg for treating intestinal schistosomiasis in the Philippines, Mauritania, Tanzania and Brazil. PLoS Negl Trop Dis 5(6):e1165
24. Cioli D, Pica-Mattoccia L, Archer S (1995) Antischistosomal drugs: past, present... and future? Pharmacol Ther 68:35–85
25. Cioli D, Pica-Mattoccia L, Basso A et al (2014) Schistosomiasis control: praziquantel forever? Mol Biochem Parasitol 195:23–29
26. Sun Q, Mao R, Wang D et al (2016) The cytotoxicity study of praziquantel enantiomers. Drug Des Devel Ther 10:2061
27. Almeida AP, Mendes T, Ferreira P et al (2018) Comparative proteomics reveals characteristic proteins on praziquantel-resistance in *Schistosoma mansoni*. BioRxiv, 314724
28. World Health Organization (2006) Preventive chemotherapy in human helminthiasis. Coordinated use of antihelminthic drugs in control interventions: a manual for health professionals and programme managers. https://apps.who.int/iris/bitstream/handle/10665/43545/9241547103_eng.pdf?sequence=1. Accessed 14 May 2019
29. World Health Organization (2013) Sustaining the drive to overcome the global impact of neglected tropical diseases. Second WHO report on neglected tropical diseases. http://apps.who.int/iris/bitstream/10665/80245/1/WHO_HTM_NTD_2013.2_eng.pdf. Accessed 14 May 2019
30. Cioli D, Pica-Mattoccia L (2003) Praziquantel. Parasitol Res 90:3–9
31. Ferrari ML, Coelho PM, Antunes CM et al (2003) Efficacy of oxamniquine and praziquantel in the treatment of *Schistosoma mansoni* infection: a controlled trial. Bull World Health Organ 81(3):190–196
32. Utzinger J, N'Goran EK, N'Dri A et al (2000) Oral artemether for prevention of *Schistosoma mansoni* infection: randomized controlled trial. Lancet 355 (9212):1320–1325
33. Araujo N, Kohn A, Katz N (1991) Activity of the artemether in experimental *Schisosomiasis mansoni*. Mem Inst Oswaldo Cruz 86(Suppl 2):185–188
34. Issoko MS, Dabo A, Traore H et al (2009) Efficacy of artesunate + sulfamethoxypyrazine/pyrimethamine versus praziquantel in the treatment of *Schistosoma haematobium* in children. PLoS One 4(10):e6732
35. Hou XY, McManus DP, Gray DJ et al (2008) A randomized double-blind placebo-controlled trial of safety and efficacy of combined praziquantel and artemether treatment for acute *Schistosomiasis japonica* in China. Bull World Health Organ 86(10):788–795
36. Utzinger J, Tanner M, Keiser J (2010) ACTs for schistosomiasis: do they act? Lancet Infect Dis 10(9):579–581
37. Gray DJ, Ross AG, Li YS (2011) Diagnosis and management of schistosomiasis. BMJ 342:2651
38. Ross AG, McManus DP, Farrar J et al (2012) Neuroschistosomiasis. J Neurol 259 (1):22–32
39. Fowler R, Lee C, Keystone JS et al (1999) The role of corticosteroids in the treatment of cerebral schistosomiasis caused by *Schistosoma mansoni*: case report and discussion. Am J Trop Med Hyg 61(1):47–50
40. Lambertucci JR, Voieta I, Silveira IS (2008) Cerebral schistosomiasis mansoni. Rev Soc Bras Med Trop 41(6):693–694
41. World Health Organization (2016) Schistosomiasis: number of people treated in 2014. Wkly Epidemiol Rec 89(4):21–28. https://

www.who.int/wer/2016/wer9105.pdf?ua=1. Accessed 15 May 2019

42. Weigelt J (2010) Structural genomics – impact on biomedicine and drug discovery. Exp Cell Res 316(8):1332–1338
43. Kumar S, Jena L, Daf S et al (2013) hpvPDB: an online proteome reserve for human papillomavirus. Genomics Inform 11(4):289–291
44. Gaulton A, Bellis LJ, Bento AP et al (2012) ChEMBL: a large-scale bioactivity database for drug discovery. Nucleic Acids Res 40 (1):1100–1107
45. Benson DA, Cavanaugh M, Clark K et al (2013) GenBank. Nucleic Acids Res 41:36–42
46. Wilson GL, Lill MA (2011) Integrating structure-based and ligand-based approaches for computational drug design. Future Med Chem 3(6):735–750
47. Blaney J (2012) A very short history of structure-based design: how did we get here and where do we need to go? J Comput Aided Mol Des 26(1):13–14
48. Lionta E, Spyrou G, Vassilatis DK et al (2014) Structure-based virtual screening for drug discovery: principles, applications and recent advances. Curr Top Med Chem 14 (16):1923–1938
49. Valasani KR, Vangavaragu JR, Day VW et al (2014) Structure based design, synthesis, pharmacophore modeling, virtual screening, and molecular docking studies for identification of novel cyclophilin D inhibitors. J Chem Inf Model 54(3):902–912
50. Berman HM, Westbrook J, Feng Z et al (2000) The protein data bank. Nucleic Acids Res 28(1):235–242
51. Leach AR, Gillet VJ, Lewis RA et al (2010) Three-dimensional pharmacophore methods in drug discovery. J Med Chem 53 (2):539–558
52. Mittal A, Paliwal S, Sharma M et al (2014) Pharmacophore based virtual screening, molecular docking and biological evaluation to identify novel PDE5 inhibitors with vasodilatory activity. Bioorg Med Chem Lett 24 (14):3137–3141
53. Mavromoustakos T, Durdagi S, Koukoulitsa C et al (2011) Strategies in the rational drug design. Curr Med Chem 18(17):2517–2530
54. Irwin JJ, Sterling T, Mysinger MM et al (2012) ZINC: a free tool to discover chemistry for biology. J Chem Inf Model 52 (7):1757–1768
55. Wang Y, Xiao J, Suzek TO (2009) PubChem: a public information system for analyzing bioactivities of small molecules. Nucleic Acids Res 37(Suppl. 2):623–633
56. Berriman M, Haas BJ, LoVerde PT et al (2009) The genome of the blood fluke *Schistosoma mansoni*. Nature 460(7253):352
57. Brindley PJ, Pearce EJ (2007) Genetic manipulation of schistosomes. Int J Parasitol 37 (5):465–473
58. Young ND, Jex AR, Li B et al (2012) Whole-genome sequence of *Schistosoma haematobium*. Nat Genet 44(2):221
59. Zerlotini A, Heiges M, Wang H et al (2008) SchistoDB: a *Schistosoma mansoni* genome resource. Nucleic Acids Res 37(suppl 1):579–582
60. Rojo-Arreola L, Long T, Asarnow D et al (2014) Chemical and genetic validation of the statin drug target to treat the helminth disease, schistosomiasis. PLoS One 9(1): e87594
61. Marek M, Kannan S, Hauser AT et al (2013) Structural basis for the inhibition of histone deacetylase 8 (HDAC8), a key epigenetic player in the blood fluke *Schistosoma mansoni*. PLoS Pathog 9:e1003645
62. Song L, Li J, Xie S et al (2012) Thioredoxin glutathione reductase as a novel drug target: evidence from Schistosoma japonicum. PLoS One 7(2):e31456
63. Sharma M, Khanna S, Bulusu G et al (2010) Corrigendum to comparative modeling of thioredoxin glutathione reductase from *Schistosoma mansoni*: a multifunctional target for antischistosomal drug discovery. J Mol Graph Model 27:665–675
64. Angelucci F, Miele AE, Boumis G et al (2008) Glutathione reductase and thioredoxin reductase at the crossroad: the structure of *Schistosoma mansoni* thioredoxin glutathione reductase. Proteins 72(3):936–945
65. Kuntz AN, Davioud-Charvet E, Sayed AA, Califf LL et al (2007) Thioredoxin glutathione reductase from *Schistosoma mansoni*: an essential parasite enzyme and a key drug target. PLoS Med 4(6):e206
66. Sayed AA, Simeonov A, Thomas CJ et al (2008) Identification of oxadiazoles as new drug leads for the control of schistosomiasis. Nat Med 14(4):407
67. Li T, Ziniel PD, He PQ et al (2015) High-throughput screening against thioredoxin glutathione reductase identifies novel inhibitors with potential therapeutic value for schistosomiasis. Infect Dis Poverty 4(1):40
68. Angelucci F, Sayed AA, Williams DL (2009) Inhibition of *Schistosoma mansoni* thioredoxin-glutathione reductase by

auranofin structural and kinetic aspects. J Biol Chem 284(42):28977–28985

69. Bannister AJ, Kouzarides T (2011) Regulation of chromatin by histone modifications. Cell Res 21(3):381–395
70. Oger F, Dubois F, Caby S et al (2008) The class I histone deacetylases of the platyhelminth parasite *Schistosoma mansoni*. Biochem Biophys Res Commun 377:1079–1084
71. Dubois F, Caby S, Oger F et al (2009) Histone deacetylase inhibitors induce apoptosis, histone hyperacetylation and up-regulation of gene transcription in *Schistosoma mansoni*. Mol Biochem Parasitol 168:7–15
72. Serrão VHB, Pereira HD, de Souza JRT et al (2017) *Schistosoma mansoni* purine and pyrimidine biosynthesis: structures and kinetic experiments in the search for the best therapeutic target. Curr Pharm Des 23 (45):6967–6983
73. Zeraik AE, Serrão VHB, Romanello L et al (2017) *Schistosoma mansoni* displays an adenine phosphoribosyltransferase preferentially expressed in mature female gonads and vitelaria. Mol Biochem Parasitol 214:82–86
74. El Kouni MH (2017) Pyrimidine metabolism in schistosomes: a comparison with other parasites and the search for potential chemotherapeutic targets. Comp Biochem Physiol B Biochem Mol Biol 213:55–80
75. Serrão VH, Romanello L, Cassago A et al (2017) Structure and kinetics assays of recombinant *Schistosoma mansoni* dihydrofolate reductase. Acta Trop 170:190–196
76. Reis RA, Calil FA, Feliciano PR et al (2017) The dihydroorotate dehydrogenases: past and present. Arch Biochem Biophys 632:175–191
77. Bader B, Knecht W, Fries M et al (1998) Expression, purification, and characterization of histidine-tagged rat and human flavoenzyme dihydroorotate dehydrogenase. Protein Expr Purif 13(3):414–422
78. Hortua MT, Huynh MH, Garavito MF et al (2012) Biochemical and molecular characterization of the pyrimidine biosynthetic enzyme dihydroorotate dehydrogenase from toxoplasma gondii. Mol Biochem Parasitol 184 (2):71–81
79. Vyas VK, Ghate M (2011) Recent developments in the medicinal chemistry and therapeutic potential of dihydroorotate dehydrogenase (DHODH) inhibitors. Mini Rev Med Chem 11(12):1039–1055
80. Loeffler M, Carrey EA, Zameitat E (2015) Orotic acid, more than just an intermediate of pyrimidine de novo synthesis. J Genet Genomics 42(5):207–219
81. Religa AA, Waters AP (2012) Sirtuins of parasitic protozoa: in search of function (s). Mol Biochem Parasitol 185(2):71–88
82. Vergnes B, Sereno D, Madjidian-Sereno N et al (2002) Cytoplasmic SIR2 homologue overexpression promotes survival of Leishmania parasites by preventing programmed cell death. Gene 296(1–2):139–150
83. Lara E, Mai A, Calvanese V et al (2009) Salermide, a Sirtuin inhibitor with a strong cancer-specific proapoptotic effect. Oncogene 28(6):781
84. Nonato MC, de Pádua RA, David JS, RAG R et al (2019) Structural basis for the design of selective inhibitors for *Schistosoma mansoni* dihydroorotate dehydrogenase. Biochimie 158:180–190
85. Smith AM, Dalton JP, Clough KA et al (1994) Adult *Schistosoma mansoni* express cathepsin L proteinase activity. Mol Biochem Parasitol 67(1):11–19
86. Brady CP, Dowd AJ, Brindley PJ et al (1999) Recombinant expression and localization of *Schistosoma mansoni* cathepsin L1 support its role in the degradation of host haemoglobin. Infect Immun 67:368–374
87. Dalton JP, Smith AM, Clough KA et al (1995) Digestion of haemoglobin by schistosomes: 35 years on. Parasitol Today 11:299–303
88. Dalton JP, Clough KA, Jones MK et al (1996) Characterization of the cathepsin-like cysteine proteinases of *Schistosoma mansoni*. Infect Immun 64:1328–1334
89. McKerrow JH, Doenhoff MJ (1988) Schistosome proteases. Parasitol Today 4:334–340
90. Clark DE, Picket SD (2000) Computational methods for the prediction of 'drug-likeness'. Drug Discov Today 5(2):49–58
91. Scott DE, Bayly AR, Abell C et al (2016) Small molecules, big targets: drug discovery faces the protein–protein interaction challenge. Nat Rev Drug Discov 15(8):533
92. Gaba S, Jamal S, Scaria V, Drug Discovery Consortium (2014) Cheminformatics models for inhibitors of *Schistosoma mansoni* thioredoxin glutathione reductase. Sci World J 2014:957107
93. Kannan S, Melesina J, Hauser AT et al (2014) Discovery of inhibitors of *Schistosoma mansoni* HDAC8 by combining homology modeling, virtual screening, and in vitro validation. J Chem Inf Model 54 (10):3005–3019
94. Melo-Filho CC, Dantas RF, Braga RC et al (2016) QSAR-driven discovery of novel chemical scaffolds active against *Schistosoma*

mansoni. J Chem Inf Model 56 (7):1357–1372

95. Otarigho B (2019) Structural, functional and docking analysis against *Schistosoma mansoni* dihydroorotate dehydrogenase for potential chemotherapeutic drugs. F1000 Res 8:651
96. Simoben C, Robaa D, Chakrabarti A et al (2018) A novel class of *Schistosoma mansoni* histone deacetylase 8 (HDAC8) inhibitors identified by structure-based virtual screening and in vitro testing. Molecules 23(3):566
97. Singh R, Pandey PN (2015) Molecular docking and molecular dynamics study on SmHDAC1 to identify potential lead compounds against Schistosomiasis. Mol Biol Rep 42(3):689–698
98. Singh R, Singh S, Pandey PN (2016) In-silico analysis of sirt2 from *Schistosoma mansoni*: structures, conformations, and interactions with inhibitors. J Biomol Struct Dyn 34 (5):1042–1051
99. Singh R, Yadav BS, Singh S et al (2016) In-silico screening of *Schistosoma mansoni* Sirtuin1 inhibitors for prioritization of drug candidates. Springerplus 5(1):286
100. Zafar A, Ahmad S, Rizvi A et al (2015) Novel non-peptide inhibitors against SmCL1 of *Schistosoma mansoni*: in silico elucidation, implications and evaluation via knowledge-based drug discovery. PLoS One 10(5): e0123996

Chapter 3

Immunofluorescent Localization of Proteins in *Schistosoma mansoni*

Rashika El Ridi and Hatem Tallima

Abstract

Immunofluorescence allows the detection, visualization, and localization of proteins by using the ability of antibodies to firmly bind to specific antigens. Proteins must be accessible to thorough interaction with the specific antibodies. Different immune evasion mechanisms of parasites are directed to hamper or prevent access of antibodies to critical proteins or virulence factors. The blood fluke *Schistosoma mansoni* would not survive a day in the host blood capillaries if antibodies were able to readily bind to proteins located at the surface and mediate its attrition and demise by the complement system and/or the FcγR- or FcαR-bearing leukocytes. The worm surface is the area of parasite-host interaction and the route to critical nutrients, but is selectively permeable, allowing access of nutrient molecules but not host antibodies. Gentle procedures, which, however, are not commonly in use in vivo, are required to increase the permeability of the parasite outer membrane shield to just allow access of specific antibodies and identify and localize the proteins at the apical surface. Robust methods involving acetone, methanol, and Triton X-100 treatment lead to disintegration of the dual lipid bilayer cover with exposure of the proteins located in the tegument beneath. Internal proteins may not be accessed except following cryostat or paraffin sectioning. Accordingly, vaccine-induced specific antibodies to the apical surface or tegument proteins are unable to harm intact parasites. Specific antibodies to surface membrane proteins may only add to the action of administered or endo schistosomicides via acceleration of killing and interference with repair of severely and lightly impacted parasites, respectively. Therefore, careful immunofluorescent localization of *S. mansoni* proteins is important for devising the different control strategies against infection.

Key words *Schistosoma mansoni*, Immunofluorescence, Sphingomyelin, Hydrogen bond barrier, Apical membrane antigens, Tegumental antigens, Cytosolic antigens

1 Introduction

Schistosoma mansoni (kingdom Animalia, phylum Platyhelminthes, class Trematoda, Schistosomatidae) are blood parasites of man and other mammals, and differently from other trematodes, adults have separate sexes [1]. Eggs able to exit the host via stool need contact with fresh water to hatch and release the miracidium. This free-living stage is infective to specific compatible snail intermediate hosts, most commonly of the genus *Biomphalaria*. Within the

David J. Timson (ed.), *Schistosoma mansoni: Methods and Protocols*, Methods in Molecular Biology, vol. 2151, https://doi.org/10.1007/978-1-0716-0635-3_3, © Springer Science+Business Media, LLC, part of Springer Nature 2020 snail, the miracidium differentiates into a mother sporocyst that develops asexually to eventually release thousands of cercariae. The cercaria is the short-lived, free-swimming stage that is infective for susceptible mammalian definitive hosts [2]. Snail *S. mansoni* endoparasite stages and free-living cercaria do not show sexual differentiation [1] and are covered by a syncytial tegument. The innermost membrane of the tegument is a conventional basal membrane, whereas the outer membrane overlaying the syncytium consists of a single trilaminate lipid bilayer [3–5]. Upon host skin invasion, cercariae must reside in the blood- and lymph-free epidermis until physiological and biochemical changes lead to exchange of the outer trilaminate lipid bilayer by a heptalaminated double lipid bilayer, each of which is composed of outer and inner leaflets [6–8]. The larvae, now termed schistosomula, reach the dermis, locate and penetrate blood capillaries, and migrate to the lungs and then the liver. The blood stream is their abode for their development, copulation, and sexual reproduction, i.e., their entire life, hence the name blood flukes. The dual outer membrane is an evident adaptation to the intravascular life style, with migrating, juvenile, and adult parasites constantly surrounded by a multitude of host immune cells and effectors [2, 6–9]. This unusual double lipid bilayer protects the parasite from all immunological attacks, insults, and even the slightest harassment. In support, surface membrane proteins of cercariae and epidermal-stage ($3 \leq 24$-h-old in vitro) schistosomula are accessed by antibodies and readily visualized and identified by surface membrane immunofluorescence (IF). Conversely, healthy and intact ex vivo dermal-, lung-, and liver-stage schistosomula and adult worms are entirely negative in membrane IF whichever antiserum is used [10–16].

Entirely unscathed by host antibodies and immune cells, *S. mansoni* worms in copula prefer the lower mesenteric veins, and the female, further protected in the male gynecophoric canal or ventral schist (hence the prefix schist), daily lays hundreds of eggs for several years. Using the lateral spine and proteolytic enzymes, eggs traverse the blood capillary and intestine wall to exit via stool and continue the life cycle in the intermediate host snail. The egg egress journey leads to colon inflammatory granulomatous responses with congestion, edema, necrosis, fibrosis, nodules and polyp formation, hemorrhage, and ulceration, the characteristics of intestinal schistosomiasis [17]. The eggs that fail to exit drift and are eventually trapped in several organs, but the majority is driven by the blood stream to settle via the portal tracts in the liver. As in the intestine wall, antigens released through the egg microscopic pores induce intense immune inflammatory responses, lesions, nodules, and fibrosis in the liver but also the central nervous system, eyes, genital organs, and skin [17, 18].

Schistosoma mansoni thus elicits serious disease. Due to poverty and government inefficiency in providing clean water, adequate sanitary conditions, and health education, *Schistosoma mansoni* affects approximately 67 million people, with 200 million, namely, children, vulnerable to infection. The disease is prevalent in Brazil, Venezuela, Suriname, the Caribbean, and 50 countries in the Middle East and sub-Saharan Africa [19, 20]. Diagnosis depending on egg detection in stool is the gold standard, but is unreliable in light infection and in chronic disease due to the changes alluded above in the intestine [17–20]. Accordingly, innumerable patients are denied treatment with the only commercially available drug, praziquantel (PZQ), which is anyhow offered to only 35.6% of the target population [17–21]. Major limitations to schistosomiasis control with PZQ have been its limited availability, deterring taste and side effects, inefficacy in heavy infection, and inability to prevent reinfection [17–23]. Vaccination is the only viable solution for control of the infection, the disease, reinfection, and transmission. A safe and efficacious vaccine remains an unmet need, predominantly because of the erraticism in immunofluorescent localization of target protein antigens in developing and adult *S. mansoni* [15, 24].

Host-specific antibodies and immune cell-bearing receptors for the Fc portion of IgG (FcγR) or IgA (FcαR) antibodies must interact with proteins in the apical lipid bilayer of developing and adult parasites to induce their demise. However, these proteins are normally hidden, shielded, and inaccessible in healthy intact parasites. We have hypothesized and provided evidence for the existence of a barrier at the host-parasite interface that is impermeable to molecules of >400 Da and presumably made of apical leaflet sphingomyelin (SM)-based hydrogen bonds with surrounding water molecules [15, 24–29]. In support, methylene blue (320 Da) and neutral red (289 Da) but not trypan blue (873 Da) can access intact healthy adult *S. mansoni* [30]. To access, localize, and identify *S. mansoni* truly surface membrane proteins using antibodies and IF, it is necessary to disrupt the outer SM-based hydrogen bond belt and increase the protein mobility to facilitate their visualization [15, 24]. Gentle procedures involving changes in pH and serum concentrations are required and sufficient. Yet, most authors resort to drastic methods of parasite exposure to methanol, acetone, Triton X-100, and fixation with formaldehyde for hours or even days. Following these procedures, the double lipid bilayer disintegrates, and proteins visualized are those within the syncytial tegument beneath the double lipid bilayer and not at the host-parasite interactive surface [31–34]. The outer dual lipid bilayer membrane overlays the syncytium and thus shields the tegumental antigens from host antibodies and effector leukocytes. Indeed, the proteins at the apical surface, or within the tegument, of intact, healthy parasites are not different from the internal molecules regarding accessibility to host immunological effectors and may not serve alone as candidate vaccines.

Exposure of proteins located at the parasite apical surface, and even the tegument, to antibody and immune cell binding could, however, occur in vitro and in vivo in parasites impacted by contact with innate or administered schistosomicides such as arachidonic acid, PZQ, artemether, and oxamniquine [35–39]. Identification of *S. mansoni* surface membrane antigens is necessary for production of vaccines that may synergize with chemotherapeutic treatment via antibody interaction with transiently exposed molecules and interference with repair of the outer surface membranes. Accordingly, it is of importance to delineate the procedures required for immunofluorescent visualization and identification of the larval and adult *S. mansoni* proteins at the apical layer, syncytial tegument, and interior of the parasite.

2 Materials

2.1 Animals

Female CD-1 mice and Syrian hamsters (*Mesocricetus auratus*) are raised at the Schistosome Biological Materials Supply Program, Theodore Bilharz Research Institute (SBSP/TBRI), Giza, Egypt, until 6 week old and maintained throughout experimentation at the animal facility of the Zoology Department, Faculty of Science, Cairo University. Hamsters are anesthetized, with its abdomen shaved, wet with sterile deionized water, and then exposed to 200 or 400 cercariae in 100 μL deionized water, protected from spilling by a sterile steel ring, as described previously [40, 41].

2.2 Parasites

1. *Miracidia* are released from eggs, as described in Subheading 3.3.1.
2. *Cercariae* of an Egyptian strain of *S. mansoni* are obtained from SBSP/TBRI and used immediately after shedding from *Biomphalaria alexandrina* snails.
3. *Lung-stage schistosomula* are obtained 6 to 8 days after percutaneous infection of hamsters with 400 cercariae of *S. mansoni*, as described in Subheading 3.3.3.
4. *Juvenile and adult male and female worms* are retrieved by hepatic and mesenteric vessel perfusion of hamsters, 3 and 6–7 weeks, respectively after infection with 200 cercariae of *S. mansoni* as previously described [40, 41].

2.3 Media

1. Roswell Park Memorial Institute medium-1640 (RPMI).
2. Antibiotic mixture: 10,000 U/mL penicillin-10,000 μg/mL streptomycin.
3. 100 mM Sodium pyruvate.
4. 200 mM L-glutamine.
5. Fetal calf serum (FCS) (*see* **Note 1**).

6. Dulbecco's phosphate-buffered saline, pH. 7.2 (D-PBS): autoclave 8 g sodium chloride, 0.2 g potassium chloride, 1.15 g dibasic sodium hydrogen phosphate, 0.2 g monobasic sodium dihydrogen phosphate in 1 L deionized water.
7. 30% (w/v) sucrose in D-PBS.

2.4 Parasite Preparation Reagents and Equipment

1. RPMI medium/antibiotics/10% (v/v) FCS.
2. 5000 U/mL heparin.
3. Meticulously clean, sterile glass beakers.
4. Conical 13 mL and 50 mL centrifuge tubes and 96-well flat-bottomed microplates.
5. Nitex 132-nylon mesh filtering screen.
6. Lysis buffer: 0.16 M ammonium chloride/0.017 M Tris, pH. 8.0.
7. Percoll (Pharmacia). Thoroughly mix 4 mL Percoll with 6 mL RPMI medium and 5 mL Percoll with 5 mL RPMI medium to prepare 40% and 50% Percoll in RPMI, respectively.
8. CO_2 incubator.
9. Centrifuge with accessories for 15 mL and 50 mL tubes and microplates.
10. Inverted ultraviolet (UV) microscope.

2.5 Immunofluorescence

1. Sodium azide (NaN_3): 0.1% (w/v) in RPMI medium (*see* **Note 2**).
2. Formaldehyde (37% w/v in H_2O). To prepare 0.1% (v/v): 0.27 mL/100 mL RPMI medium, pH 7.5.
3. Para-formaldehyde: 4% (w/v) in D-PBS, pH 7.5.
4. Arachidonic acid: 1.0 mM in FCS-free RPMI medium.
5. Triton X-100: 0.2% (v/v) in D-PBS.
6. Acetone (ice cold).
7. Paraffin wax-embedded sections.
8. Cryostat sections.
9. Mouse antisera.
10. Anti-mouse IgG (Fab-specific) fluorescein isothiocyanate (FITC)-labeled goat immunoglobulin $F(ab')_2$ fragment of affinity isolated antibody, adsorbed with bovine, horse, and human serum proteins (Sigma).

3 Methods

3.1 Preparation of Media

1. To a 500 mL bottle of RPMI medium, add 5 mL of penicillin–streptomycin (final concentration 100 U/mL penicillin and 100 μg/mL streptomycin), 5 mL sodium pyruvate, and 5 mL L-glutamine stock (final concentration 1.0 and 2.0 mM, respectively) (*see* **Note 3**).
2. For parasite incubation, add FCS to 10% (v/v) final concentration.
3. For ex vivo lung-stage schistosomula extraction, further add heparin to a final concentration of 10 U/mL.
4. For hepatic and mesenteric vessel perfusion of juvenile and adult worms, autoclave 0.85% (w/v) NaCl/1.5% (w/v) sodium citrate in deionized water.

3.2 Preparation of Antisera

1. Prepare the protein to be localized in *S. mansoni* in a pure, recombinant, or multiple antigen peptide construct.
2. Randomly distribute 10 parasite-free healthy mice in two groups of 5 mice each.
3. Immunize 5 mice subcutaneously or intramuscularly three times with 3 week interval with 100 μL alum (13% aluminum hydroxide gel, Sigma) containing 0 (adjuvant control) or 10 μg immunogen thoroughly mixed with the alum.
4. One week later, collect blood from each mouse, centrifuge twice at 400 × *g* for 20 min, and aliquot the supernatant (serum) before freezing at −20 °C.
5. Select the most specific antiserum and appropriate adjuvant control serum by enzyme-linked immunosorbent assay (ELISA) as described [42] (*see* **Note 4**).
6. In some experiments, check for the rigor of the protein localization by IF via using control and test sera previously absorbed with 25 μg/μL target molecule, overnight at room temperature, as described [15].
7. Indirect membrane or cytoplasmic IF is most convenient. Co-localization of two protein targets can readily be accomplished by direct IF (*see* **Note 5**).

3.3 Parasite Preparation

3.3.1 Miracidia

1. Expose hamsters to 400 cercariae of *S. mansoni*.
2. Seven weeks later, collect liver.
3. Mince liver with a blade to 0.1 mm^3 pieces.
4. Homogenize liver slices in 1.5% (w/v) NaCl solution to a fine consistency.

5. Centrifuge the homogenate at 100 × *g* in a 50 mL tube, pour out supernatant, and suspend the sediment in 45 mL ice-cold deionized water. Allow the homogenate to settle by gravity in the dark.
6. Wrap the tube with aluminum foil, and expose the top to direct light source for 60 min.
7. Collect the photo tactic miracidia free from tissue debris from the water upper top layer using glass Pasteur pipette.
8. Note that live healthy miracidia run in deionized water at great speed. Record viability, integrity, and number by 10× light microscopy.

3.3.2 Tailless Cercariae

1. Cercariae are the infective stage for humans. Wear gloves. Any spill should be immediately covered with 70% (v/v) ethanol or preferably 10% (v/v) Betadine.
2. Sediment cercarial suspension in water by maintaining the tube on ice for 30 min.
3. Transfer the water supernatant into a disposable container of 10% (v/v) Betadine solution, and replace with RPMI medium.
4. Vortex the cercarial suspension three times, 30 s each, as agitation elicits detachment of the cercarial tails from the bodies.
5. Pass the cercarial suspension five times using the barrel and needle of a 5 mL syringe. Dispose the syringe in 10% (v/v) Betadine.
6. Carefully over layer the cercarial suspension, drop by drop, onto 50% (v/v) Percoll in RPMI medium. Avoid mixing the cercarial suspension and the Percoll underneath. Centrifuge at 400 × *g* for 15 min using swinging bucket rotor.
7. Tailless cercariae pellet to the bottom of the tube, while tails remain in the supernatant. Discard the supernatant into disposable vessel containing 10% (v/v) Betadine.
8. Wash the tailless cercariae twice in RPMI medium by centrifugation at 100 × *g* for 5 min.
9. Record purity, viability, and number by 10× light microscopy.

3.3.3 Ex Vivo Lung-Stage Larvae

1. Infect hamsters with 400 cercariae of *S. mansoni*.
2. Six or seven days later, euthanize hamsters.
3. Dissect the hamsters to expose the thoracic cavity.
4. Perfuse lung by injecting 10 mL RPMI medium/10 U/mL heparin into the right ventricle of the heart. Transfer the lung pieces of each hamster in thoroughly clean glass beakers. Cut the lung pieces with clean scissors to an extremely fine consistency.

5. To allow the lung-stage schistosomula to crawl from the blood capillaries to the surrounding medium, add 10 mL RPMI medium/antibiotics/10% FCS/10 U/mL heparin, and incubate at 37 °C for 3–4 h, under 5% CO_2 (final pH approximately 7.2) and 0% CO_2 (final pH 7.5–7.8) in parallel.
6. Thoroughly mix the suspension with sterile plastic Pasteur pipette, and sieve it through a sterile Nitex 132 membrane into 50 mL tube using RPMI medium.
7. Pellet the lung cells and larvae at 400 × *g* for 10 min. Discard the supernatant.
8. Suspend the pellet in 10 mL ice-cold lysis buffer, and incubate at 10 °C for 10 min to allow lysis of erythrocytes while preserving the integrity of schistosomula surface membrane.
9. Centrifuge the suspension at 400 × *g* for 10 min. Discard the supernatant. Suspend the pellet in 5 mL RPMI medium, and examine the larvae for viability, contractility, and number. Larvae and cells may be pooled from different animals or animal groups at this stage.
10. Schistosomula are preferably further isolated free from lung cells by carefully over layering 5 mL suspension onto 5 mL 40% (v/v) Percoll in RPMI medium in 13 mL tubes and centrifugation at 400 × *g* for 20 min.
11. Pure schistosomula pellet in the bottom of the tubes. Wash twice in RPMI medium to eliminate residual Percoll. Inspect the larvae for viability, motility, and number. Schistosomula generally appear elongated, contractile, and remarkably transparent (Fig. 1, compare upper and lower panels) [10, 24, 41, 43, 44].

3.3.4 Juvenile and Adult Worms

1. Infect hamsters with 200 cercariae of *S. mansoni*.
2. Collect juvenile worms 21–24 days post infection (PI), and adult worms 6–7 weeks PI, by hepatic and mesenteric vessel perfusion as described previously [40, 41, 45, 46].
3. Extreme care should be exercised in handling the worms. For apical surface membrane protein localization, only use intact and healthy worms that would not allow entry of 0.4% (w/v) trypan blue (approximately 873 Da) as described previously [30] (*see* **Note 6**).

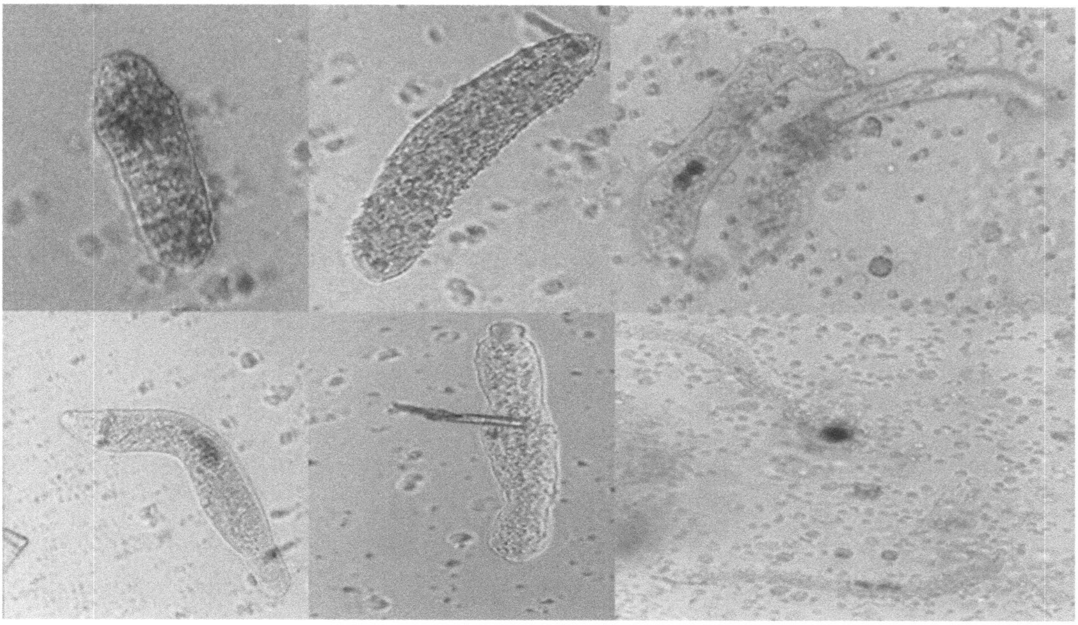

Fig. 1 Lung-stage schistosomula extracted from lung pieces at pH 7.6, fixed with 0.1% formaldehyde, and inspected by light microscopy. Lung-stage schistosomula, which appear distorted, granular, and opaque (upper panel), should not be considered in evaluation of positive or negative IF. Schistosomula used for assessment of protein localization by IF should appear elongated and remarkably transparent under light microscopy (lower panel)

3.4 Proteins Immunofluorescent Localization

3.4.1 Apical Surface Membrane Proteins

1. 1. *Miracidia and cercariae* can be fixed for 5 min or less in 0.1% (w/v) sodium azide or 1% (v/v) paraformaldehyde in FCS-free RPMI medium to inhibit movement and processed for IF localization of proteins at the apical surface membrane.
2. *Schistosomula* extracted and isolated from lung pieces and cells at pH ≤ 7.3, appearing live, healthy, elongated, contractile, and transparent, will be uniformly negative in IF assays, even following fixation in 1% paraformaldehyde or 0.1% formaldehyde, due to the inability of antibodies to access molecules on the parasite outer lipid bilayer.
3. Use lung-stage larvae extracted in the absence of CO_2, or incubate healthy ex vivo lung-stage schistosomula for 12 h in FCS-free RPMI medium in the entire absence of CO_2 to allow medium to reach pH 7.5–7.6. Optimal activation of parasite tegument-associated neutral sphingomyelinase (nSMase) at pH ≥ 7.5 elicits hydrolysis of some sphingomyelin (SM) molecules at the apical double lipid bilayer leaflet and, consequently, partial disruption of the SM-based hydrogen bond barrier.
4. Do not fix the larvae in 0.1% (w/v) NaN_3. It is necessary to fix the larvae for 10 min in 0.1% (v/v) formaldehyde [24]. Methylene hydrate (HO–CH_2–OH) polymers may now access and

increase the mobility of apical protein molecules, allowing their visualization by IF [24].

5. *Juvenile and adult worms* freshly perfused, healthy, and intact will be consistently negative in IF whichever antiserum is used, and localization of surface membrane proteins is not possible and is likely artefactual (*see* **Note** 7).
6. It is necessary to disrupt the protective SM-based hydrogen bond barrier, without impairing the integrity of the outer double lipid bilayer membrane, which conceals the proteins in the syncytial tegument underneath.
7. Fix live healthy, intact, contractile adult worms OVERNIGHT at 4 °C in 50 volumes 1% (v/v) paraformaldehyde in D-PBS (pH 7.5) [48], OR incubate the worms in 1.0 mM arachidonic acid (nSMase activator) in FCS-free RPMI medium for 1 h, fix in 0.1% (v/v) formaldehyde for 15 min [38], then wash three times in RPMI medium.

3.4.2 Tegumental Surface Membrane Proteins

1. Incubate free-living parasite stages, lung-stage schistosomula, and juvenile or adult worms in ice-cold acetone for 5 (larval stages) or 30 (juvenile and adult worm) min, followed by 0.2% (v/v) Triton X-100 for 15 min, for disintegration and full permeabilization of the double lipid bilayer covering, respectively [5, 25].
2. Wash the parasites three times in FCS-free medium; then fix with 1% (v/v) paraformaldehyde for 15 min.

3.4.3 Internal Proteins

1. Fix larval, juvenile, and adult worms in 4% (w/v) paraformaldehyde overnight at room temperature. Wash and store in D-PBS at 4 °C.
2. Prepare paraffin or cryostat 5 μm sections of the parasites depending on the sensitivity of the target protein to heat, following the procedures detailed previously [49, 50]. Of note, worms destined to cryostat sectioning must be infiltrated with a cryoprotectant before freezing and, accordingly, should be incubated overnight in 30% (w/v) sucrose in D-PBS. The worms should sink at the bottom of the tube when totally cryopreserved.
3. Deparaffinize paraffin sections in xylol; incubate successively in 96%, 80%, and 50% (v/v) ethanol in deionized water and then D-PBS.
4. Overlay the cryostat and paraffin sections with 0.2% (v/v) Triton X-100 in RPMI medium for 15 min for full cell permeabilization [49, 50].
5. Wash the sections three times in RPMI medium.

3.4.4 Immunofluorescence Assay

1. Equal number of parasites is distributed in wells of 96-well flat-bottomed assay plates (Costar), and parasite sections are incubated for 1 h at room temperature with RPMI medium supplemented with 5% (v/v) FCS to block nonspecific labeling sites and then control, and anti-target molecule serum is added to a predetermined concentration or 1:100-dilution. Incubation is overnight at room temperature in a humidified chamber.
2. Parasites and parasite sections are washed four times in RPMI medium 5% (v/v) FCS, incubated with 1:100-diluted anti-mouse IgG (Fab-specific) fluorescein isothiocyanate (FITC)-labeled goat immunoglobulin $F(ab')_2$ fragment of affinity isolated antibody, adsorbed with bovine, horse, and human serum proteins (Sigma), washed four times in RPMI medium, and inspected by alternate light and UV microscopy.
3. Neglect whole-mount immunofluorescence labeling of larvae appearing granular, opaque, distorted, and entirely fluorescent (Figs. 1 and 2, upper panel). True positivity is indicated by a

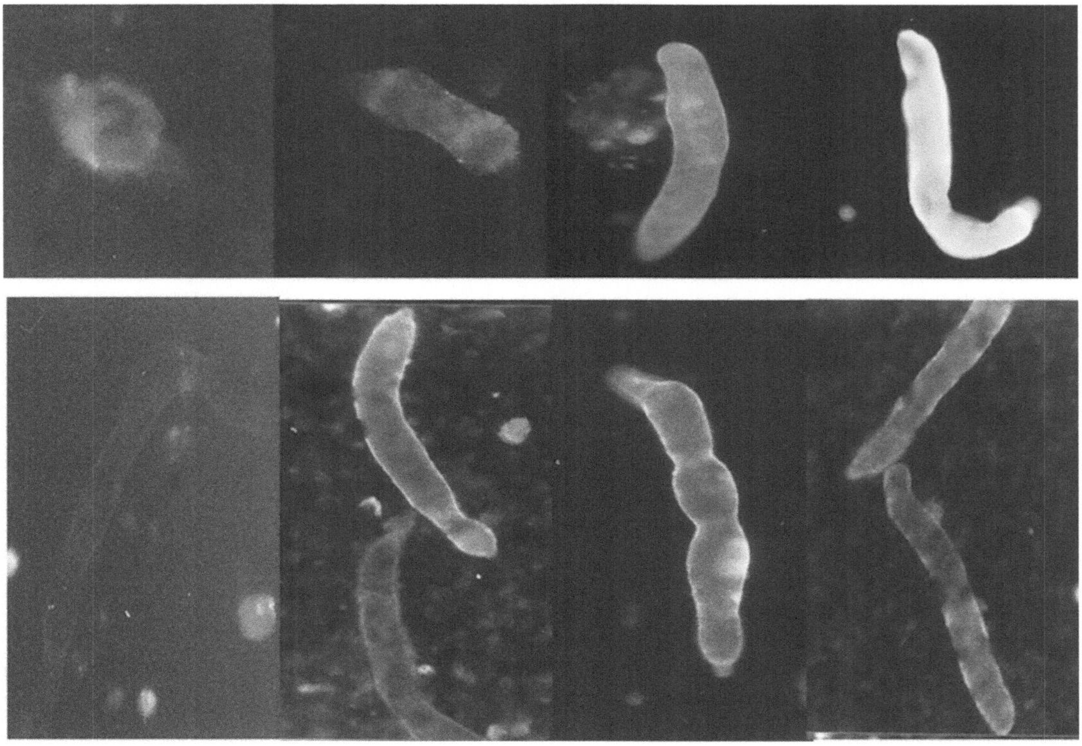

Fig. 2 Typical immunofluorescent localization of proteins at the apical surface of lung-stage schistosomula extracted from lung pieces at pH 7.6 and fixed with 0.1% formaldehyde. Larvae appearing granular, opaque, distorted, and entirely weakly or strongly fluorescent (upper panel) should be discarded from evaluation. A representative negative schistosomulum is shown in the left of lower panel, while the three other figures of the lower panel show different levels of continuous or speckled positive reactivity

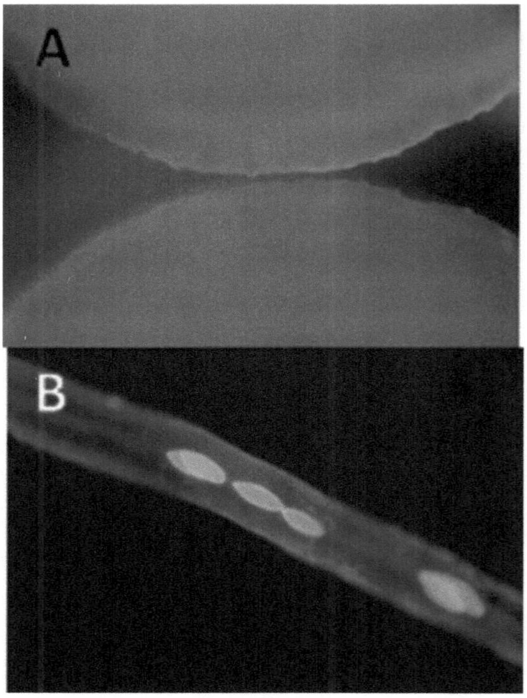

Fig. 3 Immunofluorescent localization of proteins at the surface of adult worms incubated with 1 mM arachidonic acid and fixed with 0.1% formaldehyde. (**a**) Shows positive and negative reactivity in male worms and (**b**) surface membrane IF reactivity and egg autofluorescence in female worm

continuous or speckled fluorescence at the worm rim (Fig. 2, lower panel; Fig. 3a) [15, 24, 43, 44].

4. Beware of autofluorescence of the eggshell within adult female worms (Fig. 3b).
5. Calculate percentage of truly positive worms incubated with control and test serum via alternate light and UV microscopy.
6. Confirm protein localization in sections using appropriate control sera.

4 Notes

1. Distribute stock penicillin-streptomycin, sodium pyruvate, L-glutamine, and FCS, preferably in 5 mL aliquots in sterile tubes before freezing at −20 °C.
2. Sodium azide is a powerful, acutely toxic metabolic inhibitor and must be used with caution. Due to its explosion hazard, NaN_3 should never be discarded in metal vessels or down the drain. All waste must be collected in a sealable compatible container and disposed as hazardous waste as per local,

national, and institutional hazardous waste guidelines (e.g., *https://www.chemistry.mcmaster.ca/mcnulty/SodiumAzide-SOP.pdf*).

3. Never add amphotericin B (Fungizone) or serotonin in culture medium destined to IF studies or any other substance that may intercalate in surface membranes and impair their integrity and function and autofluorescence or contain fluorescent impurities.
4. Highly specific antibodies, preferably affinity purified, with minimal or preferably no cross-reactivity to non-target parasite proteins are extremely important for artefact-free IF protein labeling.
5. For protein targets co-localization, the two different specific antibodies may be labeled with fluorescein isothiocyanate (FITC) or tetramethylrhodamine isothiocyanate (TRITC), respectively, and used simultaneously. FITC has excitation and emission spectrum peak wavelengths of approximately 495 nm/519 nm, giving it a green color, while TRITC has excitation and emission spectrum peak wavelengths of approximately 557 nm/576 nm, giving it an orange-red color.
6. To avoid artefacts, larvae and worms should not suffer at any time the slightest exposure to solvents, detergents, chaotropic agents, high or low osmotic pressure or temperature, shearing, or vortexing.
7. Numerous *S. mansoni* proteins believed to be located at the apical lipid layer of the worm were found to not be a true surface molecule and are located in the tegument membrane compartments within the tegumental matrix [31, 47].

References

1. Grossman AI, Short RB, Cain GD (1981) Karyotype evolution and sex chromosome differentiation in Schistosomes (Trematoda, Schistosomatidae). Chromosoma 84:413–430
2. Walker AJ (2011) Insights into the functional biology of schistosomes. Parasit Vectors 4:203
3. Pan SC (1996) *Schistosoma mansoni*: the ultrastructure of larval morphogenesis in *Biomphalaria glabrata* and of associated host-parasite interactions. Jpn J Med Sci Biol 49:129–149
4. Dorsey CH, Cousin CE, Lewis FA, Stirewalt MA (2002) Ultrastructure of the *Schistosoma mansoni* cercaria. Micron 33:279–323. Review
5. Collins JJ 3rd, King RS, Cogswell A, Williams DL, Newmark PA (2011) An atlas for *Schistosoma mansoni* organs and life-cycle stages using cell type-specific markers and confocal microscopy. PLoS Negl Trop Dis 5:e1009
6. Hockley DJ, McLaren DJ (1973) *Schistosoma mansoni*: changes in the outer membrane of the tegument during development from cercaria to adult worm. Int J Parasitol 3:13–25
7. Mclaren DJ, Hockley DJ (1977) Blood flukes have a double outer membrane. Nature 269:147–149
8. Foley M, Kusel JR, Garland PB (1988) Changes in the organization of the surface membrane upon transformation of cercariae to schistosomula of the helminth parasite *Schistosoma mansoni*. Parasitology 96:85–97
9. Gobert GN, Chai M, McManus DP (2007) Biology of the schistosome lung-stage schistosomulum. Parasitology 134:453–460
10. Dean DA (1977) Decreased binding of cytotoxic antibody by developing *Schistosoma mansoni*. Evidence for a surface change

independent of host antigen adsorption and membrane turnover. J Parasitol 63:418–426
11. Pearce EJ, Basch PF, Sher A (1986) Evidence that the reduced surface antigenicity of developing *Schistosoma mansoni* schistosomula is due to antigen shedding rather than host molecule acquisition. Parasite Immunol 8:79–94
12. Pearce EJ, James SL (1986) Post lung-stage schistosomula of *Schistosoma mansoni* exhibit transient susceptibility to macrophage-mediated cytotoxicity in vitro that may relate to late phase killing in vivo. Parasite Immunol 8:513–527
13. Dalton JP, Lewis SA, Aronstein WS, Strand M (1987) *Schistosoma mansoni*: immunogenic glycoproteins of the cercarial glycocalyx. Exp Parasitol 63:215–226
14. Dalton JP, Tom TD, Strand M (1987) Cloning of a cDNA encoding a surface antigen of *Schistosoma mansoni* schistosomula recognized by sera of vaccinated mice. Proc Natl Acad Sci U S A 84:4268–4272
15. Tallima H, El Ridi R (2008) *Schistosoma mansoni* glyceraldehyde 3-phosphate dehydrogenase is a lung-stage schistosomula surface membrane antigen. Folia Parasitol (Praha) 55:180–186
16. McWilliam HE, Driguez P, Piedrafita D, Maupin KA, Haab BB, McManus DP, Meeusen EN (2013) The developing schistosome worms elicit distinct immune responses in different tissue regions. Immunol Cell Biol 91:477–485
17. Elbaz T, Esmat G (2013) Hepatic and intestinal schistosomiasis: review. J Adv Res 4:445–452
18. Barsoum RS, Esmat G, El-Baz T (2013) Human schistosomiasis: clinical perspective: review. J Adv Res 4:433–444
19. Adenowo AF, Oyinloye BE, Ogunyinka BI, Kappo AP (2015) Impact of human schistosomiasis in sub-Saharan Africa. Braz J Infect Dis 19:196–205
20. Zoni AC, Catalá L, Ault SK (2016) Schistosomiasis prevalence and intensity of infection in Latin America and the Caribbean countries, 1942–2014: a systematic review in the context of a regional elimination goal. PLoS Negl Trop Dis 10:e0004493
21. World Health Organization (2018) Schistosomiasis. http://www.who.int/news-room/fact-sheets/detail/schistosomiasis
22. Coulibaly JT, Ouattara M, Barda B, Utzinger J, N'Goran EK, Keiser J (2018) A rapid appraisal of factors influencing praziquantel treatment compliance in two communities endemic for schistosomiasis in Côte d'Ivoire. Trop Med Infect Dis 3:69
23. Adriko M, Faust CL, Carruthers LV, Moses A, Tukahebwa EM, Lamberton PHL (2018) Low praziquantel treatment coverage for *Schistosoma mansoni* in Mayuge district, Uganda, due to the absence of treatment opportunities, rather than systematic non-compliance. Trop Med Infect Dis 3:111
24. El Ridi R, Tallima H (2006) Equilibrium in lung schistosomula sphingomyelin breakdown and biosynthesis allows very small molecules, but not antibody, to access proteins at the host-parasite interface. J Parasitol 92:730–737
25. Tallima H, Al-Halbosiy MF, El Ridi R (2011) Enzymatic activity and immunolocalization of *Schistosoma mansoni* and *Schistosoma haematobium* neutral sphingomyelinase. Mol Biochem Parasitol 178:23–28
26. Migliardo F, Tallima H, El Ridi R (2014) Is there a sphingomyelin-based hydrogen bond barrier at the mammalian host-schistosome parasite interface? Cell Biochem Biophys 68:359–367
27. Migliardo F, Tallima H, El Ridi R (2014) Rigidity and resistance of larval- and adult schistosomes-medium interface. Biochem Biophys Res Commun 446:255–260
28. De Jesus JW, DA Cunha Melo JR, Baba EH, Coelho PM, Kusel JR (2015) The skin migratory stage of the schistosomulum of *Schistosoma mansoni* has a surface showing greater permeability and activity in membrane internalisation than other forms of skin or mechanical schistosomula. Parasitology 142:1143–1151
29. Slotte JP (2016) The importance of hydrogen bonding in sphingomyelin's membrane interactions with co-lipids. Biochim Biophys Acta 1858:304–310
30. Mitsui Y, Kato K (2018) Application of non-fluorescent dyes to assess the antischistosomal effect of antimalarial drugs on *Schistosoma mansoni* adult worms. Jpn J Infect Dis 71:382–387
31. Tran MH, Pearson MS, Bethony JM, Smyth DJ, Jones MK, Duke M, Don TA, McManus DP, Correa-Oliveira R, Loukas A (2006) Tetraspanins on the surface of *Schistosoma mansoni* are protective antigens against schistosomiasis. Nat Med 12:835–840
32. Molehin AJ, Sennoune SR, Zhang W, Rojo JU, Siddiqui AJ, Herrera KA et al (2017) Cross-species prophylactic efficacy of Sm-p80-based vaccine and intracellular localization of Sm-p80/Sm-p80 ortholog proteins during development in *Schistosoma mansoni*, *Schistosoma japonicum*, and *Schistosoma haematobium*. Parasitol Res 116:3175–3188

33. Wang Q, Da'dara AA, Skelly PJ (2017) The human blood parasite *Schistosoma mansoni* expresses extracellular tegumental calpains that cleave the blood clotting protein fibronectin. Sci Rep 7:12912
34. McKenzie M, Kirk RS, Walker AJ (2018) Glucose uptake in the human pathogen *Schistosoma mansoni* is regulated through Akt/protein kinase B signaling. J Infect Dis 218:152–164
35. Doenhoff MJ, Modha J, Lambertucci JR (1988) Anti-schistosome chemotherapy enhanced by antibodies specific for a parasite esterase. Immunology 65:507–510
36. Fallon PG, Cooper RO, Probert AJ, Doenhoff MJ (1992) Immune-dependent chemotherapy of schistosomiasis. Parasitology 105:S41–S48. Review
37. Fallon PG, Doenhoff MJ (1995) Active immunization of mice with *Schistosoma mansoni* worm membrane antigens enhances efficacy of praziquantel. Parasite Immunol 17:261–268
38. El Ridi R, Tallima H, Salah M, Aboueldahab M, Fahmy OM, Al-Halbosiy MF, Mahmoud SS (2012) Efficacy and mechanism of action of arachidonic acid in the treatment of hamsters infected with *Schistosoma mansoni* or *Schistosoma haematobium*. Int J Antimicrob Agents 39:232–239
39. Reimers N, Homann A, Höschler B, Langhans K, Wilson RA, Pierrot C, Khalife J, Grevelding CG, Chalmers IW, Yazdanbakhsh M, Hoffmann KF, Hokke CH, Haas H, Schramm G (2015) Drug-induced exposure of *Schistosoma mansoni* antigens SmCD59a and SmKK7. PLoS Negl Trop Dis 9:e0003593
40. Smithers SR, Terry RJ (1965) The infection of laboratory hosts with cercariae of *Schistosoma mansoni* and the recovery of the adult worms. Parasitology 55:695–700
41. Tucker MS, Karunaratne LB, Lewis FA, Freitas TC, Liang YS (2013) Schistosomiasis. Curr Protoc Immunol 103:19.1.1–19.1.58
42. Konstantinou GN (2017) Enzyme-linked immunosorbent assay (ELISA). Methods Mol Biol 1592:79–94
43. El Ridi R, Tallima H, Mohamed SH, Montash M (2004) Depletion of *Schistosoma mansoni* lung-stage schistosomula cholesterol by methyl-beta-cyclodextrin dramatically increases specific antibody binding to surface membrane antigens. J Parasitol 90:727–732
44. Tallima H, Salah M, El-Ridi R (2005) In vitro and in vivo effects of unsaturated fatty acids on *Schistosoma mansoni* and *S. haematobium* lung-stage larvae. J Parasitol 91:1094–1102
45. Hackett F (1993) The culture of *Schistosoma mansoni* and production of life cycle stages. Methods Mol Biol 21:89–99
46. Mann VH, Morales ME, Rinaldi G, Brindley PJ (2010) Culture for genetic manipulation of developmental stages of *Schistosoma mansoni*. Parasitology 137:451–462
47. Schulte L, Lovas E, Green K, Mulvenna J, Gobert GN, Morgan G, Jones MK (2013) Tetraspanin-2 localisation in high pressure frozen and freeze-substituted *Schistosoma mansoni* adult males reveals its distribution in membranes of tegumentary vesicles. Int J Parasitol 43:785–793
48. Sepulveda J, Tremblay JM, DeGnore JP, Skelly PJ, Shoemaker CB (2010) *Schistosoma mansoni* host-exposed surface antigens characterized by sera and recombinant antibodies from schistosomiasis-resistant rats. Int J Parasitol 40:1407–1417
49. Vitha S, Osteryoung KW (2011) Immunofluorescence microscopy for localization of Arabidopsis chloroplast proteins. Methods Mol Biol 774:33–58
50. Alkema M, Goumans MJ, Kruithof BPT (2019) Immunofluorescent visualization of BMP signaling activation on paraffin-embedded tissue sections. Methods Mol Biol 1891:191–200

Chapter 4

Sensitive Fluorescence In Situ Hybridization on Semithin Sections of Adult *Schistosoma mansoni* Using DIG-Labeled RNA Probes

Lenka Ulrychová, Martin Horn, and Jan Dvořák

Abstract

In situ hybridization is a tool for evaluation of gene expression within tissues or single cells. This protocol describes optimized sensitive fluorescence detection of gene transcripts (mRNAs) in semithin sections of *Schistosoma mansoni* adult worms using specifically designed and labeled RNA probes. Due to improved methodologies in tissue preservation, sectioning, amplification of fluorescent signal, and prehybridization tissue treatment, it is possible to detect transcripts in the fine structures of schistosomes. The protocol is sensitive enough to detect very low abundance targets. This procedure is optimized for tissues derived from *S. mansoni* adult worms; however, it can be successfully applied to other trematode species.

Key words *Schistosoma mansoni*, Adult worm, Tissue section, In situ hybridization, RNA probe, Digoxigenin labeling, Fluorescence RNA detection

1 Introduction

In situ hybridization (ISH) is a technique that enables the localization and visualization of specific nucleic acids in the cells, tissues, or whole organisms. It is based on annealing of labelled RNA probes to complementary sequences in the sample. ISH was introduced more than 50 years ago [1]; however, this technique underwent enormous development during the time. It is mostly because different types of tissue and/or cells require optimization not only for the hybridization itself but also for subsequent visualization. Critical steps include the tissue pretreatment that is highly specific for different tissue types. Therefore, the effective in situ hybridization protocol must be always optimized.

Several ISH protocols have been developed for schistosome tissues—whole mount in situ hybridization (WISH), a protocol that enables to detect gene transcripts within the whole *S. mansoni* adults [2] and ISH performed on their semithin sections

David J. Timson (ed.), *Schistosoma mansoni: Methods and Protocols*, Methods in Molecular Biology, vol. 2151, https://doi.org/10.1007/978-1-0716-0635-3_4,

[3]. In both cases, the visualization of transcripts was performed chromogenically. Chromogens are widely used due to convenient and nontoxic properties. Despite the good overview of mRNA localization within the organism, they do not show satisfactory details within the tissue structures.

In this protocol, we describe fluorescence RNA ISH (FISH) in semithin sections of adult *Schistosoma mansoni*. We employed digoxigenin-labeled (DIG) probes, which are detected using anti-DIG antibody connected with the fluorescent tyramide signal amplification (TSA) that amplifies signal and allows probe visualization. Our FISH improves original methodology based on previously established protocols [2, 3]: we modified several steps including *S. mansoni* adult tissue fixation and prehybridization treatment. Moreover, this protocol is universal for trematode species as shown for *Eudiplozoon nipponicum* [4].

2 Materials

To avoid RNase contamination and RNA degradation, work under RNase-free conditions during the whole procedure. It is also recommended to autoclave glassware and use sterilized RNase-free plastic tubes and tips. Prepare all buffers using diethyl pyrocarbonate-treated water (DEPC-H_2O), and autoclave them. To prepare DEPC-H_2O, mix 1 mL of DEPC with 1 L of Milli-Q water, cover, and mix it in the fume hood overnight. Autoclave twice to deactivate DEPC, and store in a dark place at 4 °C. The majority of buffers and solutions can be prepared in advance and stored for a month; those, which must be prepared fresh prior the use, are marked with asterisks (*).

2.1 Probe Generation

1. cDNA from adult *S. mansoni* prepared as described in Stefanic et al. [5].
2. Proofreading polymerase (e.g., *Pfu* DNA polymerase).
3. pGEM-T Easy Vector (Promega).
4. A pair of gene-specific primers designed for amplification of the defined size of the target sequence (500–1500 bases).
5. Equipment and reagents for DNA electrophoresis.
6. DNA gel extraction kit.
7. DIG RNA Labeling Kit (SP6/T7).
8. Restriction endonucleases for pGEM-T Easy Vector and their recommended buffers; see a vector map and check the sequence of your target to avoid unwanted internal restriction.
9. 0.2 M EDTA.
10. PCR clean-up kit.

11. LB/ampicillin plates: 1%(w/v) tryptone, 0.5%(w/v) yeast extract, 0.5%(w/v) NaCl. Adjust pH with 1 M NaOH to 7.5, add 1.5%(w/v) agar, and autoclave. Add ampicillin to a final concentration of 50 μg/mL before pouring plates.
12. LB/ampicillin medium: 1%(w/v) tryptone, 0.5%(w/v) yeast extract, 0.5%(w/v) NaCl. Adjust pH with 1 M NaOH to 7.5, and autoclave. Add ampicillin to a final concentration of 100 μg/mL before use.
13. Kit for plasmid DNA isolation.

2.2 Tissue Preparation for In Situ Hybridization

1. RPMI 160 medium or medium M-199.
2. *S. mansoni* adult worms recovered from mice hepatic portal system by transcardial perfusion with medium (*see* **Note 1**).
3. 10× PBS: 80 mM sodium phosphate, 20 mM potassium phosphate, pH 7.4, 1.37 M sodium chloride, 27 mM potassium chloride.
4. 1× PBS: dilute ten times 10× PBS in DEPC-H_2O and autoclave.
5. * 4%(v/v) formaldehyde in 1× PBS.
6. * Ethanol solutions (25%, 50%, 70%, 90%, 96% v/v): these solutions should be prepared on the day of use.
7. Chromotrope 2R.
8. X-tra Adhesive glass slides (Leica).
9. Cover glass.
10. Microtome.
11. * 0.2 M HCl; this solution should be prepared immediately before use.
12. Antigen retrieval buffer: 10 mM sodium citrate buffer, pH 6.0.
13. 0.2%(w/v) glycine.
14. * Cold 20%(v/v) acetic acid: cool down DEPC-H_2O on ice and add glacial acetic acid.
15. 20%(v/v) glycerol.
16. 20× SSC: 0.3 M sodium citrate, pH 7.0, 3 M sodium chloride.
17. * 2× SSC: prepare by dilution of 20× SSC; this solution should be prepared on the day of use.

2.3 RNA In Situ Hybridization

1. 20× SSC: 0.3 M sodium citrate, pH 7.0, 3 M sodium chloride.
2. 10× PBS stock solution: 80 mM sodium phosphate, 20 mM potassium phosphate, pH 7.4, 1.37 M sodium chloride, 27 mM potassium chloride.
3. Torula yeast RNA (Sigma-Aldrich).

4. Deionized formamide: mix 100 g of formamide with 10 g of mixed bed resin and incubated overnight and rotate. Prepare aliquots of 1 mL and store in −80 °C. Use always new aliquot for preparation of hybridization buffer.
5. * Hybridization mix A: in a RNase-free tube, combine 625 μL of 20× SSC, 300 μL of 10× PBS, 2.5 mg of torula yeast RNA, and RNA probe. The final concentration of the probe in hybridization mixture must be always tested experimentally; for method described in this protocol, start with probe concentrations 0.625, 1.25, and 5 ng/mL. Fill with DEPC-H_2O to 2.5 mL. This solution must be prepared shortly before use.
6. * Hybridization mix B: combine 1.25 mL of deionize formamide, 250 μL of 10%(v/v) Tween20, and 250 mg of dextran sulfate. Fill with DEPC-H_2O to 2.5 mL. This solution must be prepared on the day of use.
7. 2× SSC/0.1%(v/v) Tween20: prepared by dilution of 20× SSC and 10%(v/v) Tween20 in DEPC-H_2O; this solution should be prepared on the day of use. Do not autoclave. Preheat the solution to the hybridization temperature before use.
8. * SSC buffers: prepare 2× SSC, 1× SSC, 0.5× SSC, and 0.1× SSC buffers by diluting 20× SSC in DEPC-H_2O; these solutions should be prepared on the day of use.
9. MAB buffer: 100 mM maleic acid, pH 7.5, 0.15 M NaCl.
10. * 2% or 4% blocking solutions: prepare 2% or 4%(v/v) heat-inactivated horse serum in MAB buffer; these solutions should be prepared immediately before use.
11. Humidity chamber.

2.4 Probe Detection

1. Anti-DIG-HRP antibodies: anti-digoxigenin antibody conjugated with the horseradish peroxidase (HRP).
2. TSA Plus Cyanine 5 System, for 50–150 Slides (Perkin Elmer).
3. TNT buffer: 100 mM Tris HCl, pH 7.5, 0.15 M NaCl, 0.05% (v/v) Tween20.
4. ProLong Diamond Antifade Mountant with DAPI (Thermo Fisher).

3 Methods

The RNA in situ hybridization procedure described here (Fig. 1) covers the following steps: (1) synthesis of DIG-labeled RNA probe, (2) fixation of adult schistosoma worms, (3) preparation of slides with schistosoma sections, (4) in situ hybridization procedure, and (5) fluorescent visualization and signal amplification of

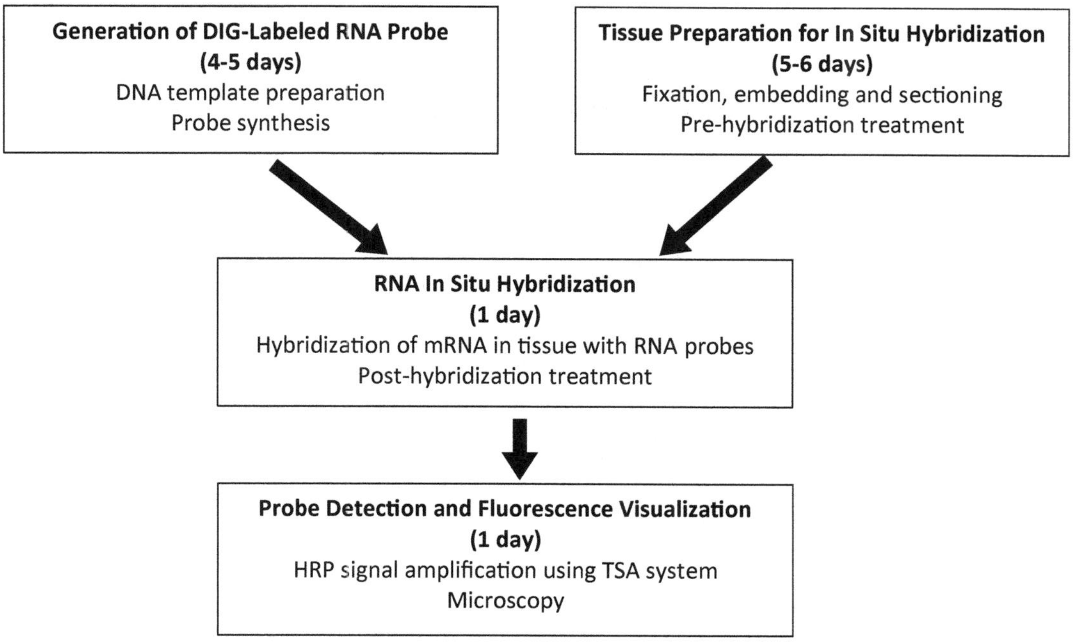

Fig. 1 Schematic overview of the protocol for fluorescence RNA in situ hybridization in *S. mansoni*

the probes hybridized to desired mRNA within the tissue. Figure 2 shows an example of the mRNA detection using protocol described here. It shows visualization of digestive protease *S. mansoni*cathepsin B1 (SmCB1) [6, 7]. If not otherwise noted, perform all steps at room temperature (*see* **Note 2**).

3.1 Generation of DIG-Labeled RNA Probe

3.1.1 DNA Template Preparation

1. Design primer pair specific for the sequence of desired region (500 to 1500 bases) of the target.
2. Amplify the desired DNA template by the polymerase chain reaction (PCR) using *S. mansoni* adult cDNA, gene-specific primers, and a proofreading polymerase according to the manufacturer's protocols.
3. Separate resulting PCR product using an agarose gel, excise the band, extract, and purify the DNA using a commercial DNA gel-extraction kit according to the manufacturer's instructions.
4. Clone the DNA into the pGEM-T Easy Vector according to the manufacturer's protocol.
5. Transform competent *E. coli* cells by pGEM-T Easy vector containing the template, select-positive transformants on LB/ampicillin plates. Screen colonies by method of your choice for an insert.
6. Pick colonies from the transformation plate, and grow them separately in 5 mL of LB/ampicillin at 37 °C overnight.

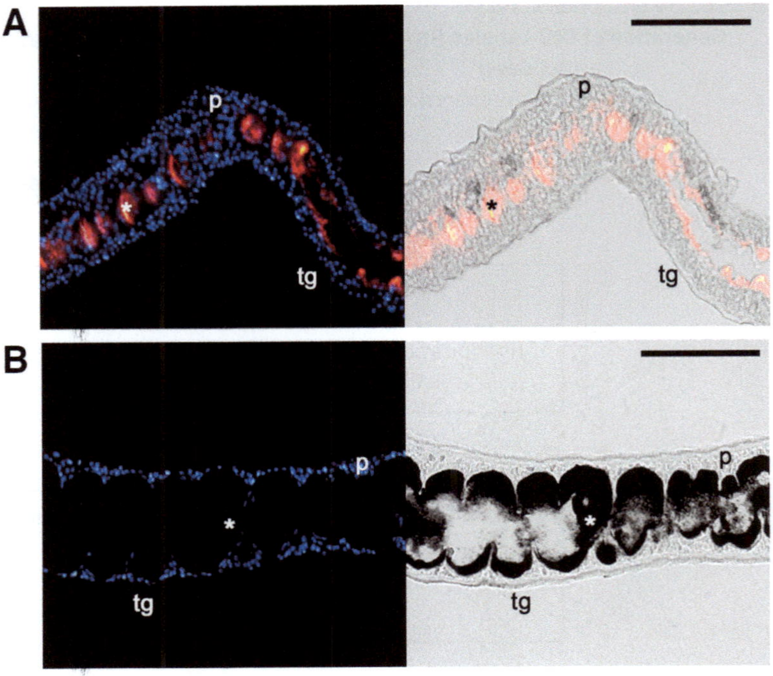

Fig. 2 RNA in situ hybridization of digestive protease *S. mansoni*cathepsin B1 (SmCB1) transcripts in adult females. Detection of SmCB1 transcripts was performed by RNA in situ hybridization using the protocol described here on semithin sections (6 μm) of *S. mansoni* adult females. (**a**) Detection of SmCB1 mRNA (red) using the anti-sense RNA probe showing transcripts in the gut. (**b**) RNA in situ hybridization procedure performed without the probe. Result shows no unspecific bindings and clear background in the tissue. DAPI was used to label nuclei (blue). *p* parenchyma, *tg*tegument, * gut. The scale bare is 100 μm

7. Harvest the cells and isolate plasmid using commercial kit (according to the manufacturer's instructions).
8. Verify the insert in the p-GEM plasmid by DNA sequencing and determine its orientation. The target gene sequence can be inserted during cloning into the pGEM-T Easy plasmid in both direction—under the T7 or SP6 promoter (see pGEM-T Easy manufacture's protocol).

3.1.2 Probe Synthesis

To generate DIG-labeled probes, employ standard in vitro transcription using DIG RNA labeling kit (SP6/T7) that produces DIG-labeled single-stranded RNA probes. Either SP6 or T7 RNA polymerase transcribes these probes in vitro from the template DNA in the presence of DIG-UTP. It is crucial to keep strict RNA-free conditions during the whole procedure.

1. Linearize the vector (the pGEM-T Easy containing DNA template from **step 8** Subheading 3.1.1) with restriction endonuclease according to the pGEM-T Easy map, DNA sequence, and its orientation in the plasmid.

2. Purify the linearized vector using a PCR cleaning kit according to the manufacturer protocol. Elute DNA with DEPC-H_2O.
3. Add 1 μg of the linearized vector to the RNase-free vial, place it on ice, and add 2 μL of 10x NTP labeling mixture (containing DIG-UTP), 2 μL of transcription buffer, 1 μL of protector RNase inhibitor, and T7 or SP6 RNA polymerase (depending on the orientation of the DNA template in the pGEM-T Easy vector).
4. Incubate the reaction at 37 °C for 2 h.
5. Add 2 μL of DNase I to remove DNA template and incubate at 37 °C for 15 min.
6. Terminate the reaction by adding 2 μL 0.2 M EDTA.
7. Analyze the transcripts by formaldehyde agarose gel electrophoresis.

3.2 Tissue Preparation for In Situ Hybridization

3.2.1 Formaldehyde Fixation, Paraffin Embedding, and Sectioning

The following incubation steps must be performed under RNA-free conditions and, if possible, with gentle shaking.

1. After perfusion (*see* **Note 1**), transfer adult schistosoma worm pairs to a Petri dish, and chill them on ice.
2. Separate males and females by gentle prodding with a small brush.
3. Wash separated worms three times with 1× PBS, and place them into two different containers.
4. Meanwhile, cover the beaker with 4% formaldehyde (the fixation solution), and heat the solution to the boiling point while stirring. The boiling point is reached when the fixation solution starts to condensate on the lid.
5. Transfer worms into the preheated fixation solution, leave them to cool down to 25 °C, and incubate for 90 min.
6. Dehydrate the worms by incubating in an ethanol series of increasing ethanol concentration (25%, 50%, and 70% v/v ethanol) for 5 min each.
7. Incubate the worms in 90%(v/v) ethanol containing Chromotrope 2R for 5 min.
8. Continue worm dehydration by incubation in 96%(v/v) ethanol and twice in 100% ethanol for 5 min.
9. Incubate the worms in methyl benzoate (*see* **Note 3**), males for 45 min and females for 20 min.
10. Wash the worms twice with benzene for 5 min each.
11. Transfer the worms into hot paraffin (60–65 °C), and incubate at 60 °C for 1 h, repeat twice.
12. Embed the worms in paraffin blocks and let them harden.

13. Prepare semithin sections (6–8 μm) of the tissue on microtome, and apply them onto X-tra adhesive glass slides.
14. Carefully select slides with high-quality tissue sections under a light microscope (*see* **Note 4**).

3.2.2 Prehybridization Treatment

1. Deparaffinize sections with fixed tissue (from **step 13** in Subheading 3.2.1) by two 5 min washes in xylene followed by two 5 min washes in 100% ethanol.
2. Rehydrate the section by incubation for 5 min in an ethanol series of decreasing ethanol concentration (96%, 90%, 70%, 50%, 25% v/v).
3. Wash the sections in DEPC-H_2O for 5 min.
4. Incubate the sections in 0.2 M HCl for 20 min (*see* **Note 5**).
5. Incubate the sections in the antigen retrieval buffer preheated in the boiling water bath for 15 min (*see* **Note 6**).
6. Let the sections cool down for 30 min at 25 °C.
7. Incubate the sections in 0.2%(w/v) glycine for 5 min.
8. Rinse the sections in cold 20%(v/v) acetic acid for 15 s (*see* **Note 7**).
9. Wash the sections in 1× PBS for 5 min.
10. Incubate the section in 20%(v/v) glycerol for 15 min.
11. Shortly rinse the sections with 2× SSC.
12. Heat the slides with sections on the metal plate preheated to 70 °C for 10 min.

3.3 RNA In Situ Hybridization

3.3.1 Hybridization

1. Heat 100 μL of the hybridization solution A to 70 °C to denature RNA structures, then shortly cool down on ice, and immediately mix with 160 μL of the hybridization solution B to generate the hybridization buffer (*see* **Notes 8** and **9**).
2. Very quickly take the slides from the metal desk preheated to 70 °C (from **step 12** in Subheading 3.2.2), shortly cool down on a precooled metal desk (−80 °C), and immediately apply 150 μL of the hybridization on them.
3. Cover slides with a cover slip and hybridize the tissue with probes in the humidity chamber at 42 °C for 16 h (for probes of length 500–1500 nucleotides).

3.3.2 Posthybridization Treatment

It is recommended to perform the following steps with gentle agitation. Be very accurate with the washing times (*see* **Note 10**).

1. After hybridization (**step 3** in Subheading 3.3.1), wash the slides for 15 min in 2× SSC/0.1%(v/v) Tween20 prewarmed to hybridization temperature (42 °C).

2. Wash the slides in 1× SSC buffer for 15 min, followed by 15 min washes in 0.5× SSC buffer and 0.1× SSC buffer.
3. Incubate the slides in MAB buffer for 5 min.
4. Incubate the slides in 4% blocking solution for 30 min.

3.4 Probe Detection and Fluorescence Visualization

The target mRNA hybridized with the DIG-labeled RNA probe (Subheading 3.3) is recognized by anti-DIG antibodies conjugated to HRP. Activity of HRP is visualized fluorescently using TSA Plus Cyanine 5 System (TSA). HRP catalyzes conversion of fluorescently labeled tyramide (a component of TSA system) to its active form that covalently react with the tyrosine residues in the tissue laying in very close proximity of HRP. This leads in significantly increased fluorescent signal compared to the conventional immunodetection. TSA assay was performed according to the manufacturer protocol. However, dilution of TSA components and incubation times must be experimentally tested for a particular tissue. Here, we describe the steps suitable for semithin sections of *S. mansoni* adults. Generally, it is recommended to perform the following steps in the dark.

1. Apply anti-DIG-HRP antibodies diluted 1:500 in 2% blocking solution on the slides (from **step 4** in Subheading 3.3.2), cover with a cover slip, and incubate them in a humidity chamber at 37 °C for 2 h.
2. Wash the slides two times in MAB buffer for 10 min.
3. Wash the slides three times in TNT buffer for 5 min each.
4. Immediately before use, prepare TSA Working Solution (*see* **Note 11**): dilute TSA Plus Stock Solution 1:50 with 1× Amplification Diluent.
5. Apply 150 μL of the TSA Plus Working Solution on the slide, cover with coverslip, and incubate for 5 min.
6. Wash the slides three times in TNT buffer for 5 min.
7. Wash the slides one time in DEPC-H_2O for 5 min.
8. Mount the sections in ProLong Diamond Antifade Mountant (with DAPI).
9. Subject the samples to the appropriate fluorescent microscopy technique.

4 Notes

1. *S. mansoni* is routinely maintained in the laboratory by cycling between the intermediate snail host (*Biomphalaria glabrata*) and definitive host ICR (CD-1) mice as described previously

[8, 9]. Adult worms are recovered from mice hepatic portal system by transcardial perfusion with RPMI 160 medium 6 weeks post infection.

2. Any failure in the protocol describing tissue fixation, prehybridization, and hybridization treatment such as change in incubation times or reagent concentrations may impair signal amplification and/or increase the background signal.
3. Methyl benzoate and benzene are more gentle and rapid clearing agents compared to xylene. Methyl benzoate does not harden the tissue, and both chemicals are penetrating the tissue significantly faster than xylene. Work in a fume hood all time.
4. Store slides containing paraffin-embedded sections up to 6 months in −80 °C.
5. This step helps in tissue clearing and improves the signal (e.g., reduce nonspecific binding). 0.2 M HCl irreversibly quenches the endogenous peroxidases which produce significant background when not inhibited properly [10].
6. Do not leave *S. mansoni* semithin sections in a boiling water bath with the antigen retrieval buffer for more than 15 min. Prolonged incubation time increases the risk of tissue loss from the slide, but do not increase the signal. This step is highly recommended especially for paraffin-embedded tissues.
7. This step improves cell permeabilization allowing access of the probe and the antibodies. It also reduces nonspecific binding of probes, minimizing the background [11].
8. It is important to preheat the hybridization mix A to denature RNA probes to 70 °C. Do not forget to chill the preheated hybridization mix A on ice to prevent RNA reannealing. On the other hand, never preheat the hybridization mix B or the hybridization buffer (hybridization mix A + B) to 70 °C because of chemical instability. The final composition of hybridization buffer should be 5× SSC, 1×PBS, 0.1% (w/v) torula yeast RNA, 50% (v/v) deionized formamide, 1% (v/v) Tween20, and 10% (w/v) dextran sulfate (Mw 4000).
9. It is recommended to run a negative control—sections that undergo the whole procedures but in the absence of the RNA probe.
10. Strictly follow the incubation time and temperature of stringency washes. These steps wash out the nonspecific or repetitive bindings.
11. TSA Working Solution must be always prepared freshly, and the incubation should be performed in the dark. Always optimize the incubation time for your sample. Dilution of TSA Working Solution and incubation time must be always tested experimentally for an individual probe and tissue.

Acknowledgments

This work was supported by the grant NV18-05-00345 from the Ministry of Health of the Czech Republic; grant 18-14167S from the Czech Science Foundation; Operational Programme Research, Development, and Education; the Call International Mobility of Researchers—MSCA—IF CZ.02.2.69/0.0/0.0/17_050/0008014; and the institutional project RVO 61388963.

References

1. Gall JG, Pardue ML (1969) Formation and detection of RNA-DNA hybrid molecules in cytological preparations. Proc Natl Acad Sci U S A 63(2):378–383
2. Cogswell AA, Collins JJ 3rd, Newmark PA, Williams DL (2011) Whole mount in situ hybridization methodology for *Schistosoma mansoni*. Mol Biochem Parasitol 178 (1-2):46–50. https://doi.org/10.1016/j.molbiopara.2011.03.001
3. Quack T, Knobloch J, Beckmann S, Vicogne J, Dissous C, Grevelding CG (2009) The formin-homology protein SmDia interacts with the Src kinase SmTK and the GTPase SmRho1 in the gonads of *Schistosoma mansoni*. PLoS One 4 (9):e6998. https://doi.org/10.1371/journal.pone.0006998
4. Jedlickova L, Dvorakova H, Dvorak J, Kasny M, Ulrychova L, Vorel J, Zarsky V, Mikes L (2018) Cysteine peptidases of Eudiplozoon nipponicum: a broad repertoire of structurally assorted cathepsins L in contrast to the scarcity of cathepsins B in an invasive species of haematophagous monogenean of common carp. Parasit Vectors 11(1):142. https://doi.org/10.1186/s13071-018-2666-2
5. Stefanic S, Dvorak J, Horn M, Braschi S, Sojka D, Ruelas DS, Suzuki B, Lim KC, Hopkins SD, McKerrow JH, Caffrey CR (2010) RNA interference in *Schistosoma mansoni* Schistosomula: selectivity, sensitivity and operation for larger-scale screening. PLoS Negl Trop Dis 4(10):e850. https://doi.org/10.1371/journal.pntd.0000850
6. Sajid M, McKerrow JH, Hansell E, Mathieu MA, Lucas KD, Hsieh I, Greenbaum D, Bogyo M, Salter JP, Lim KC, Franklin C, Kim JH, Caffrey CR (2003) Functional expression and characterization of *Schistosoma mansoni* cathepsin B and its trans-activation by an endogenous asparaginyl endopeptidase. Mol Biochem Parasitol 131(1):65–75
7. Jilkova A, Horn M, Rezacova P, Maresova L, Fajtova P, Brynda J, Vondrasek J, McKerrow JH, Caffrey CR, Mares M (2014) Activation route of the *Schistosoma mansoni* Cathepsin B1 drug target: structural map with a glycosaminoglycan switch. Structure 22 (12):1786–1798. https://doi.org/10.1016/j.str.2014.09.015
8. Leontovyc A, Ulrychova L, O'Donoghue AJ, Vondrasek J, Maresova L, Hubalek M, Fajtova P, Chanova M, Jiang Z, Craik CS, Caffrey CR, Mares M, Dvorak J, Horn M (2018) SmSP2: a serine protease secreted by the blood fluke pathogen *Schistosoma mansoni* with anti-hemostatic properties. PLoS Negl Trop Dis 12 (4):e0006446. https://doi.org/10.1371/journal.pntd.0006446
9. Dvorak J, Fajtova P, Ulrychova L, Leontovyc A, Rojo-Arreola L, Suzuki BM, Horn M, Mares M, Craik CS, Caffrey CR, O'Donoghue AJ (2016) Excretion/secretion products from *Schistosoma mansoni* adults, eggs and schistosomula have unique peptidase specificity profiles. Biochimie 122:99–109. https://doi.org/10.1016/j.biochi.2015.09.025
10. Liu G, Amin S, Okuhama NN, Liao G, Mingle LA (2006) A quantitative evaluation of peroxidase inhibitors for tyramide signal amplification mediated cytochemistry and histochemistry. Histochem Cell Biol 126 (2):283–291. https://doi.org/10.1007/s00418-006-0161-x
11. Hyytinen E, Visakorpi T, Kallioniemi A, Kallioniemi OP, Isola JJ (1994) Improved technique for analysis of formalin-fixed, paraffin-embedded tumors by fluorescence in situ hybridization. Cytometry 16(2):93–99. https://doi.org/10.1002/cyto.990160202

Chapter 5

Collection of Excretory/Secretory Products from Individual Developmental Stages of the Blood Fluke *Schistosoma mansoni*

Adrian Leontovyč, Lenka Ulrychová, Martin Horn, and Jan Dvořák

Abstract

Individual developmental stages of blood fluke *Schistosoma mansoni* excrete or secrete a different set of molecules. Here we describe optimized protocols for collection of excretory/secretory products (E/S products) from cercariae, schistosomula, adult worms, and eggs. These E/S products are essential for successful parasitism functioning at the host-parasite interface, enabling invasion into the host and contributing to the survival of the parasite by modulation of host physiology and immune responses. Collection of sufficient amounts of E/S products is required for detailed research of these processes.

Key words *Schistosoma mansoni*, Developmental stages, Excretory/secretory products

1 Introduction

Blood flukes of the genus *Schistosoma* are highly adapted to the human host. Their larvae (cercariae) invade the host via skin, transform into schistosomula (larval migratory worms), and based on schistosome species migrate through the lungs to the blood vessels of mesentery or urinary bladder where they reach maturity. Adult schistosomes can reside for decades in the host vascular system as male-female pairs producing hundreds of eggs per day. Entrapped eggs in the host tissues induce inflammatory processes that significantly contribute to disease morbidity [1].

Schistosomes excrete or secrete a broad range of molecules from their bodies into the surrounding environment of the host. Composition of these excretory/secretory products (E/S products) greatly differs among individual developmental stages [2–5]. E/S products of cercariae are mostly produced by pre- and post-acetabular glands, and their main function is to penetrate the host skin and to downregulate the host's immune response during the invasion [6–9]. The molecules secreted by schistosomula and adults

David J. Timson (ed.), *Schistosoma mansoni: Methods and Protocols*, Methods in Molecular Biology, vol. 2151, https://doi.org/10.1007/978-1-0716-0635-3_5,

during the life cycle are important for parasite survival in the host. They have the capability to modulate immune response and regulate the host hemostatic system [10, 11]. Schistosome eggs need to bypass host tissues in order to be released with feces or urine to the water environment. Mature eggs secrete bioactive molecules, which modulate immune responses and induce inflammatory processes and cytotoxicity to the surrounding cells enabling egg passage through the tissue [12, 13]. Some eggs are entrapped in tissues, and their E/S products cause tissue fibrosis [14, 15].

This chapter describes collection of E/S products from *Schistosoma mansoni* developmental stages that infect or reside in the human host: cercariae, schistosomula, adults, and their eggs. Obtaining E/S products is necessary for further studies on the role of various molecules secreted by schistosomes into the host. These protocols were originally designed to study activities of *Schistosoma* peptidases in E/S products [3, 16, 17] and are adjusted for a general collection of protein content from E/S products of schistosomes, and with certain adaptation, of other flukes.

2 Materials

2.1 Biological Material

1. Developmental stages of *S. mansoni*: cercariae, in vitro transformed schistosomula, adults, and eggs. Maintenance of *S. mansoni* life cycle and preparation of individual developmental stages are described in Subheading 3.

2.2 Media

1. Penicillin-Streptomycin (P/S) stock solution: 10000 U/ml penicillin, 10 mg/ml streptomycin. Final dilution in medium 1:100.
2. Fungizone stock solution: 250 μg/ml Amphotericin B, 205 μg/ml sodium deoxycholate. Final dilution in medium 1:100.
3. Fetal bovine serum (FBS) (*see* **Note 1**).
4. RPMI 1640 medium.
5. M-199 medium with Earle's salts.
6. Basch medium 169: Prepared as described previously [18, 19]. To make 100 ml of Basch medium, combine:

   0.1 g lactalbumin Hydrolysate

   0.1 g glucose.

   50 μl 1 mM hypoxanthine.

   100 μl 1 mM serotonin.

   200 μl 10 mg/ml insulin.

   100 μl 1 mM hydrocortisone.

   100 μl 0.2 mM triiodothyronine.

500 μl 100x MEM vitamins (Thermo Fisher, 11120052).

5 ml Schneider's Drosophila medium.

1 ml 1 M HEPES pH 7.4.

Make up to 100 ml with Basal Medium Eagle (BME).

Check pH; if necessary, adjust to 7.4 using 1 M NaOH or 1 M HCl.

Filter sterilize, store up to 1 month in 4 °C.

7. Medium A: M-199, P/S.
8. Medium B: M-199, P/S, Fungizone, 10% FBS.
9. Medium C: M-199, P/S, 5% FBS.
10. Medium D: Basch medium 169, P/S, Fungizone, 10% FBS.
11. Medium E: Basch medium 169, P/S, 5% FBS.

2.3 Chemicals and Buffers

1. Porcine trypsin (Sigma-Aldrich T4799).
2. Phosphate buffered saline (PBS): 10 mM sodium phosphate, 1.8 mM potassium phosphate, pH = 7.4, 137 mM NaCl, 2.7 mM KCl.
3. 2× PBS: 20 mM sodium phosphate, 3.6 mM potassium phosphate, pH = 7.4, 274 mM NaCl, 5.4 mM KCl.
4. Dissolving buffer for cercariae secretion vesicles: 50 mM sodium citrate, 100 mM sodium phosphate, pH 6.0, 0.1% Triton X-100.
5. Dechlorinated water.

2.4 Plastics and Other Equipment

1. 1.5 ml tubes.
2. 50 ml tubes.
3. 6-well plates.
4. Pasteur pipettes.
5. Stainless steel sieves with mesh openings of 420 μm, 180 μm, 105 μm, and 45 μm.
6. 10 ml and 20 ml syringes.
7. 22G needle.
8. Syringe filters (pore size 0.22 μm).
9. Centrifugal concentrators (e.g., Amicon® Ultracel centrifugal unit with cut-off 3000 kDa from Millipore).
10. CO_2 incubator.
11. Laminar flow box.

3 Methods

3.1 Maintenance of S. mansoni Life Cycle

S. mansoni (a Puerto Rican isolate) is routinely maintained in the laboratory by cycling between the intermediate snail host, *Biomphalaria glabrata*, and outbred ICR mice as definitive hosts. Infective larvae (cercariae) are released after light stimulation from infected snails and placed in fresh water (*see* Subheading 3.2.1). Adult mice are infected by exposure to approximately 300 cercariae or by subcutaneous injection of 200 in vitro transformed schistosomula (*see* Subheadings 3.2 and 3.3). Mice are euthanized by an intraperitoneal injection of ketamine (Narkamon 5%—1.2 ml/kg body weight) and xylazine (Rometar 2%—0.6 ml/kg body weight) 6 weeks post-infection, and the worms are recovered from the portal vein by transcardial perfusion with RPMI 1640 medium or M-199 medium (*see* Subheading 3.4.1). Eggs are isolated from the dissected mice livers (*see* Subheading 3.5.1). Schistosomula are prepared in vitro by mechanical transformation of cercariae (*see* Subheading 3.3.1).

3.2 Cercariae

3.2.1 Shedding of Cercariae

1. Before shedding, place snails infected with *S. mansoni* in the dark for 24–48 h.
2. Spray snails intensively with water to remove contaminations.
3. Place snails in a clean beaker with dechlorinated water (approximately 40 ml of water per 20 snails). Snails should be completely covered with water.
4. Place the beaker with snails under the light for 1 h.
5. Collect the water containing cercariae, pass it through a sieve to remove snail feces, then transfer it to a 50 ml tube, and place on ice for 30 min or until cercariae settle.

3.2.2 Collection of E/S Products from Cercariae During Transformation Process

1. Wash cercariae three times with M-199 medium. Cercariae will pellet on ice in 10–15 min.
2. Resuspend cercariae in 5 ml of M-199 medium after the third wash.
3. Passage the suspension of cercariae through the 22G needle 25 times using a 10 ml syringe (*see* **Note 2**). This will initiate the transformation process including secretion of content from acetabular glands.
4. Check a small drop of the suspension under a microscope to see the efficiency of tail shearing.
5. Place transforming cercariae into a 6-well plates (approximately 2000 cercariae per well) in 5 ml of Basch medium 169 with P/S, and incubate at 37 °C in 5% CO_2 for 90 min to induce spontaneous secretion of acetabular gland contents.

6. Secretion vesicles produced by transforming cercariae will pellet at the bottom of the wells. Carefully remove medium containing transforming cercariae, and wash the vesicles five times with the same medium.
7. Transfer vesicles into a 1.5 ml tube, remove medium, and add 200 μl of dissolving buffer (50 mM sodium citrate, 100 mM sodium phosphate, pH 6.0, 0.1% Triton X-100).
8. Incubate at room temperature for 5 min.
9. Centrifuge dissolved vesicles at 10 000 × *g* at 4 °C for 5 min and collect the supernatant (*see* **Note 3**). This step serves for removing any residual cercariae.
10. Buffer-exchange the collected medium with at least 40 ml of ice-cold PBS, and concentrate using a centrifugal concentrator in precooled centrifuge at 0 °C. One 6-well plate should typically give the final volume of approximately 2 ml.
11. Make 100 μl aliquots of E/S products and store in −80 °C (*see* **Note 4**).

3.3 Schistosomula

3.3.1 Transformation of Cercariae to Schistosomula

1. Prepare cercariae (*see* Subheading 3.2.1).
2. Wash cercariae three times with M-199 medium. Cercariae will pellet on ice in 10–15 min.
3. Resuspend cercariae in 5 ml of M-199 after the third wash.
4. Passage the suspension of cercariae through the 22G needle 25 times using a 10 ml syringe.
5. Check a small drop of the suspension under microscope to see the efficiency of tail shearing.
6. Place the suspension into a Petri dish and add M-199 medium to the half of the dish and gently swirl. The cercariae transforming into schistosomula will concentrate in the middle of the dish while the tails will go to the sides.
7. Take out the transforming cercariae from the middle to a fresh sterile tube, and repeat swirling until there are still transforming cercariae concentrating in the middle of the dish.

3.3.2 Collection of E/S Products from In Vitro Cultivated Schistosomula

1. Work in a flow box and maintain sterility.
2. Wash transforming cercariae three times in medium D.
3. Incubate transforming cercariae in medium D at 37 °C in 5% CO_2 for 1 h.
4. Wash ten times with medium E, then transfer transforming cercariae into a 6-well plate (approximately 2000 transforming cercariae per a well), add 5 ml of medium E, and incubate at 37 °C in 5% CO_2 for 5 days to allow complete transformation from cercariae into schistosomula and remove contaminating cercarial products.

5. Wash ten times with medium A to remove the FBS.
6. Transfer schistosomula to a new 6-well plate, and incubate in medium A (*see* **Note 5**) at 37 °C in 5% CO_2 for 16 h.
7. Collect medium and filter it through a 0.22 μm pore size syringe filter. Keep it on ice.
8. Buffer-exchange the collected medium with at least 40 ml of ice-cold PBS, and concentrate using a centrifugal concentrator in precooled centrifuge at 0 °C. One 6-well plate should typically give the final volume of approximately 2 ml.
9. Make 100 μl aliquots of E/S products and store in −80 °C (*see* **Note 4**).

3.4 Adults

3.4.1 Isolation of Adult Worms from Mice

1. Euthanize mice using intraperitoneal injection of ketamine (Narkamon 5%—1.2 ml/kg body weight) and xylazine (Rometar 2%—0.6 ml/kg body weight) 6 weeks post-infection.
2. Dissect the mouse so that its abdominal and thoracic cavities are opened.
3. Cut the ribs on the left half of the thoracic cavity.
4. Cut the hepatic portal vein.
5. Insert a needle into left ventricle and pump RPMI 1640 medium through the needle. Worms will be perfused from the cut of the hepatic portal vein.
6. Collect the perfusate into a container and then transfer perfused worms into a 50 ml tube with fresh medium.

3.4.2 Collection of E/S Products from Adults

1. Work in a flow box and maintain sterility.
2. Wash worms three times in medium B.
3. Incubate worms in medium B at 37 °C in 5% CO_2 for 1 h.
4. Wash ten times with medium C, then transfer them into a 6-well plate, and add medium C to the final volume of 5 ml. Incubate at 37 °C in a 5% CO_2 overnight. FBS deactivates serine proteases from the host in this step.
5. Wash ten times with medium A to remove FBS, and then incubate in medium A at 37 °C in a 5% CO_2 for 16 h (*see* **Note 5**).
6. Collect medium and filter it through a 0.22 μm pore size syringe filter. Keep it on ice.
7. Buffer-exchange the collected medium with at least 40 ml of ice-cold PBS, and concentrate using a centrifugal concentrator at 0 °C in precooled centrifuge. One 6-well plate should typically give the final volume of approximately 2 ml.
8. Make 100 μl aliquots of E/S products and store in −80 °C (*see* **Note 4**).

3.5 Eggs

3.5.1 Isolation of Schistosome Eggs from Mouse Liver

1. Subcutaneously inject mice with 200 cercariae.
2. After 6 weeks post-infection, euthanize mice using intraperitoneal injection of ketamine (Narkamon 5%—1.2 ml/kg body weight) and xylazine (Rometar 2%—0.6 ml/kg body weight).
3. Dissect mouse liver (*see* **Note 6**) and blend it in 100 ml of 2× PBS to a smooth homogenate; then, add 2× PBS to the final volume of 1 l (this volume should be enough for up to six livers).
4. Add trypsin to the final concentration of 0.7% (w/v).
5. Incubate for 2 h in 37 °C with shaking at 120 rpm.
6. Pass the homogenate subsequently through the sieves with 425 μm, 180 μm, 125 μm, and 45 μm aperture. Rinse the sieves well with 2× PBS (*see* **Note 7**). The eggs are captured on the 45 μm sieve.
7. Rinse the 45 μm sieve with 2× PBS from a wash bottle, and collect eggs to a 50 ml tube using a Pasteur pipette (*see* **Note 8**).

3.5.2 Collection of E/S Products from Eggs

1. Move to a flow-box and maintain sterility.
2. Centrifuge the eggs at 300 × *g* for 3 min, remove the supernatant, and resuspend the eggs in medium A. Repeat this step ten times to wash the eggs properly.
3. Transfer the eggs using a Pasteur pipette to new sterile 50 ml tube, and resuspend them in 10 ml of medium B. Incubate eggs at 37 °C in a 5% CO_2 for 1 h.
4. Wash ten times with medium C, then transfer eggs to a 6-well plate (approximately 5000 eggs per well), add 5 ml of medium C, and incubate at 37 °C in a 5% CO_2 overnight. FBS in medium will inhibit trypsin used during egg isolation and host serine proteases from liver tissue.
5. Wash ten times with medium A to remove FBS.
6. Incubate eggs in medium A (*see* **Note 5**) at 37 °C in a 5% CO_2 for 16 h.
7. Collect medium and filter it through a 0.22 μm pore size syringe filter. Keep it on ice.
8. Buffer-exchange the collected medium with at least 40 ml of ice-cold PBS, and concentrate using a centrifugal concentrator at 4 °C in precooled centrifuge. One 6-well plate should typically give the final volume of approximately 2 ml.
9. Make 100 μl aliquots of E/S products and store in −80 °C (*see* **Note 4**).

4 Notes

1. Since FBS varies in quality from company to company, perform batch testing to select your FBS supplier for in vitro cultivation of schistosomes.
2. When transforming cercariae by passaging through a needle, keep always a little bit of air in the syringe. This prevents crushing the cercariae with a plunger.
3. When handling E/S products, keep them always on the ice, and work as fast as possible to minimize autodegradation.
4. You can freeze aliquots of E/S products in a liquid nitrogen, and then move them into a −80 °C freezer to minimize autodegradation.
5. If you plan to analyze E/S products using mass spectrometry, you can use DMEM medium instead of M-199. DMEM contains less amino acids and peptides which could interfere with mass spectrometry analysis.
6. You can leave mouse liver for 24 h in fridge before isolation of eggs. This may help to soften the liver tissue and increase egg yield in the case of higher liver fibrotization.
7. Be patient when rinsing the sieves during isolation of the eggs. Eggs are sticky and you can lose many on the sieves, if you do not rinse them well. In addition, insufficient liver digestion leads to significant yield reduction.
8. It is possible to separate mature and immature eggs using a Percoll gradient centrifugation as described previously [20].

Acknowledgments

This work was supported by the project 1404317 from the Charles University Grant Agency; grants 19-17269S and 18-14167S from the Czech Science Foundation; Operational Programme Research, Development, and Education; the Call International Mobility of Researchers—MSCA—IF CZ.02.2.69/0.0/0.0/17_050/0008014; and the institutional project RVO 61388963.

References

1. Gryseels B, Polman K, Clerinx J, Kestens L (2006) Human schistosomiasis. Lancet 368 (9541):1106–1118
2. Call JL, Pilcher JB, Freeman GL Jr, Tsang VC (1995) Serum-free culturing of adult *Schistosoma mansoni* in dialysis bags for the production of excretory/secretory antigens. J Parasitol 81(5):742–746
3. Dvorak J, Fajtova P, Ulrychova L, Leontovyc A, Rojo-Arreola L, Suzuki BM, Horn M, Mares M, Craik CS, Caffrey CR, O'Donoghue AJ (2016) Excretion/secretion products from *Schistosoma mansoni* adults,

eggs and schistosomula have unique peptidase specificity profiles. Biochimie 122:99–109. https://doi.org/10.1016/j.biochi.2015.09.025

4. Kawanaka M, Carter CE (1992) Schistosoma japonicum: excretory-secretory products of the eggs during miracidial development. Exp Parasitol 74(2):143–150
5. Yoshino TP, Lodes MJ, Rege AA, Chappell CL (1993) Proteinase activity in miracidia, transformation excretory-secretory products, and primary sporocysts of *Schistosoma mansoni*. J Parasitol 79(1):23–31
6. Haas W, Diekhoff D, Koch K, Schmalfuss G, Loy C (1997) *Schistosoma mansoni* cercariae: stimulation of acetabular gland secretion is adapted to the chemical composition of mammalian skin. J Parasitol 83(6):1079–1085
7. McKerrow JH, Doenhoff MJ (1988) Schistosome proteases. Parasitol Today 4 (12):334–340
8. Dvorak J, Mashiyama ST, Braschi S, Sajid M, Knudsen GM, Hansell E, Lim KC, Hsieh I, Bahgat M, Mackenzie B, Medzihradszky KF, Babbitt PC, Caffrey CR, McKerrow JH (2008) Differential use of protease families for invasion by schistosome cercariae. Biochimie 90 (2):345–358. https://doi.org/10.1016/j.biochi.2007.08.013
9. Ingram J, Knudsen G, Lim KC, Hansell E, Sakanari J, McKerrow J (2011) Proteomic analysis of human skin treated with larval Schistosome peptidases reveals distinct invasion strategies among species of blood flukes. PLoS Negl Trop Dis 5(9):E1337. https://doi.org/10.1371/Journal.Pntd.0001337
10. Mebius MM, van Genderen PJ, Urbanus RT, Tielens AG, de Groot PG, van Hellemond JJ (2013) Interference with the host haemostatic system by schistosomes. PLoS Pathog 9(12): e1003781. https://doi.org/10.1371/journal.ppat.1003781
11. Dvorak J, Horn M (2018) Serine proteases in schistosomes and other trematodes. Int J Parasitol 48(5):333–344. https://doi.org/10.1016/j.ijpara.2018.01.001
12. Cass CL, Johnson JR, Califf LL, Xu T, Hernandez HJ, Stadecker MJ, Yates JR 3rd, Williams DL (2007) Proteomic analysis of *Schistosoma mansoni* egg secretions. Mol Biochem Parasitol 155(2):84–93. https://doi.org/10.1016/j.molbiopara.2007.06.002
13. De Marco VC, Potriquet J, You H, McManus DP, Mulvenna J, Jones MK (2019) Qualitative and quantitative proteomic analyses of Schistosoma japonicum eggs and egg-derived secretory-excretory proteins. Parasit Vectors 12 (1):173. https://doi.org/10.1186/s13071-019-3403-1
14. Costain AH, MacDonald AS, Smits HH (2018) Schistosome egg migration: mechanisms, pathogenesis and host immune responses. Front Immunol 9:3042. https://doi.org/10.3389/fimmu.2018.03042
15. Sombetzki M, Fuchs CD, Fickert P, Osterreicher CH, Mueller M, Claudel T, Loebermann M, Engelmann R, Langner C, Sahin E, Schwinge D, Guenther ND, Schramm C, Mueller-Hilke B, Reisinger EC, Trauner M (2015) 24-nor-ursodeoxycholic acid ameliorates inflammatory response and liver fibrosis in a murine model of hepatic schistosomiasis. J Hepatol 62(4):871–878. https://doi.org/10.1016/j.jhep.2014.11.020
16. Horn M, Fajtova P, Arreola LR, Ulrychova L, Bartosova-Sojkova P, Franta Z, Protasio AV, Opavsky D, Vondrasek J, McKerrow JH, Mares M, Caffrey CR, Dvorak J (2014) Trypsin- and chymotrypsin-like serine proteases in *Schistosoma mansoni*—"The undiscovered Country". PLoS Negl Trop Dis 8(3). https://doi.org/10.1371/journal.pntd.0002766
17. Leontovyc A, Ulrychova L, O'Donoghue AJ, Vondrasek J, Maresova L, Hubalek M, Fajtova P, Chanova M, Jiang Z, Craik CS, Caffrey CR, Mares M, Dvorak J, Horn M (2018) SmSP2: a serine protease secreted by the blood fluke pathogen *Schistosoma mansoni* with anti-hemostatic properties. PLoS Negl Trop Dis 12 (4):e0006446. https://doi.org/10.1371/journal.pntd.0006446
18. Basch PF (1981) Cultivation of *Schistosoma mansoni* in vitro. I. Establishment of cultures from cercariae and development until pairing. J Parasitol 67(2):179–185
19. Tucker MS, Karunaratne LB, Lewis FA, Freitas TC, Liang YS (2013) Schistosomiasis. Curr Protoc Immunol 103:Unit 19 11. https://doi.org/10.1002/0471142735.im1901s103
20. Ashton PD, Harrop R, Shah B, Wilson RA (2001) The schistosome egg: development and secretions. Parasitology 122 (Pt 3):329–338

Chapter 6

Labeling of the *Schistosoma mansoni* Tegument

George Wendt and James J. Collins

Abstract

Schistosomes are deadly pathogens responsible for the neglected tropical disease schistosomiasis. The parasite's virulence is aided by a skin-like tissue called the tegument. The study of the tegument is hampered by a lack of tools suitable for visualizing the tissue. Here we describe a novel methodology employing fluorophore-conjugated dextrans that allows specific fluorescent labeling of the tegument and that is compatible with downstream fluorescence-labeling techniques including phalloidin labeling, RNA FISH, and immunofluorescence.

Key words Schistosome, Tegument, Fluorescence, Dextran, Phalloidin, Fluorescence in situ hybridization, Immunofluorescence

1 Introduction

Schistosomes are parasitic flatworms that cause the neglected tropical disease schistosomiasis, a devastating disease that infects more than 230 million people causing wide-ranging pathology including severe liver, kidney, and urogenital disease, as well as death, but its greatest impact is likely on the impoverished children who lack access to treatment and prevention. If not treated, chronically infected children can develop cognitive impairment and disability that will follow them for the rest of their lives, propagating a vicious cycle where poverty drives infection that drives more poverty [1]. Indeed, without treatment, the parasites can live inside the host's blood for decades at a time, aided by their skin-like tegument. The tegument is a syncytial tissue that covers the entire surface of the parasite, supported by thousands of cell bodies that connect to the surface via small cytoplasmic stalks that intercalate between the parasite's muscle layers (Fig. 1). Although it has long been thought to be important for diverse tasks such as immune evasion [2] and metabolism [3], we have a very poor understanding of the molecular details of these processes. The study of the tegument is complicated by the fact that up until now the only way to

David J. Timson (ed.), *Schistosoma mansoni: Methods and Protocols*, Methods in Molecular Biology, vol. 2151,
https://doi.org/10.1007/978-1-0716-0635-3_6,

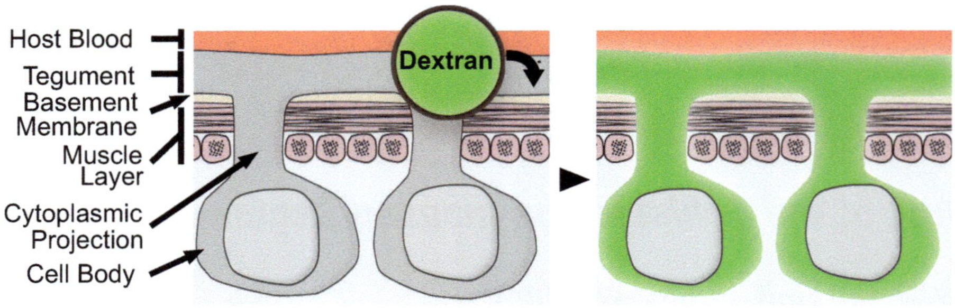

Fig. 1 Anatomical cartoon of the tegument illustrating dextran labeling. When performed as described here, dextran will specifically label the tegument of the parasite, allowing fluorescent visualization of the tissue (Adapted from Wendt et al. [5])

visualize the tegument at the cellular level has been by electron microscopy, which precludes using other methods such as RNA FISH or immunofluorescence. Though robust techniques for studying various other tissues in schistosomes already exist [4], the lack of techniques that specifically label the tegument has hampered the study of this important and fascinating tissue. We recently described a novel method for labeling the tegument using a fluorescently labeled high molecular weight dextran [5] (Fig. 1). This method is compatible with other commonly used methods in the study of schistosome biology including phalloidin labeling, RNA FISH, and immunofluorescence.

2 Materials

2.1 General

1. Small incubation basket (100 μm mesh).
2. Intavis Bioanalytical Instruments 12.340.
3. 48-well culture plates.
4. 24-well culture plates.
5. Aluminum foil for covering tubes/plates to protect from light.
6. Tricaine methanesulfonate.
7. PBS: pH 7.40.
8. 20x SSC: 3 M NaCl, 0.3 M trisodium citrate, pH 7.00 in DEPC H_2O.
9. Premounting solution: 80%(v/v) glycerol in PBS, pH 7.40.
10. Vectashield: Vector Laboratories H-1000.

2.2 Dextran Labeling

1. Dextran labeling solution: 5 mg/mL fluorophore-conjugated dextran (Dextran, TAMRA and biotin, 10,000 MW, Lysine Fixable (mini-Ruby): LifeTech) dissolved in double-distilled H_2O. Make 100 μL aliquots and store at −20 °C in the dark for up to 6 months (*see* **Note 1**).

2. PBSTx: 0.3%(v/v) triton x-100 in PBS.
3. Fixing solution: 4%(v/v) formaldehyde in PBSTx.

2.3 Phalloidin Labeling

1. Phalloidin blocking solution: 1%(v/v) BSA in PBSTx.
2. Phalloidin working solution: 4%(v/v) fluor-conjugated phalloidin (e.g., Alexa Fluor 488 Phalloidin: LifeTechnologies) in phalloidin blocking solution (*see* **Note 2**).
3. 1 mg/mL 4′,6-Diamidino-2-phenylindole dihydrochloride (DAPI) in double-distilled H_2O.

2.4 Fluorescence In Situ Hybridization (FISH)/ Immunofluorescence

1. 20 mg/mL proteinase K.
2. Prehybridization solution: 5× SSC, 1 mg/mL yeast RNA, 50% (v/v) deionized formamide, 1% (v/v) tween-20 in DEPC H_2O.
3. Hybridization solution: 5× SSC, 1 mg/mL yeast RNA, 50% (v/v) deionized formamide, 10% (v/v) dextran sulfate, 1% (v/v) tween-20 in DEPC H_2O.
4. Stringency wash solution 1: 3.5× SSC, 25% (v/v) formamide, 0.5% (v/v) Tween-20, 0.05%(v/v) Triton X-100 in DEPC H_2O.
5. Stringency wash solution 2: 2× SSC, 0.5% (v/v) Tween-20 in DEPC H_2O.
6. Stringency wash solution 3: 0.2× SSC, 0.5% (v/v) Tween-20 in DEPC H_2O.
7. TNT: 0.1 M tris, 0.15 M NaCl, pH 7.50 in double-distilled H_2O, 0.1%(v/v) Tween-20.
8. FISH blocking solution: 5% (v/v) heat-inactivated horse serum, 0.5% western blocking reagent (Roche) v/v in TNT.
9. Tyramide labeling buffer: 2 M NaCl, 0.1 M boric acid, pH 8.5. Sterile filtered, store at 4 °C.
10. Tyramide labeling solution: per 1 mL tyramide labeling buffer: 1 μL 4-Iodophenylboronic acid (4-IPBA), 1 μL 3% H_2O_2, 2 μL fluorophore-conjugated tyramide (*see* **Notes 3** and **4**).

3 Methods

Unless otherwise noted, all steps are to be performed at room temperature with gentle agitation. Steps that take place at 52 °C should be carried out in a hybridization oven with gentle agitation.

3.1 Dextran Labeling

1. Collect 6-week post-exposure parasites from infected mice as described in [6].

2. Separate the male parasites from the female parasites by incubating in 0.05% tricaine methanesulfonate in Basch media 169 [7] for 1 min while gently pipetting.
3. Discard female parasites; collect ~50 male parasites at the bottom of a 1 5 mL conical tube.
4. Remove as much of the media as possible by using a p1000 tip to drag the parasites up to the side of the tube. Removing as much media as possible is absolutely critical for the labeling process (*see* **Note 5**).
5. Add 100 μL of dextran labeling solution to the parasites. Vortex constantly for 3 min, ensuring that worms are always covered in dextran labeling solution. This may involve forcing worms to the bottom of the tube periodically (*see* **Note 6**).
6. Douse parasites with 10 mL of fixing solution. Invert tube several times, allow the parasites to settle for approximately 30 s, and then remove and replace the fixing solution with 10 mL of fresh fixing solution.
7. Wrap the 15 mL conical with aluminum foil to protect the labeled parasites from light.
8. Fix for 4 h in the dark.
9. Rinse worms for 10 min in 10 mL PBSTx in the dark.
10. For PHALLOIDIN: proceed directly to Subheading 3.2 (*see* **Note 7**).
11. For ISH/IMMUNOFLUORESCENCE: replace PBSTx with 10 mL methanol; store at −20 °C for up to 1 year protected from light.

3.2 Dextran-Phalloidin Labeling

1. Transfer between 10 and 20 freshly fixed dextran-labeled parasites to a 1.5 mL microcentrifuge tube.
2. Rinse for 10 min in 1 mL of PBSTx in the dark.
3. Block for 1 h in 1 mL of phalloidin blocking solution in the dark.
4. Replace phalloidin block solution with 1 mL phalloidin working solution.
5. Incubate for 8–16 h at 4 °C in the dark.
6. Rinse for 10 min in 1 mL of PBSTx in the dark. Repeat three times.
7. Incubate for 1 h in 1 mL of PBSTx with 1 μg/mL of DAPI in the dark.
8. Transfer worms to one well of a 24-well culture plate.
9. Incubate in 300 μL of premounting solution, and store at 4 °C until ready to mount.

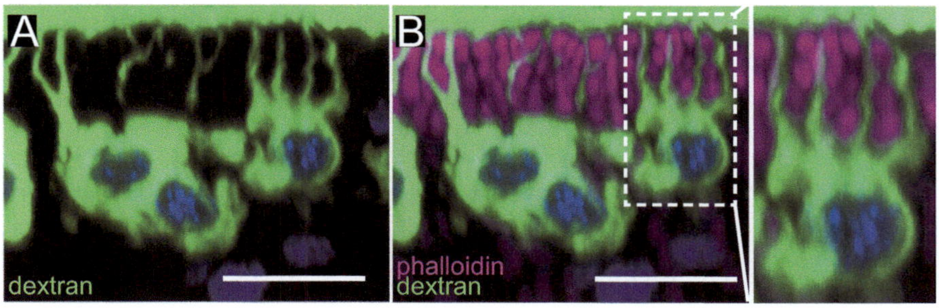

Fig. 2 Dextran labeling in conjunction with phalloidin labeling. Dextran labeling can be performed prior to phalloidin labeling, allowing for concurrent labeling of both the tegument and the underlying muscle. This allows for unambiguous identification of cytoplasmic stalks traveling from tegument cell bodies up to the parasite's surface (**a**), intercalating between muscle fibers (**b**). Scale bar is 10 μm (Adapted from Wendt et al. [5])

10. Mount worms on slide with Vectashield, image with confocal microscope.

See Fig. 2 for example of dextran-phalloidin labeled parasites imaged with a confocal microscope.

3.3 Dextran-FISH Labeling

1. Transfer up to 20 dextran-labeled parasites from methanol storage into a 1.5 mL microcentrifuge tube, and incubate in 1 mL of 50%(v/v) methanol in PBSTx for 10 min in the dark.
2. Rinse for 10 min in 1 mL of PBSTx in the dark.
3. Treat with proteinase K: 2 μg/mL in 1 mL of PBSTx for 30 min in the dark (*see* **Note 8**).
4. Post-fix for 10 min in 1 mL of fixing solution in the dark.
5. Rinse for 10 min in 1 mL of 50% prehybridization solution v/v in PBSTx in the dark.
6. Transfer up to 10 parasites into one small incubation basket.
7. Transfer basket to 48-well plate with 300 μL of prehybridization solution per well (*see* **Note 9**). Cover top of wells with parafilm to prevent evaporation of liquids.
8. Incubate for 2 h at 52 °C in the dark.
9. Transfer basket to a new well with 300 μL of hybridization solution with between 10 and 200 ng/mL of labeled riboprobe (*see* **Note 10**).
10. Incubate for 12–18 h at 52 °C in the dark.
11. Transfer basket to a new well with 300 μL of stringency wash solution 1.
12. Incubate for 30 min at 52 °C in the dark.
13. Repeat **steps 11** and **12** once.

14. Transfer basket to a new well with 300 μL of stringency wash solution 2.
15. Incubate for 30 min at 52 °C in the dark.
16. Repeat **steps 14** and **15** once.
17. Transfer basket to a new well with 300 μL of stringency wash solution 3.
18. Incubate for 30 min at 52 °C in the dark.
19. Repeat **steps 17** and **18** once.
20. Transfer basket to a new well with 300 μL of TNT.
21. Incubate for 10 min in the dark.
22. Repeat **steps 20** and **21** once.
23. Transfer basket to a new well containing 300 μL FISH blocking solution.
24. Incubate for 2 h in the dark.
25. Transfer baskets to a new well containing 300 μL of FISH blocking solution with a 1:1000 dilution of peroxidase-conjugated antibody against your probe (*see* **Note 11**).
26. Incubate for 8–16 h at 4 °C in the dark.
27. Transfer baskets to a new well containing 300 μL of TNT.
28. Incubate for 20 min in the dark.
29. Repeat **steps 27** and **28** seven times.
30. Transfer baskets to a new well containing 300 μL tyramide labeling solution (*see* **Note 12**).
31. Incubate for 10 min in the dark.
32. Transfer baskets to a new well containing 300 μL of TNT.
33. Incubate for 10 min in the dark.
34. Repeat **steps 32** and **33** five times.
35. Transfer baskets to a new well containing 30 0 μL of TNT with 1 μg/mL DAPI.
36. Incubate for 1 h in the dark.
37. Transfer worms from basket into 24-well plate, incubate in 300 μL of premounting solution, and store at 4 °C until ready to mount.
38. Mount worms on slide with Vectashield, image with confocal microscope.

See Fig. 3 for example of dextran-FISH labeled parasites imaged with a confocal microscope.

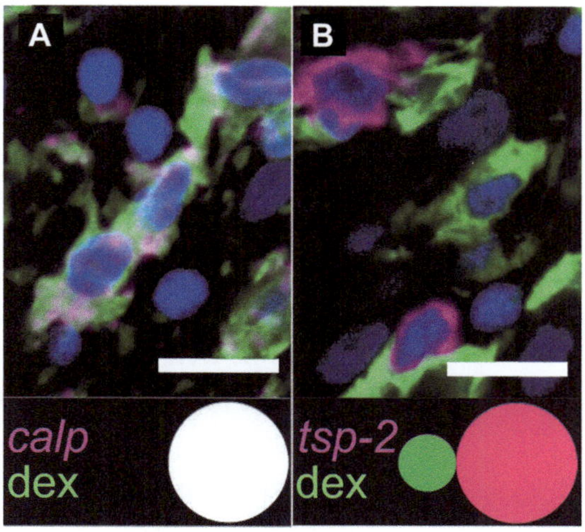

Fig. 3 Dextran labeling in conjunction with FISH. Dextran labeling of the tegument can be performed along with FISH to determine whether a given gene is expressed in the tegument. (**a**) Dextran colocalizes with the tegument marker *calp*. (**b**) Dextran does not colocalize with the tegument progenitor cell marker *tsp-2*. Venn diagram below image is a graphical representation of proportion of overlap between dextran signal and FISH signal. Scale bar is 10 μm (Adapted from Wendt et al. [5])

3.4 Dextran--Immunofluorescence Labeling

1. Transfer up to 20 dextran-labeled parasites from methanol storage into a 1.5 mL microcentrifuge tube, and incubate in 1 mL of 50% methanol v/v in PBSTx for 10 min in the dark.
2. Rinse for 10 min in 1 mL of PBSTx in the dark.
3. Digest with proteinase K: 2 μg/mL in 1 mL of PBSTx for 30 min in the dark.
4. Post-fix with 1 mL of fixing solution for 10 min in the dark.
5. Rinse for 10 min in 1 mL of PBSTx in the dark. Repeat once.
6. Transfer up to 10 parasites into 1 small incubation basket.
7. Transfer basket to 1 well of a 48-well plate with 300 μL of FISH blocking solution.
8. Incubate for 1 h in the dark.
9. Transfer baskets to a new well with 300 μL of FISH blocking solution with 1:1000 dilution of primary antibody (*see* **Note 13**).
10. Incubate for 8–16 h at 4 °C in the dark.
11. Transfer basket to a new well containing 300 μL of TNT.
12. Incubate for 20 min in the dark.
13. Repeat **steps 11** and **12** seven times.

14. Transfer basket to a new well containing 30 0 μL of FISH blocking solution with 1:1000 dilution of secondary antibody (*see* **Note 14**).
15. Incubate for 8–16 h at 4 °C in the dark.
16. Transfer basket to a new well containing 300 μL of TNT.
17. Incubate for 10 min in the dark.
18. Repeat **steps 16** and **17** five times.
19. Transfer basket to a new well containing 300 μL of TNT with 1 μg/mL DAPI.
20. Incubate for 1 h in the dark.
21. Transfer worms from basket into 24-well plate, incubate in 300 μL of premounting solution, and store at 4 °C until ready to mount.
22. Mount worms on slide with Vectashield, image with confocal microscope.

See Fig. 4 for example of dextran-FISH-immunofluorescence labeled parasites imaged with a confocal microscope.

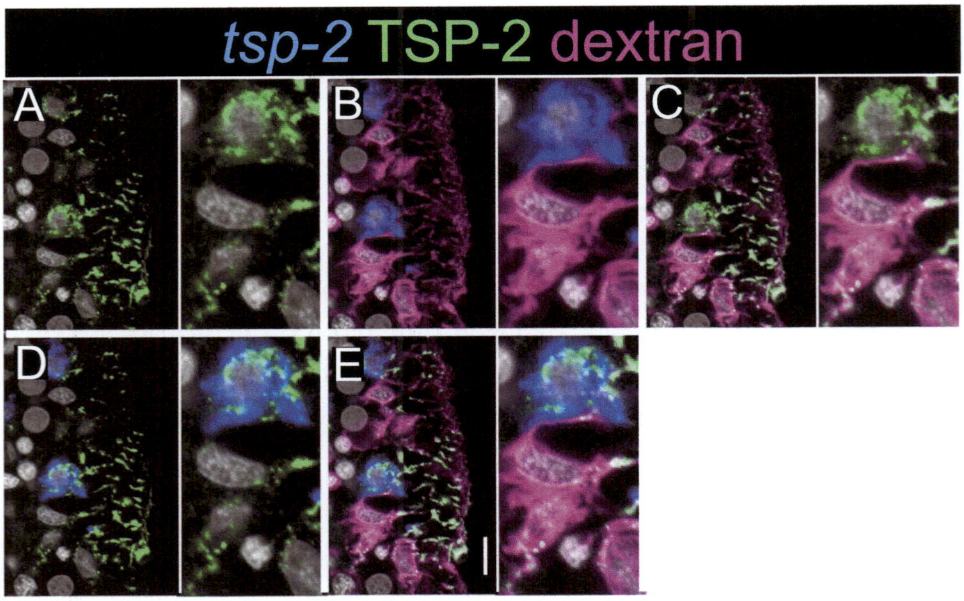

Fig. 4 Dextran labeling in conjunction with FISH and immunofluorescence. Using dextran, FISH, and immunofluorescence, we can demonstrate that TSP-2 protein is found in *tsp-2*$^+$ progenitor cells as well as in *tsp-2*$^-$ tegument cells. (**a**) TSP-2 protein labeled by immunofluorescence. Right panel is higher magnification. (**b**) *tsp-2* mRNA labeled by FISH overlaid with tegument labeled by dextran. Right panel is higher magnification. (**c**) TSP-2 protein labeled by immunofluorescence overlaid with tegument labeled by dextran. Right panel is higher magnification. (**d**) *tsp-2* mRNA labeled by FISH overlaid with TSP-2 protein labeled by immunofluorescence. Right panel is higher magnification. (**e**) *tsp-2* mRNA labeled by FISH overlaid with both TSP-2 protein labeled by immunofluorescence and tegument labeled by dextran. Right panel is higher magnification. Scale bar is 10 μm (Adapted from Wendt et al. [5])

4 Notes

1. Other colors of fluorophore-conjugated dextran can also be used as long as they do not overlap with fluorophores of other labeling steps.
2. Other colors of fluorophore-conjugated phalloidin can also be used as long as they do not overlap with fluorophores of other labeling steps.
3. Allow the tyramide labeling buffer to come to room temperature before making the tyramide labeling solution.
4. Any fluorophore-conjugated tyramide can be used, e.g., Alexa Fluor 488 tyramide reagent (Invitrogen B40953), Alexa Fluor 546 tyramide reagent (Invitrogen B40954), or Alexa Fluor Alexa Fluor 647 tyramide reagent (Invitrogen B40958).
5. Dextran labeling of the tegument appears to be dependent upon the tonicity of the solution. Therefore, removing *as much residual culture media as possible* is absolutely essential for efficient labeling. Leftover media will increase tonicity and decrease labeling efficiency.
6. The length of time spent in the dextran labeling solution directly affects the strength and specificity of the labeling. Less than 3 min will typically result in weak labeling, whereas more than 3 min will sometimes result in dextran signal bleeding into tissues past the tegument such as muscle. That being said, age/strain of parasite, batch-to-batch variation of dextran, and other factors could potentially affect the efficiency of labeling. As such, it may be necessary to test and compare a small range of labeling times (e.g., 2, 3, and 4 min).
7. Do not dehydrate dextran-labeled parasites in methanol prior to phalloidin labeling. Doing so will destroy the native conformation of F-actin and prevent phalloidin labeling from working.
8. FISH efficiency is dramatically improved by treating with proteinase K. Proteinase k treatment, however, negatively impacts the strength of dextran labeling. As such, it may be necessary to optimize the concentration of proteinase K and the duration of digestion. Digesting for 30 min with 2 μg/mL of proteinase K is a good starting point. Always prepare proteinase K immediately before use.
9. When transferring incubation baskets between wells, be sure to blot the basket on paper towel to remove the current contents of the basket and to jostle baskets up and down after entering each new well to ensure the contents of the new well enter the interior of the basket.

10. Ideal probe concentration depends on physical properties of the probe as well as expression levels of the gene. Start at 100 ng/mL of probe and then adjust up or down if necessary. *See* [8] for more details about ISH and riboprobe synthesis.
11. Antibody concentrations should be empirically determined. For antibodies commonly used in FISH protocols such as Anti-Digoxigenin-POD, Fab fragments, from sheep (MilliporeSigma 11207733910) and Anti-Fluorescein-POD, Fab fragments, from sheep (MilliporeSigma 11426346910), a 1:1000 dilution is a good starting point.
12. Tyramide labeling solution should be prepared immediately prior to use. Some probes benefit from more than 10 min of incubation, but increased labeling time also causes an increase in background. Optimization is recommended, but labeling for 10 min is a good starting point.
13. Primary antibody concentrations should be empirically determined.
14. Secondary antibody concentrations should be empirically determined. Any secondary appropriate for immunofluorescence may be used, e.g., Goat anti-Mouse IgG (H + L) Cross-Adsorbed Secondary Antibody, and Alexa Fluor 488 (ThermoFisher A-11001).

References

1. Colley DG, Bustinduy AL, Secor WE, King CH (2014) Human schistosomiasis. Lancet (London, England) 383(9936):2253–2264. https://doi.org/10.1016/s0140-6736(13) 61949-2
2. Skelly PJ, Wilson RA (2006) Making sense of the schistosome surface. Adv Parasitol 63:185–284. https://doi.org/10.1016/S0065-308X(06) 63003-0
3. Faghiri Z, Camargo SM, Huggel K, Forster IC, Ndegwa D, Verrey F, Skelly PJ (2010) The tegument of the human parasitic worm *Schistosoma mansoni* as an excretory organ: the surface aquaporin SmAQP is a lactate transporter. PLoS One 5(5):e10451. https://doi.org/10.1371/journal.pone.0010451
4. Collins JJ III, King RS, Cogswell A, Williams DL, Newmark PA (2011) An atlas for *Schistosoma mansoni* organs and life-cycle stages using cell type-specific markers and confocal microscopy. PLoS Negl Trop Dis 5(3):e1009. https://doi.org/10.1371/journal.pntd.0001009
5. Wendt GR, Collins JN, Pei J, Pearson MS, Bennett HM, Loukas A, Berriman M, Grishin NV, Collins JJ (2018) Flatworm-specific transcriptional regulators promote the specification of tegumental progenitors in *Schistosoma mansoni*. Elife 7. https://doi.org/10.7554/eLife.33221
6. Lewis F (2001) Schistosomiasis. Curr Protoc Immunol. Chapter 19:Unit 19 11. https://doi.org/10.1002/0471142735.im1901s28
7. Basch PF (1981) Cultivation of *Schistosoma mansoni in vitro*. I. Establishment of cultures from cercariae and development until pairing. J Parasitol 67(2):179–185
8. Cogswell AA, Collins JJ III, Newmark PA, Williams DL (2011) Whole mount in situ hybridization methodology for *Schistosoma mansoni*. Mol Biochem Parasitol 178(1-2):46–50. https://doi.org/10.1016/j.molbiopara.2011.03.001

Chapter 7

A Procedure for Analyzing the Proteomic Profile of *Schistosoma mansoni* Cercariae

Dilhan Perera, Makan Golizeh, and Momar Ndao

Abstract

Schistosomiasis is one of the most important helminthic parasitic infections in the world, with over 700 million people at risk of infection. Species of *Schistosoma* have a complex life cycle involving the infection of freshwater snails before infecting their mammalian definitive host. Taking about 130,000 lives per annum, *S. mansoni* is the major cause of intestinal schistosomiasis worldwide. Within *Biomphalaria glabrata* snails, asexual replication of the parasite gives rise to cercariae larvae. Cercariae actively penetrate the host's skin to complete their life cycle and eventually transform into adult worms. If left untreated, intestinal schistosomiasis can lead to peripheral destruction of the portal vein system, gastric hemorrhage from esophageal varices, as well as hepatic failure. Mass spectrometry (MS) is the method of choice for proteomics analysis. The bottom-up proteomics approach—also known as "shotgun proteomics"—typically includes a protein extraction and solubilization step followed by proteolytic digestion and tandem MS (MS/MS) analysis. Proteins are later identified by peptide de novo sequencing upon MS and MS/MS spectra of digest peptides. In this chapter, we introduce an analytical workflow for proteome profiling of *S. mansoni* cercariae using bottom-up proteomics. The cercariae were isolated and lysed. Proteins were then extracted, enzymatically digested, and subjected to liquid chromatography-tandem mass spectrometry (LC-MS/MS) analysis. Proteins were identified using MaxQuant software. Cercariae are the first life stage of the parasite *S. mansoni* which humans encounter, and conducting proteomic analysis on this life cycle stage can shed light on possible drug or vaccine candidates to help disable the parasite's ability to infect or arm the immune system for parasite clearance.

Key words Shotgun proteomics, Bottom-up proteomics, Tandem MS, De novo sequencing, Tegument, Drug target identification, Schistosomiasis

1 Introduction

Schistosomiasis (or Bilharzia) is one of the most important helminthic parasitic infections in the world [1], with over 700 million people at risk of infection [2]. With many species able to infect humans, the three most common agents are *Schistosoma haematobium*, *Schistosoma mansoni*, and *Schistosoma japonicum*. Species of *Schistosoma* have a complex life cycle involving the infection of freshwater snails before infecting their mammalian definitive host.

David J. Timson (ed.), *Schistosoma mansoni: Methods and Protocols*, Methods in Molecular Biology, vol. 2151, https://doi.org/10.1007/978-1-0716-0635-3_7,

S. mansoni can be found in tropical and subtropical regions causing mental and physical growth retardation in children [3], anemia, megasyndromes, and other chronic pathologies in infected patients [4]. Taking approximately 130,000 lives per annum [5], *S. mansoni* is the major cause of intestinal schistosomiasis worldwide.

S. mansoni eggs deposited into freshwater will hatch miracidia larvae which are the free-swimming infectious agent of the aquatic snail intermediate host. Within *Biomphalaria glabrata* snails, asexual replication gives rise to cercaria larvae. Cercariae are the infectious agents of mammals, such as humans, and they actively penetrate host skin for parasitic entry. While burrowing, cercariae transform into schistosomulae and remain in the skin tissues for about 2 days making its way into circulation. Schistosomulae travel first to the lung and then back to circulation before arriving at the liver where they will migrate into juvenile worms. Juvenile worms will mature into adult worms; males and females will pair and remain in copula as they migrate to the mesenteric veins. As mature worms mate, females will lay up to 300 eggs per day. Eggs excreted in the feces will propagate the life cycle; however, many will become trapped in host tissues causing major pathology. If left untreated, intestinal schistosomiasis can lead to peripheral destruction of the portal vein system, gastric hemorrhage from esophageal varices, as well as hepatic failure [6].

Currently, the first-line treatment for schistosomiasis is praziquantel. This antihelminthic is widely used due to its efficacy of worm clearance by paralysis. Although effective, chemotherapy does not reverse egg deposition or protect from reinfection. The importance of new drugs to treat infection is growing as praziquantel drug resistance has already been shown in vitro [7, 8]. In addition, a vaccine preventing infection could be ideal to help stop the transmission and propagation of the parasite. Drug and vaccine candidates are being sought through a variety of methods including library screening, affinity chromatography, biochemical suppression assays, proteomic analysis, as well as many others [9].

Mass spectrometry (MS) is the method of choice for protein identification and the core technique in proteomics analysis. The bottom-up proteomics approach—also known as "shotgun proteomics"—typically includes a protein extraction and solubilization step followed by proteolytic digestion and tandem MS (MS/MS) analysis. To reduce sample complexity, proteins are either separated by gel electrophoresis prior to digestion ("gel-based approach") or first digested into peptides which are then separated by high-performance liquid chromatography (LC) ("gel-free approach"). Protein identification is performed using a protein search software via peptide de novo sequencing based on the MS and MS/MS spectra of digest peptides.

During MS analysis, peptides are ionized by laser (matrix-assisted laser desorption ionization; MALDI) or by being sprayed into a high-voltage capillary electrode (electrospray ionization, ESI). Generated ions are separated in an electric or magnetic field based on their mass-to-charge (m/z) ratios and detected upon hitting an ion-counting detector. Certain ions are then selected and sent to an activation cell where they are fragmented into smaller ions (i.e., amino acid residues) resulting in a MS/MS spectrum. A protein search software then combines data from MS and MS/MS spectra (peptide exact masses and fragmentation patterns) to generate tentative peptide sequences known as "peptide-spectrum matches" or PSMs. Confident PSMs will be searched against a protein sequence database, and matches will be reported in association with their corresponding proteins only if they pass false-discovery rate (FDR) assessment [10]. In bottom-up proteomics, peptide ions are mainly separated on time-of-flight (TOF) or orbital ion trap (OT) mass spectrometers. TOF-based and OT-based platforms have different performance characteristics although they offer substantial complementarity [11].

To maximize proteome coverage, proteins must be fully accessible for digestion. Using an appropriate solubilizing agent can largely improve the outcome of a proteomics analysis. A solubilizing agent disrupts protein non-primary structures and cancels protein-protein and/or protein-lipid interactions. Most solubilizing agents used in proteomics fall under one of the three categories of detergents, chaotropes, and organic solvents [12]. Another technique to improve protein sequence coverage is breaking intercysteine disulfide bonds using a reducing agent, such as dithiothreitol or 2-mercaptoethanol. However, to protect newly made free thiol groups from (re)oxidation, they are usually immediately reacted with an alkylating agent, such as 2-iodoacetamide or 2-iodoacetic acid.

In this chapter, we introduce an analytical workflow for proteome profiling of *S. mansoni* cercariae using gel-free bottom-up proteomics. Cercariae were isolated and lysed. Proteins were then extracted, enzymatically digested, cleaned-up using solid-phase extraction, and then subjected to liquid chromatography-tandem mass spectrometry (LC-MS/MS) analysis. Proteins were identified against the Trematoda subset of UniProt protein database at 1% false-discovery rate using the freely available MaxQuant software [13].

Cercariae are the first life stage of the parasite *S. mansoni* which humans encounter. By conducting proteomic analysis on the products expressed by this life cycle stage, it is possible to identify possible drug or vaccine targets to help disable the parasite's ability to infect or arm the immune system for parasite clearance. This method will describe how to analyze the proteome of whole cercaria larvae of the species *Schistosoma mansoni* using LC-MS/MS.

2 Materials and Equipment

2.1 Snails and Chemical Reagents

1. *Biomphalaria glabrata* snails infected with the *Schistosoma mansoni*.
2. Ammonium hydroxide (ammonia) solution.
3. Ammonium bicarbonate.
4. Dithiothreitol.
5. Formic acid.
6. Iodoacetamide.
7. Protease inhibitor mixture (tablet or cocktail).
8. Protein quantitation kit.
9. Sequencing grade-modified trypsin.
10. Sodium chloride.
11. Sodium dodecyl sulfate.
12. Trichloroacetic acid.
13. Triton X-100.

2.2 Solvents

1. Acetone.
2. Acetonitrile.
3. Methanol.

2.3 Solutions

1. Lysis solution: 1% (w/v) sodium dodecyl sulfate, 1% (v/v) Triton X-100, 0.5% (w/v) sodium chloride containing 1% (w/v) protease inhibitor (*see* **Note 1**).
2. Protein digestion solution: 0.1 M ammonium bicarbonate pH 8.0.
3. Reduction solution: 0.1 M dithiothreitol in protein digestion solution.
4. Alkylation solution: 0.25 M iodoacetamide in protein digestion solution (*see* **Note 2**).
5. Protein precipitation solution: 10% (w/v) trichloroacetic acid in 90% (v/v) acetone (*see* **Note 3**).
6. pH adjustment solution: 5% ammonium hydroxide (*see* **Note 4**).
7. pH adjustment solution: 1% formic acid (*see* **Note 4**).

2.4 Equipment

1. LC-MS/MS instrument.
2. C18 reverse-phase high-performance liquid chromatography (HPLC) column. Recommended column features: Dimension 2.1 × 150 mm. Particle size 1.7 μm. Pore size 100 Å.
3. C18 reverse-phase solid-phase extraction column or cartridge.

4. 100 μm cell strainer.
5. Centrifuge for 2 mL conical tubes.
6. Centrifuge for 50 mL conical tubes.
7. Heat block.
8. Incandescent light source.
9. pH meter, pH strip, or pH paper.
10. Ultrasound bath.
11. Vacuum evaporator.
12. Vortex mixer.

3 Methods

3.1 Parasite Shedding

1. Place up to 15–20 *Biomphalaria glabrata* snails into 100 mL beakers with 15 mL carbon-filtered water that was allowed to aerate for 16 h (*see* **Notes 5** and **6**).
2. Put beakers under direct light from an incandescent light bulb, 30–35 cm from the source, and expose snails to light for 60–90 min.
3. Return snails to their tanks and *be cautious with water as it contains live, infectious cercaria* (*see* **Note** 7).
4. Using a Stripette, collect infectious water and filter through a 100 μm filter into 50 mL conical tubes.

3.2 Protein Extraction

1. Rest conical tubes containing filtered cercaria on ice for 10 min allowing cercariae to settle at the bottom of the tubes.
2. Centrifuge 2 × 50 mL conical tubes at a time at 2000 × *g*, 4 °C, for 5 min.
3. After centrifugation, aspirate water carefully as to not disturb the pellet.
4. Add cercaria water to pelleted sample, and centrifuge again, concentrating the cercariae.
5. Repeat **steps 3** and **4** until left with two tubes of cercariae.
6. Aspirate most of the water leaving about 700 μL of water in each conical tube.
7. Using a pipette, mix up and down to resuspend cercariae, and transfer the contents of both tubes to a 2 mL conical tube.
8. Centrifuge at 2000 × *g*, 4 °C, for 10 min to pellet cercariae.
9. Aspirate water carefully as to not disturb the pellet.
10. Resuspend cercariae in 500 μL lysis solution.

11. Using an ultrasound homogenizer, disrupt cercariae in lysis solution, using 3 × 10 s durations at low to moderate amplitude or until fully suspended.
12. Let sample rest on ice for 10 min.
13. Centrifuge sample at 15,000 × *g*, 4 °C, for 15 min.
14. Transfer supernatant containing proteins to a clean 2 mL conical tube.
15. Add 1.5 mL cold protein precipitation solution and store sample at −20 °C overnight.
16. Centrifuge sample at 18,000 × *g*, 4 °C, for 20 min.
17. After centrifugation, a pellet should be visible. Remove supernatant, and wash pellet with 200 μL ice-cold acetone.
18. Briefly vortex sample and centrifuge again at 18,000 × *g*, 4 °C, for 5 min.
19. Repeat **steps 17** and **18**.
20. Remove supernatant, and air-dry pellet for 10 min.
21. Resuspend pellet in 104 μL protein digestion solution, using an ultrasound homogenizer with sonication 3 × 10 s duration at low to moderate amplitude or until fully suspended.

3.3 Protein Digestion

1. Use 4 μL of sample to determine protein concentration using bicinchoninic acid (BCA) assay (or any other protein quantification assay).
2. Transfer an aliquot equivalent to 0.5 mg protein to a clean 2 mL conical tube.
3. Add 10 μL reduction solution to sample (*see* **Note 8**).
4. Incubate sample at 70 °C in a heat block for 30 min.
5. Remove sample from heat block and allow it to cool for 5 min at room temperature.
6. Add 15 μL of alkylation solution to sample.
7. Incubate at room temperature protected from light for 30 min.
8. Add 10 μg sequencing grade-modified trypsin to sample.
9. Using a pH strip, ensure pH is 8.0. If pH < 8.0, then adjust pH to 8.0 with 5% ammonium hydroxide.
10. Incubate sample at 37 °C overnight (16–20 h).
11. Stop digestion using 10 μL 1%(v/v) formic acid.
12. Using a pH strip, ensure pH is 6.0. If pH > 6.0, then adjust pH to 6.0 with 1% (v/v) formic acid.
13. Clean-up sample using a C18 reverse-phase solid-phase extraction column or cartridge. Suggested clean-up procedure: Use methanol followed by the protein digestion solution to

equilibrate the column or cartridge. Load sample. Rinse with neutral water. Elute with methanol.

14. Dry sample under vacuum at 37 °C for 60–90 min.
15. Reconstitute sample in 50 μL 0.1% formic acid.
16. Store sample at −80 °C until ready for LC-MS/MS analysis.

3.4 LC-MS/MS Analysis

1. Load 20 μL sample onto a C18 reverse-phase HPLC column using loop injection.
2. Use water and acetonitrile (both containing 1% (v/v) formic acid) as HPLC mobile phases A and B. Filter mobile phases at 0.45 μm to protect column and HPLC tubing from blockage.
3. Conduct HPLC at 125 μL/min, 50 °C, in gradient mode for 60 min. Begin chromatography at 0% B, hold for 3 min, then increase to 40% B at 30 min and to 90% B at 38 min, hold at 90% B for 6 min, then decrease to 5% B, and hold for 15 min to re-equilibrate the column. To protect mass spectrometer against inorganic contaminants, use a switch valve to send the first 3 min to waste.
4. MS source parameters: Capillary voltage +4500 V. End-plate offset 500 V. Nebulizer pressure 2.5 bar (99.5% nitrogen). Dry gas flow 6.0 L/min (99.5% nitrogen at 200 °C).
5. Precursor ion (MS) scan: m/z 150–2200. Acquisition time 0.5 s in high-resolution mode.
6. Product ion (MS/MS) scan: m/z 300–2200. Acquisition time 0.06–0.25 s depending on precursor ion intensity. Collision energy 18–65 eV depending on precursor ion mass and charge. Singly charged, low-abundance, and redundant precursor ions were excluded. Total cycle time 1.5 s. *See* Fig. 1 for expected LC-MS/MS results.

3.5 Proteomics Data Analysis

1. Load LC-MS/MS analysis data file in MaxQuant software. For this study, version 1.6 of the MaxQuant software has been used.
2. Set digestion parameters to standard proline-compatible trypsin (trypsin/P) digestion allowing up to two missed cleavages.
3. Select methionine oxidation and protein N-terminal acetylation as permitted variable modifications and cysteine carbamidomethylation as the fixed modification.
4. Set first and main search tolerances to 70 and 6 mDa, respectively, for peptides with a maximum charge state of 5, minimum length of 7 residues, and maximum mass of 4600 Da at 1% false-discovery rate (PSM, protein, and site decoy levels). To account for LC retention shifts, the "match between runs" option was enabled with a match time window of 0.7 min and an alignment time window of 20 min.

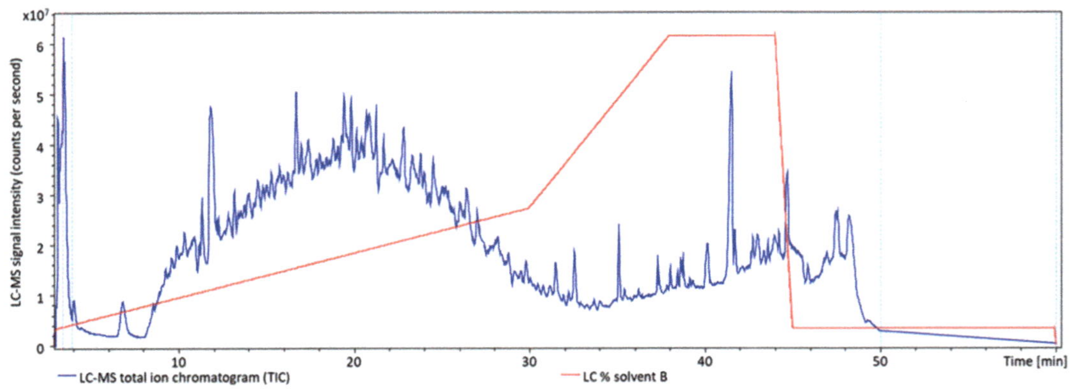

Fig. 1 Total ion chromatogram (blue) and LC % solvent B (red) from LC-MS/MS analysis of *Schistosoma mansoni* cercariae proteome. The protein extract was digested by trypsin and then separated on a C18 reverse-phase HPLC column prior to tandem mass spectrometry analysis. The peak at 3 min is associated with the internal calibrant

Table 1
Selected proteins identified by MaxQuant in *Schistosoma mansoni* cercariae

Protein name	Accession	Peptides	Unique peptides	Sequence coverage [%]	Mol. mass [kDa]	Sequence length	Score	MS/MS count
Malate dehydrogenase	G4VBJ0	10	10	54.8	36.3	341	99.8	17
Glyceraldehyde-3-phosphate dehydrogenase	G4V5I1	9	9	49.3	31.7	296	124.3	21
Cercarial elastase (S01 family)	G4M1L4	7	1	43.7	28.6	263	6.6	1
Histone H2A	G4VFE8	2	2	41.6	13.4	125	27.4	5
Putative fatty acid binding protein	G4VCN1	2	2	39.4	15.1	132	14.0	2
Fructose-bisphosphate aldolase	P53442	8	8	36.6	39.6	363	74.4	11
Elastase 2a	Q8MUV8	5	5	36.5	28.6	263	82.4	13
Superoxide dismutase [Cu-Zn]	Q01137	3	3	35.3	15.7	153	21.8	4
Uncharacterized protein	G4VG17	4	4	34.1	14.7	138	56.1	7

The protein table usually includes protein name or FASTA header, accession, number of peptides, and unique peptides identified for each protein, protein % sequence coverage, molecular weight, length in amino acid residues, protein score reflecting the level of confidence, and number of MS/MS spectra associated with each identified protein

5. Conduct the protein search against the Trematoda subset of the UniProt protein database (270,086 sequences, downloaded in February 2019). *See* Table 1 for expected MaxQuant results.

4 Notes

1. Store lysis buffer at room temperature between uses. If a precipitate forms, heat solution to 37 °C to dissolve.
2. To minimize loss of efficiency due to degradation, prepare alkylating solution shortly before use and protect it from light.
3. To maximize recovery, keep protein precipitation solution and acetone used for washing the pellet at or below 4 °C at all times.
4. All solvents and reagents must be analytical grade or higher quality.
5. Water containing cercariae is infectious to humans. Handle with care. If infectious water touches your skin, cover the area in 70% ethanol; this will kill cercariae on contact.
6. Any equipment that comes into contact with live cercariae should be cleaned with 70% ethanol or soaked in a 10% bleach solution.
7. Wear gloves and cover exposed skin when handling infectious water. If infectious water touches your skin, cover the area in 70% ethanol; this will kill cercariae on contact.
8. Mix sample for at least 10 s using a vortex mixer after every addition of reagent or solution.

References

1. World Health Organization (2016). http://www.who.int/mediacentre/factsheets/fs115/en. Accessed 13 Apr 2019
2. World Health Organization (2016). https://www.who.int/schistosomiasis/disease/en. Accessed 13 Apr 2019
3. Colley DG, Bustinduy AL, Secor WE, King CH (2014) Human schistosomiasis. Lancet 383:2253–2264. https://doi.org/10.1016/S0140-6736(13)61949-2
4. Mazigo HD, Nuwaha F, Kinung'hi SM, Morona D, de Moira AP, Wilson S, Heukelbach J, Dunne DW (2012) Epidemiology and control of human schistosomiasis in Tanzania. Parasit Vectors 5:274. https://doi.org/10.1186/1756-3305-5-274
5. van der Werf MJ, de Vlas SJ, Brooker S, Looman CWN, Nagelkerke NJD, Habbema J, Dik F, Engels D (2003) "Quantification of clinical morbidity associated with schistosome infection in sub-Saharan Africa" (PDF). Acta Trop 86(2–3):125–139. https://doi.org/10.1016/S0001-706X(03)00029-9
6. Dunne DW, Vennervald BJ (2012) 11.1 Schistosomiasis. In: Warrell DA, Cox TM, Firth JD, Török E (eds) Oxford textbook of medicine: infection, 6th edn. Oxford University Press, Oxford, pp 831–841
7. Fallon PG, Doenhoff MJ (1994) Drug-resistant schistosomiasis: resistance to praziquantel and oxamniquine induced in *Schistosoma mansoni* in mice is drug specific. Am J Trop Med Hyg 51:83–88. https://doi.org/10.4269/ajtmh.1994.51.83

8. Ismail MM, Taha SA, Farghaly AM, el-Azony AS. (1994) Laboratory induced resistance to praziquantel in experimental schistosomiasis. J Egypt Soc Parasitol 24:685–695
9. Terstappen G, Schlüpen C, Raggiaschi R, Gaviraghi G (2007) Target deconvolution strategies in drug discovery. Nat Rev Drug Discov 6(11):891–903. https://doi.org/10.1038/nrd2410
10. Reiter L, Claassen M, Schrimpf SP, Jovanovic M, Schmidt A, Buhmann JM, Hengartner MO, Aebersold R (2009) Mol Cell Proteomics 8(11):2405–2417. https://doi.org/10.1074/mcp.M900317-MCP200
11. Hessling B, Büttner K, Hecker M, Becher D (2013) Global relative quantification with liquid chromatography-matrix-assisted laser desorption ionization time-of-flight (LC-MALDI-TOF)-cross-validation with LTQ-Orbitrap proves reliability and reveals complementary ionization preferences. Mol Cell Proteomics 12(10):2911–2920. https://doi.org/10.1074/mcp.M112.023457
12. Speers AE, Wu CC (2007) Chem Rev 107 (8):3687–3714. https://doi.org/10.1021/cr068286z
13. Tyanova S, Temu T, Cox J (2016) The MaxQuant computational platform for mass spectrometry-based shotgun proteomics. Nat Protoc 11:2301–2319. https://doi.org/10.1038/nprot.2016.136

Chapter 8

Proteomic Analysis of *Schistosoma mansoni* Tegumental Proteins

Mark S. Pearson, Alex Loukas, and Javier Sotillo

Abstract

The tegument (outer surface) of *Schistosoma mansoni* and other trematodes is in intimate contact with the host and plays an important role in host-parasite interactions. It is a complex structure that contains hundreds of proteins implicated in a variety of functions, although, so far, only a few proteins have been well characterized. Indeed, a few of these proteins have been shown to be effective vaccine and diagnostic candidates against *S. mansoni* and other schistosomes, and so the proteomic characterization of tegumental molecules could open new avenues for the development of novel control and surveillance strategies to combat schistosomiasis. Here, we describe the step by step isolation of tegumental proteins from the different tegument compartments using a biotinylation approach, as well as the materials and reagents needed.

Key words Schistosoma, Tegument, Proteomics, Biotin, Mass spectrometry

1 Introduction

The tegument of schistosomes and other trematodes is a complex structure covering the entire body of the worm and plays a key role in host-parasite interactions. Indeed, the molecules present in the membrane of several species of schistosomes are implicated in parasite development, nutrition, secretion, sensory reception, and immune evasion, among other functions [1–4]. Furthermore, several proteins present on the tegument from adult and larval schistosomes have been described as promising diagnostic and vaccine candidates [5–8]. Indeed, the *S. mansoni* tegumental tetraspanin *Sm*-TSP-2 is currently in phase I clinical trials as a vaccine against schistosomiasis [9].

Schistosomes have a dynamic tegument that grows from a trilaminate outer plasma membrane in cercariae to a heptalaminate membrane containing an overlying membranocalyx in adult worms [10]. In order to get a complete picture of the organization of the tegument and locate the different proteins within this structure,

David J. Timson (ed.), *Schistosoma mansoni: Methods and Protocols*, Methods in Molecular Biology, vol. 2151,
https://doi.org/10.1007/978-1-0716-0635-3_8, © Springer Science+Business Media, LLC, part of Springer Nature 2020

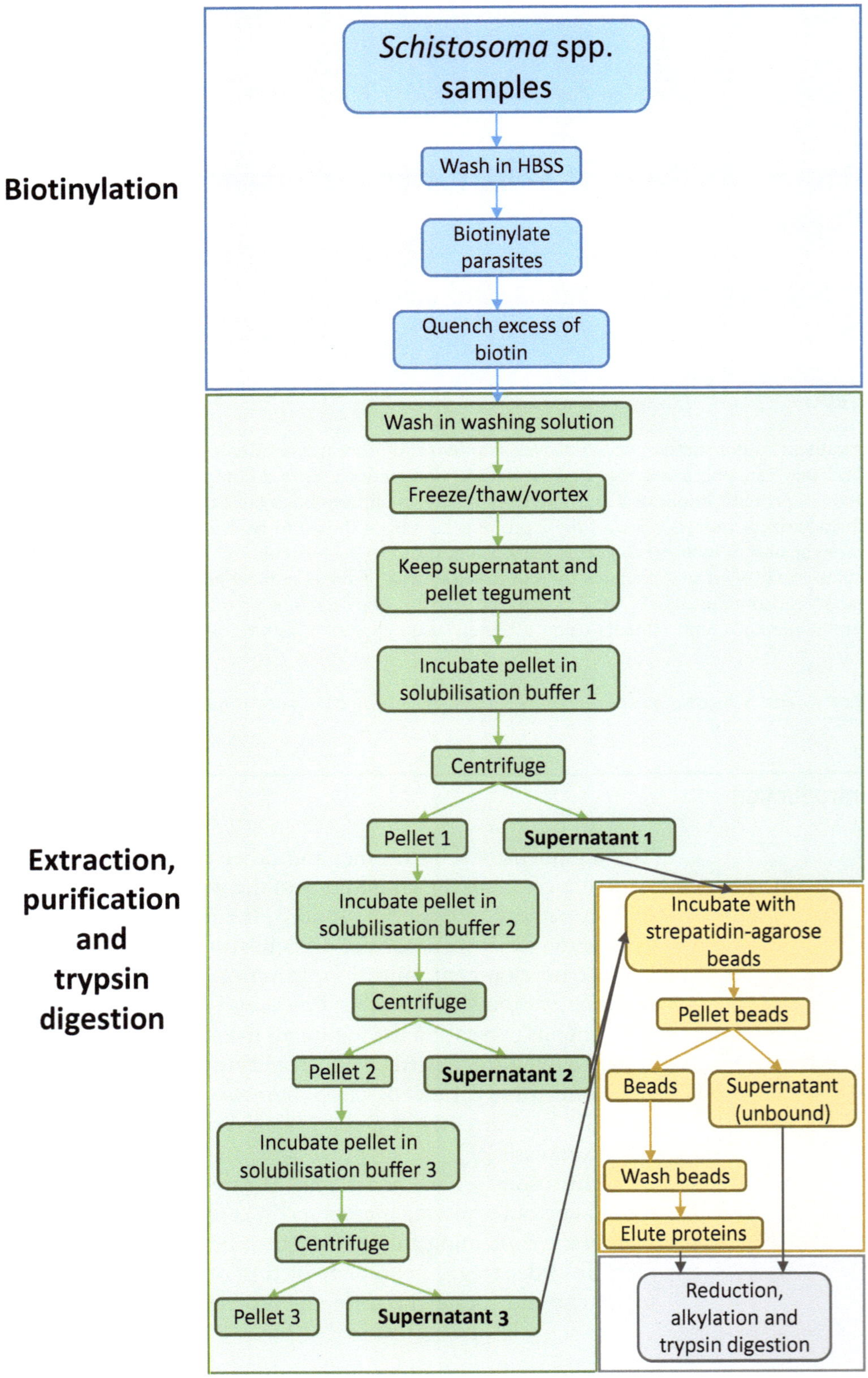

Fig. 1 Flow chart representing all the steps described in the method

sequential tegument extractions employing several buffers of increasing strength can be performed, although not all buffers are necessary and their use will depend on the researcher's needs and which parts of the tegument are to be studied. In addition, to avoid contamination with somatic proteins and other pollutant molecules, extraction techniques make use of impermeant biotinylation reagents that label the tegument of the worms, and streptavidin-agarose beads are employed to capture the labeled proteins (Fig. 1). This method has been used and validated by different researchers for adult worms and schistosomula of several schistosomatids, including *S. mansoni* and *S. japonicum* [5, 11, 12], and it could be applied to any trematode and life stage.

2 Materials

Prepare all solutions using ultrapure water (with sensitivity of 18 MΩ-cm at 25 °C) and analytical grade reagents. Prepare all reagents in a laminar flow cabinet, and filter them using a 0.45 μm filter (unless indicated otherwise).

2.1 Labelling of Tegumental Proteins

1. Hanks buffered salt solution (HBSS): 140 mM NaCl, 5 mM KCl, 1 mM $CaCl_2$, 0.4 mM $MgSO_4 \cdot 7H_2O$, 0.5 mM $MgCl_2 \cdot 6H_2O$, 0.3 mM $Na_2HPO_4 \cdot 2H_2O$, 0.4 mM KH_2PO_4, 6 mM glucose, 4 mM $NaHCO_3$, pH 7.4.
2. Biotinylation solution 1: 1 mM sulfo-*N*-succinimidobiotin (EZ-Link™ Sulfo-NHS-SS-Biotin) in HBSS. Dissolve 6.07 mg of EZ-Link Sulfo-NHS-SS-Biotin in 10 mL HBSS at room temperature. Do not filter. Keep at 4 °C.
3. Biotinylation solution 2: 1 mM sulfo-*N*-succinimidyl-6-(biotinamido) hexanoate (EZ-Link™ Sulfo-NHS-SS-Biotin) in HBSS. Dissolve 5.57 mg of EZ-Link Sulfo-NHS-LC-Biotin in 10 mL HBSS at room temperature. Do not filter. Keep at 4 °C.
4. RPMI 1640 medium with free amino acids.
5. Protease inhibitor cocktail. Prepare this solution in HBSS according to the manufacturer's instructions.

2.2 Differential Extraction of Tegumental Proteins and Processing of Biotinylated Proteins

1. Washing solution 1: 10 mM Tris–HCl, 0.84%(w/v) NaCl, pH 7.4. Store at 4 °C.
2. Washing solution 2: 10 mM Tris–HCl, pH 7.4. Store at 4 °C.
3. Extraction buffer 1: 40 mM Tris, pH 7.4. Store at room temperature.
4. Extraction buffer 2: 5 M urea, 2 M thiourea in 40 mM Tris, pH 7.4. Store at room temperature.

5. Extraction buffer 3: 0.1% (w/v) sodium dodecyl sulfate (SDS), 1.0% (v/v) Triton X-100 in 40 mM Tris, pH 7.4. Store at room temperature.
6. Immobilized streptavidin-agarose beads.
7. Bead washing solution: 0.1% (w/v) SDS, 1% (v/v) Triton X-100 in PBS.
8. Elution buffer: 2% (w/v) SDS.
9. 3 kDa spin concentrators.
10. Streptavidin-FITC.

2.3 Protein Digestion and Preparation for Liquid Chromatography-Tandem Mass Spectrometry (LC-MS/MS)

Prepare these solutions fresh each time and keep on ice. Keep the iodoacetamide protected from light.

1. 50 mM ammonium bicarbonate.
2. 100 mM dithiothreitol (DTT): Dissolve 15.4 mg in 1 mL ultrapure water.
3. 10 mM DTT: Prepare a 1/10 dilution from the 100 mM DTT solution.
4. 55 mM iodoacetamide (IAM): Dissolve 5.08 mg in 500 μL ultrapure water.
5. 200 μg/mL proteomics grade trypsin.
6. 5%(v/v) formic acid.
7. C18 sample desalting tip columns.

3 Methods

3.1 Labelling of Tegumental Proteins

1. Wash freshly obtained or cultured parasites in HBSS twice at room temperature, and remove damaged parasites with the aid of a 10× magnifying glass. The number of parasites to be used will depend on the life stage studied (*see* **Note 1**).
2. Incubate parasites with biotinylation solution 1 or 2 for 30 min at 4 °C (*see* **Notes 2** and **3**).
3. Remove biotinylation solution and wash the parasites three times in RPMI 1640 with free amino acids (*see* **Note 4**).
4. Remove excess RPMI 1640, snap-freeze parasites in liquid nitrogen, and keep at −80 °C with a minimal volume of protease inhibitors (*see* **Note 5**).
5. Keep 5–10 worms for testing of labeling efficiency.

3.2 Tegument Extraction and Processing of Biotinylated Proteins

Perform each extraction three times with extraction buffers, and pool the samples after each buffer (*see* **Notes 6** and **7**). Immediately after pooling, add protease inhibitor cocktail to a final concentration of 1× to each of the pooled samples, and store at −80 °C.

1. Thaw samples on ice.
2. Gently wash three times in cold washing solution 1 and discard supernatant.
3. Incubate parasites in washing solution 2 for 5 min on ice.
4. Vortex samples (5 × 1 s bursts) and collect supernatant. The remaining material will contain stripped worms (keep them for testing protocol efficiency).
5. Centrifuge samples at 1000 × *g* for 30 min at 4 °C and discard supernatant. Pellet will contain tegumental extract.
6. Solubilize pellet in 200 μL of extraction buffer 1 and pellet at 15,000 × *g* for 30 min at 4 °C. Repeat three times and combine supernatants to a final volume of 600 μL.
7. Solubilize pellet in 200 μL of extraction buffer 2 and pellet at 15,000 × *g* for 30 min at 4 °C. Repeat three times and combine supernatants to a final volume of 600 μL.
8. Solubilize pellet in 200 μL of extraction buffer 3 and pellet at 15,000 × *g* for 30 min at 4 °C. Repeat three times and combine supernatants to a final volume of 600 μL.
9. Incubate each of the samples after solubilization with 240 μL of streptavidin-agarose beads with gentle head-over-head mixing for 2 h at room temperature.
10. Pellet the beads at 3000 × *g* for 1 min and collect the supernatant, which will be the unbound material, and wash the beads five times with 500 μL of bead washing solution.
11. Incubate with 300 μL of elution buffer for 10 min at 95 °C followed by vortexing for 2 min to collect the biotinylated fraction (*see* **Note 8**).
12. Buffer exchange in 50 mM ammonium bicarbonate, concentrate down to 50 μL using 3 kDa spin concentrators, and quantify by BCA (*see* **Note 9**).

3.3 Analysis of Biotiniylation and Tegumental Extraction Efficiency

1. Wash labeled and stripped parasites (Subheading 3.2, **step 4**) three times with 200 μL of HBSS.
2. Incubate parasites in streptavidin-FITC for 30 min and wash twice more with HBSS (*see* **Note 10**).
3. Visualize under fluorescence microscope.

3.4 Protein Digestion and Preparation for Liquid Chromatography-Tandem Mass Spectrometry (LC-MS/MS)

Use 18.5 μg of protein in 10 μL of 50 mM ammonium bicarbonate for this procedure.

1. Add 2 μL of 50 mM ammonium bicarbonate and 3 μL of 10 mM DTT.
2. Incubate 20 min at 60 °C.

3. Add 3 μL of 50 mM ammonium bicarbonate and 2 μL of 55 mM IAM.
4. Incubate for 30 min at room temperature in darkness.
5. Add 2.5 μL of 50 mM ammonium bicarbonate and 2.5 μL of 100 mM DTT.
6. Incubate for 30 min at room temperature.
7. Add 2 μL of 200 μg /mL proteomics grade trypsin and incubate overnight at 37 °C.
8. Stop the reaction by adding 3 μL of 5%(v/v) formic acid.
9. Desalt using a C18 sample desalting tip column according to the manufacturer's instructions.
10. Dry in vacuum concentrator.

4 Notes

1. Ideally, the number of adult parasites used for proteomic analysis should be at least 500 worms. In the case of schistosomula, since they are much smaller in size, the starting material should be over 50,000 parasites.
2. In Subheading 2, we have described two impermeant biotinylation reagents (EZ-Link Sulfo-NHS-SS-Biotin and EZ-Link Sulfo-NHS-LC-Biotin). The use of the short-chain or long-chain reagent will depend on the needs of the researcher. For example, it has been shown in *S. mansoni* adult worms that the short form has a greater degree of penetration of the tegument and the long chain reagent can be useful if proteins only present within or above the plasma membrane need to be studied [5].
3. For correct biotinylation, incubate 500 adult worms or 50,000 schistosomula in 5 mL of biotinylation solution. Adjust the volume of biotinylation solution according to number of worms. It is also very important to remove potentially damaged parasites after biotinylation with the aid of a 10× magnifying glass to minimize contamination with somatic proteins.
4. When washing parasites after biotinylation, make sure to use complete RPMI 1640 with amino acids. These amino acids will quench any excess biotin and will make washing more efficient, minimizing potential contamination with non-parasitic proteins.
5. At this point, parasites can be kept at −80 °C. This is useful so parasites can be biotinylated right after they are obtained, and then all labeled samples can be processed at the same time to minimize researcher-based errors that can lead to bias in protein determination and quantification.

6. Parasites should be kept at −80 °C in a minimal volume of protease inhibitors prepared in HBSS. This volume will depend on the amount of parasites processed but should be enough to just cover the parasites.
7. Washing and incubation will depend on the amount of parasites processed. We have use 200 μL of buffer for 500 adult worms or 50,000 schistosomula.
8. The final number of fractions will be six: three fractions containing unbound (non-biotinylated proteins) and three biotinylated fractions after solubilization in solubilization buffers 1, 2, and 3.
9. For protein concentration and buffer exchange, 3 kDa filters are recommended. 10 kDa filters can also be used, although small proteins could be missed if using the latter.
10. It is very useful to include controls such as (a) unbiotinylated parasites but incubated with streptavidin-FITC, (b) biotinylated but not incubated with streptavidin-FITC, and (c) unbiotinylated parasites and not incubated with streptavidin-FITC.

Acknowledgments

This work was supported by a grant from the Health Institute Carlos III (ISCIII), Ministry of Economy and Competitiveness (Spain) (MPY 406/18), and a Miguel Servet I fellowship to JS (CP17III/00002) and a program grant (APP# 1037304) from the National Health and Medical Research Council (NHMRC, Australia) and a Senior Principal Research fellowship from NHMRC to AL (APP# 1117504).

References

1. Jones MK, Gobert GN, Zhang L, Sunderland P, McManus DP (2004) The cytoskeleton and motor proteins of human schistosomes and their roles in surface maintenance and host–parasite interactions. BioEssays 26 (7):752–765
2. Ramajo-Hernández A, Pérez-Sánchez R, Ramajo-Martín V, Oleaga A (2007) *Schistosoma bovis*: plasminogen binding in adults and the identification of plasminogen-binding proteins from the worm tegument. Exp Parasitol 115(1):83–91
3. Skelly PJ, Da'dara AA, Li X-H, Castro-Borges W, Wilson RA (2014) Schistosome feeding and regurgitation. PLoS Pathog 10 (8):e1004246
4. Skelly PJ, Alan WR (2006) Making sense of the schistosome surface. Adv Parasitol 63:185–284
5. Braschi S, Wilson RA (2006) Proteins exposed at the adult schistosome surface revealed by biotinylation. Mol Cell Proteomics 5 (2):347–356
6. Fonseca CT, Braz Figueiredo Carvalho G, Carvalho Alves C, de Melo TT (2012) Schistosoma tegument proteins in vaccine and diagnosis development: an update. J Parasitol Res 2012:54126
7. Wilson RA (2012) Proteomics at the schistosome-mammalian host interface: any prospects for diagnostics or vaccines? Parasitology 139(9):1178–1194

8. Zhang M, Fu Z, Li C, Han Y, Cao X, Han H et al (2015) Screening diagnostic candidates for schistosomiasis from tegument proteins of adult *Schistosoma japonicum* using an immunoproteomic approach. PLoS Negl Trop Dis 9 (2):e0003454
9. ClinicalTrials.gov. A phase I study of the safety, reactogenicity, and immunogenicity of Sm-TSP-2/alhydrogel® with or without GLA-AF for intestinal schistosomiasis in healthy adults
10. Hockley D, McLaren D (1971) The outer membrane of *Schistosoma mansoni*. Trans R Soc Trop Med Hyg 65(4):432
11. Mulvenna J, Moertel L, Jones MK, Nawaratna S, Lovas EM, Gobert GN et al (2010) Exposed proteins of the *Schistosoma japonicum* tegument. Int J Parasitol 40 (5):543–554
12. Sotillo J, Pearson M, Becker L, Mulvenna J, Loukas A (2015) A quantitative proteomic analysis of the tegumental proteins from *Schistosoma mansoni* schistosomula reveals novel potential therapeutic targets. Int J Parasitol 45 (8):505–516

Chapter 9

Measuring Histone Modifications in the Human Parasite *Schistosoma mansoni*

Ronaldo de Carvalho Augusto, David Roquis, Marion AL Picard, Cristian Chaparro, Celine Cosseau, and Christoph Grunau

Abstract

DNA-binding proteins play critical roles in many major processes such as development and sexual biology of *Schistosoma mansoni* and are important for the pathogenesis of schistosomiasis. Chromatin immunoprecipitation (ChIP) experiments followed by sequencing (ChIP-seq) are useful to characterize the association of genomic regions with posttranslational chemical modifications of histone proteins. Challenges in the standard ChIP protocol have motivated recent enhancements in this approach, such as reducing the number of cells required and increasing the resolution. In this chapter, we describe the latest advances made by our group in the ChIP methods to improve the standard ChIP protocol to reduce the number of input cells required and to increase the resolution and robustness of ChIP in *S. mansoni*.

Key word Native chromatin immunoprecipitation, ChIPmentation, Histone modifications, Schistosomiasis, Trematode, Development

1 Introduction

Measuring posttranslational histone modifications is particularly useful for studying chromatin profiles and potentially linking them to gene expression patterns in many organisms. For the purpose of this method chapter, we will focus on histone methylation, one of the many ways in which histones can be modified in the cell nucleus. When histone methylation occurs, one, two, or three methyl groups are transferred from *S*-adenosyl-L-methionine to lysine or arginine residues of histone proteins by histone methyltransferases (HMTs). Histone demethylases (HDMs) are enzymes that remove methyl groups from target sites. Both modifications can lead to changes in chromatin accessibility and can allow to turn

The original version of this chapter was revised. The correction to this chapter is available at https://doi.org/10.1007/978-1-0716-0635-3_20

David J. Timson (ed.), *Schistosoma mansoni: Methods and Protocols*, Methods in Molecular Biology, vol. 2151,
https://doi.org/10.1007/978-1-0716-0635-3_9, © Springer Science+Business Media, LLC, part of Springer Nature 2020,
corrected publication 2020

genes "on" or "off." Histone modifications are covalent posttranslational modifications which act by decreasing/increasing nucleosome contact (between histone-histone or histone-DNA) and, therefore, altering structure and stability of the chromatin, and/or recruiting proteins with chromatin modifying enzyme activity [1]. Specific methylation of histone 3 at lysine 4, 9, 27, 36, or 79 (H3K4me3, H3K9me3, H3K27me3, H3K36me3, respectively) and methylation of histone 4 at lysine 20 (H4K20me1) have been shown to be extremely conserved through all kingdoms and involved in many biological processes.

The human parasite *Schistosoma mansoni* is a digenetic trematode with two consecutive host. The parasite goes through at least five developmental stages during its life cycle; two of them are free living larvae that transit between the two obligatory hosts within a fresh water ecosystem. Each developmental stage is characterized by specific posttranslational histone modifications, in particular, methylations. Histone methylation also responds to environmental changes, development, sexual biology, and parasite pathogenesis and seems to be involved in an adaptive response [2–6]. Nowadays, several methods are used for mapping histone methylation on the genome, with chromatin immunoprecipitation being the one with the highest resolution. In cross-linking ChIP (X-ChIP), the chromatin is cross-linked, sonicated, or digested into small fragments, and then a specific antibody is used to immunoprecipitate the fragments containing the protein of interested. Next, reverse cross-link is needed, and the DNA is purified. Specific protein-DNA interaction at known genomic binding site sequences can be identified by quantitative PCR (ChIP-qPCR) or high-throughput sequencing (ChIP-seq). Alternatively, in native ChIP (nChIP), non-cross-linked chromatin is fragmented by enzymatic digestion and then processed in its native form. Our goal with the present book chapter is to share improved methods and immunoprecipitation protocols on *S. mansoni* samples that will be followed by high-throughput sequencing.

Native ChIP-seq (*nChIP-seq*) is a powerful technique used to determine the role of histone posttranslational modifications, often characterized by the modification's position, abundance, and type. In contrast to other chromatin immunoprecipitation methods, nChIP uses enzymatic chromatin fragmentation, leaving histone proteins in their native form and keeping the natural interaction with DNA under physiological salt conditions. Recently, *ChIPmentation*, a method that combines chromatin immunoprecipitation with sequencing library preparation by Tn5 transposase ("tagmentation"), was adapted by our group for non-model organisms, among them *S. mansoni* adult worms. Knowing that there are similarities and differences between principles of immunoprecipitation techniques, we will share here methods and protocols used by our group for experiments that will be followed by large-scale sequencing, i.e., *nChip-seq* and *ChIPmentation*.

2 Materials

2.1 Antibodies

One of the most important parts of ChIP techniques is related to the quality and specificity of the antibodies. Specificity and the strength of antibody-antigen interaction must be experimentally determined since they will affect the quality of the ChIP experiment in terms of specificity and sensitivity (Please check [7] for detailed information). Even if antibodies against histone isoforms (monoclonal or polyclonal) are sold as "ChIP-certified," they must be validated on your focal organism first. By a simple western blot, it is possible to determine if the antibody can be used or should be avoided in your experiment. Antibody that gives multiple bands must be avoided; those that do not reveal a band can potentially be used in ChIP. Keep in mind that all antibodies must be aliquoted on arrival and stored (to avoid freeze–thaw degradation every time the antibody is used) at −20 °C. Furthermore, using the correct antibody concentration can significantly improve the signal-to-background ratio of your library. In this sense, the determination of the amount of antibody required by performing a ChIP experiment is instrumental for each antibody type and concentrations. A titration is necessary to determine the optimal amount of antibody to use with your organism (please check Subheading 3.2 for detailed information).

In addition, this chapter offers the opportunity to share knowledge on antibody contamination coming from the production method. In the last 2 years, our group worked on improving the nChIP technique to reduce the number of cells needed in each sample. In our hands, nChIP had been routinely performed with a sample size of 150,000 cells [7]. However, due to recent demands (e.g., studies using stem cells and reproductive organs), an improvement in sensitivity of our procedures was required. For that, we compared nChIP with other methods proposed in the literature, such as commercially available X-ChIP and ultra-low-input native ChIP (ULI-ChIP) [8]. We obtained the best signal-to-background ratio for nChIP in all tested conditions. Our experiments were performed on the following samples: 150,000, 80,000, 20,000, 10,000, and 1000 of *S. mansoni* cells, in duplicate for each antibody (anti-H3K4me3 and anti-H3K27me3) + control, as well as without antibody (background). Although all samples passed our quality control step before sequencing, we observed sometimes high levels of nonspecific DNA for ≤20,000 cells. An unexpected source of contamination was the antibodies themselves. For example, in anti-H3K4me3 from Merck (cat. Nr. 04-745, several lots), we found high amounts of rabbit DNA. This contamination became predominant when working with less than 100,000 cells where ≥90% of the reads were rabbit DNA. Interestingly, our laboratory does not raise rabbits which means it is very unlikely that this material has been contaminated during the procedure. We suspect the contamination to come directly from the antibodies, which is plausible as many antibodies are produced in rabbits. Therefore, we have adopted in our laboratory an additional procedure for testing all antibody when working with a low number of cells. In that case, we tested the antibodies produced in rabbits by PCR using GAPDH primer (F-5′-GCCGCTTCTTCTCGT GCAG-3′; R-5′-ATGGATCATTGATGGCGACAACAT-3′). We have tested monoclonal antibodies from three suppliers before selecting the one with the lowest level of contamination. Our group reported this incident for the first time in [6].

2.2 Materials

1. Microdialysis units (Slide-a-Lyzer 3500 D cut-off, Pierce 69550).
2. 15 ml centrifugation tubes.
3. Centrifuges.
4. Stirrers.
5. Water bath or similar.
6. PCR machine.
7. Gel electrophoresis equipment.
8. 1 M KCl, autoclaved.
9. 5 M NaCl, filtered and autoclaved.
10. 1 M MgCl, autoclaved.
11. 1 M Tris/Cl pH 7.4–7.6, autoclaved.
12. 0.5 M EDTA, autoclaved.
13. 1 M $CaCl_2$, 10 ml, filtered.
14. 100 mM DTT, 1 ml (store at −20 °C).
15. Roche Complete Protease Inhibitor (ref: 11 697 498 001).
16. 2.5 M sodium butyrate (store at 4 °C).
17. 25 mM PMSF in isopropanol, 10 ml (store at −20 °C).
18. 15 U/μl Micrococccal nuclease (MNase) (USB 70196Y) in sterile 50%(v/v) glycerol, aliquot to ~10 μl and store at −20 °C.
19. Protein A—sepharose CL-4B (Sigma P3391 250 mg) (store at 4 °C).
20. Agarose gel loading buffer.
21. 20%(w/v) SDS, filtered.
22. 20 g/l glycogen solution (store at −20 °C).
23. 2%(w/v) NaN_3 in water (store at 4 °C), NaN_3 is very toxic!

Working solution	
2× base buffer	60 mM KCl, 15 mM NaCl, 5 mM $MgCl_2$, 0.1 mM EDTA, 15 mM Tris/Cl, pH 7.5 Add 2 of Roche protease inhibitor tabletts
Buffer 1	0.3 M sucrose, 5 mM $CH_3CH_2CH_2COONa$, 0.1 mM PMSF, 0.5 mM DTT, 60 mM KCl, 15 mM NaCl, 5 mM $MgCl_2$, 0.1 mM EDTA, 15 mM Tris/cl, pH 7.5
Buffer 2	0.3 M sucrose, 5 mM $CH_3CH_2CH_2COONa$, 0.1 mM PMSF, 0.5 mM DTT, 0.8% (v/v) NP40, 60 mM KCl, 15 mM NaCl, 5 mM $MgCl_2$, 0.1 mM EDTA, 15 mM Tris/Cl, pH 7.5
Buffer 3 for 3 cell samples	1.2 M sucrose, 5 mM $CH_3CH_2CH_2COONa$, 0.1 mM PMSF, 0.5 mM DTT, 60 mM KCl, 15 mM NaCl, 5 mM $MgCl_2$, 0.1 mM EDTA, 15 mM Tris/Cl, pH 7.5
MNase digestion buffer	0.3 M sucrose, 5 mM $CH_3CH_2CH_2COONa$, 0.2 mM PMSF, 4 mM $MgCl_2$, 1 mM $CaCl_2$, 50 mM Tris/Cl, pH 7.5
Dialysis buffer	5 mM $CH_3CH_2CH_2COONa$, 0.2 mM PMSF, 0.2 mM EDTA, 1 mM Tris/Cl, pH 7.5
Antibody incubation buffer	150 mM Nacl, 20 mM Tris/Cl, 20 mM sodium butyrate, 5 mM EDTA, 100 μM PMSF
Elution buffer	20 mM Tris/Cl, 50 mM NaCl, 5 mM EDTA, 20 mM sodium butyrate, 100 μM PMSF, 1% SDS

Washing Buffers	
Buffer A	50 mM Tris/Cl, 10 mM EDTA, 5 mM sodium butyrate, 75 mM NaCl
Buffer B	50 mM Tris/Cl, 10 mM EDTA, 5 mM sodium butyrate, 125 mM NaCl
Buffer C	50 mM Tris/Cl, 10 mM EDTA, 5 mM sodium butyrate, 175 mM NaCl

1. 100 mM dithiothreitol (DTT) in distilled water, aliquot to 500 μl (store at −20 °C). DTT is a reducing agent that prevents the formation of disulfide bonds in and between proteins.
2. 25 mM phenylmethanesulfonyl fluoride (PMSF) in isopropanol, 10 ml (store at −20 °C). PMSF is a serine protease inhibitor.

3. Roche Complete Protease Inhibitor tablets (ref: 11 697 498 001) (store at 4 °C).
4. 2.5 M sodium butyrate in distilled water (store at 4 °C). Sodium butyrate is a histone deacetylase (HDAC) inhibitor. Sodium butyrate solution is not very stable and should not be stored for more than 4 weeks. The product is an irritant and the solution has a nauseating odor.
5. 2%(w/v) NaN_3 in water (store at 4 °C) as preservative. Sodium azide is very toxic.
6. 15 U/μl micrococcal nuclease (MNase) (EC 3.1.31.1) (USB 70196Y) in sterile 50% (v/v) glycerol, aliquot to ~10 μl and store at −20 °C. Do not refreeze. MNase digests DNA between nucleosomes.

3 Methods

3.1 Preparation of Sepharose-Protein A

1. Weigh 250 mg Sepharose-protein A in 15 ml falcon tube (*see* **Note 1**).
2. Wash with 10 ml sterile water.
3. Centrifuge for 10 min at 4000 × *g* at 4 °C.
4. Remove supernatant.
5. Repeat washing (**steps 2–4**) four times.
6. Add sterile water to 5 ml.
7. Add NaN3 to 0.02% (w/v) and store at 4 °C.

3.2 Antibody Titration

Ideally, the amount of immunoprecipitated DNA must not depend on the amount of antibody used meaning that antibody should be in excess over the protein you want to precipitate. The inappropriate antibody concentration would lead to many difficulties in reproducibility of subsequent experiments. For that, a titration procedure is needed before ChIP experiment. In our hands, a titration is done with a constant quantity of chromatin from our sample incubated with increasing amounts of each antibodies.

The titration procedure is costly because substantial amounts of antibodies are consumed, but it is a critical step and assures the reproducibility of the results. For titration, do the following:

1. Prepare a dilution series of your chromatin in MNase buffer starting with 20–40 μg chromatin DNA for histone ChIP, and use 2–20 μg antibody (if the concentration is not known, use 2–20 μl antibody).
2. Add appropriate amounts of stock solutions to generate the antibody incubation buffer (150 mM NaCl, 20 mM sodium butyrate, 5 mM EDTA, 0.2 mM PMSF, 20 mM Tris/Cl

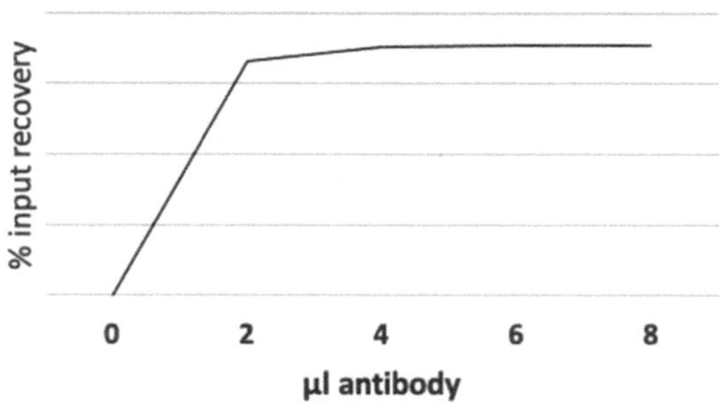

Fig. 1 Schematic representation of a titration experiment where antibody saturation was reached at around 4 µl antibody

pH 7.5) by taking into consideration the amount of S1 and S2 and their respective buffers.

3. Dilute S1 and S2 in 1 ml final volume of antibody incubation buffer.
4. Add the appropriate amount of antibody. Add an equal volume of water to the control tube without antibody.
5. Incubate overnight at 4 °C on a slowly rotating wheel.
6. Proceed starting from *Day 3* of the protocol.

Antibody saturation could be detected directly by the amount of precipitated genomic region of interest (bound) in comparison to the input (unbound fraction of control). In general, this ratio is expressed in % input recovery:

$$\%\text{input recovery} = 100 \times \text{PCR} - \text{efficiency}\ (\text{Ct} - \text{input} - \text{Ct} - \text{bound}).$$

From a certain quantity of antibody, the amount of immunoprecipitated DNA should remain constant (Fig. 1). This approach is sensitive to pipetting errors and requires careful standardization, and several technical and biological duplicates should be performed. For a successful ChIP experiment, it is instrumental to test the specificity of the antibodies and to determine the optimal amount by titration.

3.3 Day 1: Cell Lysis, Chromatin Preparation, MNase Digestion (Duration 2 h)

The following procedure has been improved in sensitivity and can now be performed with 10,000 cells of *S. mansoni*. This corresponds roughly to 100 larvae (miracidia or sporocyst 1). A single adult worm can also be used. Cells can be aliquoted and stored at −80 °C for up to 6 months or in liquid nitrogen for up to 12 months. However, it is preferable to use fresh biological material.

1. Cool down centrifuge to 4 °C.
2. Prepare a 2% (w/v) agarose gel in 0.5× TBE with 20 µl wells.

3. Preheat a water bath to exactly 37 °C.
4. Prepare 2× base buffer, buffers 1, 2, 3, MNase digestion buffer, and dialysis buffer solutions with autoclaved distilled water.
5. Put all buffer solutions on ice (except MNase buffer), which should be preheated at 37 °C in water bath.

3.3.1 Cell Lysis

1. Prepare aliquots of 10,000 to 100,000 cells, or 1 adult worm (fresh, stored at −80 °C or in liquid nitrogen).
2. Remove excess liquid (if any) and resuspend in 1 ml buffer 1, add 1 ml buffer 2 (lysis buffer), and transfer to Dounce.
3. Homogenize for 3 min with Dounce (pestle A) on ice.
4. Put on ice for 7 min.
5. Fill 8 ml of buffer 3 into a 15 ml or 50 ml centrifugation tube that resists 10,000 × *g*.
6. Overlay the 8 ml of buffer 3 with 1 ml cell suspension, so that the tubes are ready for centrifugation 15 min (sporocysts) of 10 min (adults) after buffer 2 has been added to the cells. Do not stir or shake the tube. You should have two distinct phases in the tube.
7. Slightly disturb the interface by passing a pipet tip through it 3–4 times.
8. Use two centrifugation tubes for the sporocyst samples that are in 2 ml of buffer 1+ and 2.
9. Mark tubes on the side to know on which side to look at the nuclei after centrifugation.
10. Centrifuge at 7800 × *g* for 20 min at 4 °C.
11. Carefully remove the supernatant completely (by pipetting with 10 ml pipette and then pipetting with micropipettes 1000 μl, 100 μl, and 10 μl).

3.3.2 MNase Digestion

1. Resuspend the pellet in 1 ml of MNase digestion buffer (1 ml for 10,000 cells).
2. Aliquot 500 μl of this suspension in 1.5 ml microcentrifuge tubes.
3. Add 1 μl MNase (15 U) and incubate for 4 min at 37 °C in water bath.
4. To stop the digestion, add 20 μl of0.5 M EDTA to each 500 μl MNase digestion tubes, and put them on ice.
5. Centrifuge 13,000 × *g* for 10 min at 4 °C.
6. Transfer the supernatant to a new tube (S1) and keep the pellet (P1).

7. Store S1 at −20 °C.
8. Quantify chromatin in S1 with the Qubit® Fluorometer (HS DNA assay) following the manufacturer instructions.

3.3.3 Dialysis of P1

1. Humidify a Slide-A-Lyzer with 50 μl dialysis membrane of buffer.
2. Resuspend the pellet P1 in 100 μl of dialysis buffer, and dialyze overnight at 4 °C in 50 ml dialysis buffer with gentle stirring. The Slide-A-Lyzer should be in a floating device, and we work with a magnetic spinner to do the gentle stirring.

3.4 Day 2: Incubation with the Antibody (Duration of 30 min Plus Overnight Incubation)

1. On the next day, thaw S1.
2. Transfer dialyzed sample into Eppendorf tubes.
3. Centrifuge both tubes side by side (S1 and dialyzed) at 13,000 × *g* for10 min at 4 °C.
4. Transfer the supernatant to a new tube and repeat the centrifugation twice. These triple centrifugations are critical to reduce unspecific background.
5. Supernatant of dialyzed sample is fraction S2.
6. Optional: use 50 μl of S1 and S2 for DNA extraction, centrifuge, and load 20 μl of supernatant on a 2% agarose 0.5× TBE gel (100 V, for 25 min).
7. Combine supernatants of S1 and S2 into a new tube.
8. Add appropriate amounts of stock solutions to generate the antibody incubation buffer.
9. Dissolve chromatin from S1 (and S2 if you have dialyzed) in a final volume of 1 ml of antibody incubation buffer, and adjust the concentrations of the compounds in the antibody incubation buffer using stock solutions, and sterile DNA-free water.
10. Add the amount of antibody needed (as determined in Subheading 3.2, above).
11. Incubate overnight at 4 °C on a rotating wheel.

3.5 Day 3: Precipitation and DNA Extraction (Duration 3 h)

3.5.1 Precipitation

1. Prepare 50 μl of protein A—sepharose for each tube.
2. Wash the beads with ChIP grade water in a 1.5 ml Eppendorf centrifuge 4000 × *g* at 4 °C for 5 min, remove the supernatant, and replace with an equal volume of sterile water.
3. Repeat the washing two more times. It is very important to resuspend the protein A—sepharose carefully to maintain a homogeneous solution in the stock tube.
4. Add 50 μl of protein A—sepharose to each tube.

5. Incubate for at least 4 h at 4 °C on a rotating wheel.
6. Prepare washing buffers (10 ml/tube), and cool down to 4 °C or on ice.
7. Centrifuge chromatin/antibody mixture for 10 min at 4 °C, 11,600 × *g*.
8. Keep the supernatant in a 2 ml tube. This is the unbound fraction UB.
9. Resuspend the pellet in approx. 1 ml of washing buffer A, and transfer into a 15 ml Falcon tube containing 9 ml of washing buffer A.
10. Mix for 10 min on a rotating wheel at 4 °C.
11. Centrifuge for 10 min at 3400 × *g* and 4 °C and pour off supernatant (by inversion).
12. Add 10 ml of washing buffer B, mix for 10 min on a rotating wheel at 4 °C, and centrifuge for 10 min, 4000 × *g*, at 4 °C.
13. Pour off the supernatant (by inversion).
14. Add 10 ml of washing buffer C, mix for 10 min on a rotating wheel at 4 °C, and centrifuge for 10 min at 4000 × *g*, at 4 °C.
15. Pour off supernatant (by inversion).
16. Centrifuge for 10 min, 4000 × *g*, at 4 °C.
17. Remove the remaining supernatant completely (not by inversion but by pipetting with micropipettes 1000 μl, 100 μl, and then 10 μl).
18. Resuspend the pellet in 500 μl elution buffer:
19. Transfer the suspension to a 1.5 ml microcentrifuge tube.
20. Incubate for 15 min at room temperature on a rotating wheel.
21. Centrifuge for 10 min, 11,600 × *g*, at 18 °C.
22. Transfer the supernatant to a 1.5 ml microcentrifuge tube. This is the bound fraction B.

3.5.2 DNA Extraction

1. Perform DNA purification on fractions using the QiaQuick PCR Purification Kit (Qiagen, Cat. No. 28104) or similar, following manufacturer instructions, but adding the following slight modifications.
2. **Steps 3** and **4**: Centrifugation time of 60 s. **Step 5**: Centrifuge the column in a 2 ml collection tube for 1 min, without the cap.
3. **Step** 7: Before the centrifugation, let the column stand for 2 min at room temperature. Elution volume: 50 μl.
4. In order to obtain more DNA, it is possible to proceed to a second elution in 30 μl.

5. Quantify purified DNA with accurate and sensitive DNA quantification methods following manufacturer instructions.
6. Use 1 μl of this DNA for PCR in 25 μl reactions (quantitative real-time PCR).

3.6 ChIPmentation

ChIPmentation is based on chromatin immunoprecipitation with sequencing library preparation by Tn5 transposase ('tagmentation'). ChIPmentation introduces sequencing compatible adaptors in a single-step reaction directly on bead-bound chromatin, which reduces time and cell input in comparison to nChIP. There are ChIPmentation kits available, and the method can even be automated on suitable pipetting robots or made manually. In our hands, the Diagenode Auto ChIPmentation Kit for Histones (Diagenode, Cat. No. C01011000) on an IP-Star® Compact Automated System (Diagenode, Cat. No. B03000002) gave excellent results (*see* **Note 1**). The procedure can be performed routinely in 2 days with up to 16 samples in parallel. We followed the provided user manual and will, in the following, only describe handling instructions specific to schistosomes. A single worm is sufficient for each antibody.

1. Shear the chromatin by sonication using the Bioruptor® or other sonication device. Choose the protocol which results in fragments of about 1000 bp in length, e.g., when using the Bioruptor® Pico shear for single adult worm, use 5 cycles [30 s "ON", 30 s "OFF"].
2. Library amplification varies from 16 to 19 cycles with C_t values dependent on the qPCR machine and calculation method. We observed better results when using 1/3 of slope = number of amplifications. Instead of using Ct that depends on the qPCR analysis algorithm, we recommend to display the fluorescence intensity as a function of amplification cycles and use the number of cycles for library amplification that correspond to 1/3 of the slope at the exponential phase of PCR amplification.

4 Notes

1. 250 mg protein A—sepharose swells to approximately 1 ml gel and binds about 20 mg human IgG. You will need 50 μl of the homogeneously mixed sepharose-protein A per ChIP. Protein A has strong affinity to human, mouse, and rabbit IgG. For antibodies raised in goat and sheep, sepharose-protein G or protein A/G mixtures should be used. Paramagnetic sepharose particles are available, but in our hands, background was on average 20 times higher than in a centrifugation-based separation.

2. *Brief overview of comparative and analytical approach.* We present here sequencing results of nChIP-seq and ChIPmentation for adult male worms of *S. mansoni* following the protocol described above. Antibodies were carefully tested for specificity as previously described and were used in saturating quantities (4 μl for each replicate). Two biological replicates were used to perform immunoprecipitation using antibodies against H3K4me3 (Diagenode, C15410003, Lot: A5051-001P). For each experiment, we used a control without antibody to assess nonspecific background (bound fraction) and input (unbound fraction). Inputs were used for normalization in all subsequent bioinformatics analyses. Both products were sequenced as paired-end 75 bp reads on an Illumina HiSeq 2500. All data processing was performed in parallel on a local GALAXY instance (http://bioinfo.univ-perp.fr). Read quality was verified using the FastQC toolbox (https://www.bioinformatics.babraham.ac.uk/projects/fastqc/). Sequences were aligned to the *S. mansoni* reference genome v7 with Bowtie v2.1 using parameters–end-to-end,–sensitive,–gbar 4 [9]. BAM files generated by Bowtie2 were sorted and then filtered for unique matches with samtools v1.3.1 [10]. PCR duplicates were also removed using samtools (samtools rmdup). Wiggle files generated in the process were visualized with IGV [11]. We started our analyses by characterizing the genome-wide distribution of narrow posttranslational histone modification. Our choice to compare narrow histone mark is due to the fact that they are highly specific and related to transcriptional start site (TSS), reducing the possibility of false-positive detection. In addition, it has been demonstrated that the histone mark H3K4me3 is typically restricted to narrow regions over specific functional genomic motifs being less dynamic than other marks already tested for *S. mansoni* [6]. Peak detection and comparative analysis were generated by ChromstaR, an R package that uses Hidden Markov Models (HMM) to perform computational inference of discrete combinatorial chromatin state dynamics over the whole genome [12]. BAM files were processed under the differential mode, with a false-discovery rate (FDR) cutoff of 1×10^{-4} and bin size of 150. We used the input as negative controls for comparative analyses. TSS of genes was defined as 3 bp (one base pair upstream and downstream of the +1 transcription site) based on the genome annotation file v.7. First, we analyzed the correlation between all samples from each experiment to understand how the distribution of H3K4me3, spanning the entire genome, changed from one approach to another (Fig. 2). Both approaches showed a similar profile with over 75% of peak positions being conserved throughout all samples. We conducted our analyses by comparing the genome-wide abundance of

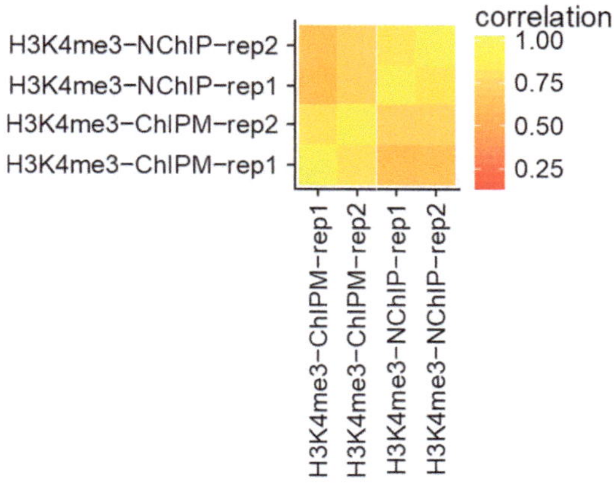

Fig. 2 Heatmap of both nChIP and ChIPmentation for H3K4me3 representing similarity of genome-wide profiles and peak positions

Table 1
Genome-wide description of frequency and number of peaks for H3K4me3 of both nChIP and ChIPmentation

	nChIP	ChIPmentation
Frequency	3.42%	3.19%
Number of peaks	6099	6098

detected peaks and their frequency for narrow H3K4me3 (Table 1). Differential frequency of H3K4me3 was lower than 0.3% with just one peak exclusively detected to a specific procedure (Table 1). Genome-wide analyses revealed high similarity between both techniques. In both techniques, H3K4me3 enrichment starts around 500 bp upstream of TSS and do not spread more than 3000 bp downstream (Fig. 3a). Wig files were produced to visualize the distribution of H3K4me3 (Fig. 3b). We highlight here the instrumental importance of good quality of antibodies to obtain high quality and deeply sequenced input DNA libraries which has fundamental implication for data analysis. We also show that nChIP-seq and ChIPmentation deliver similar results. The minimum number of cells for reliable peakcall with our nChIP technique is 10,000 cells, and we recommend using $\geq$20,000 in the routine.

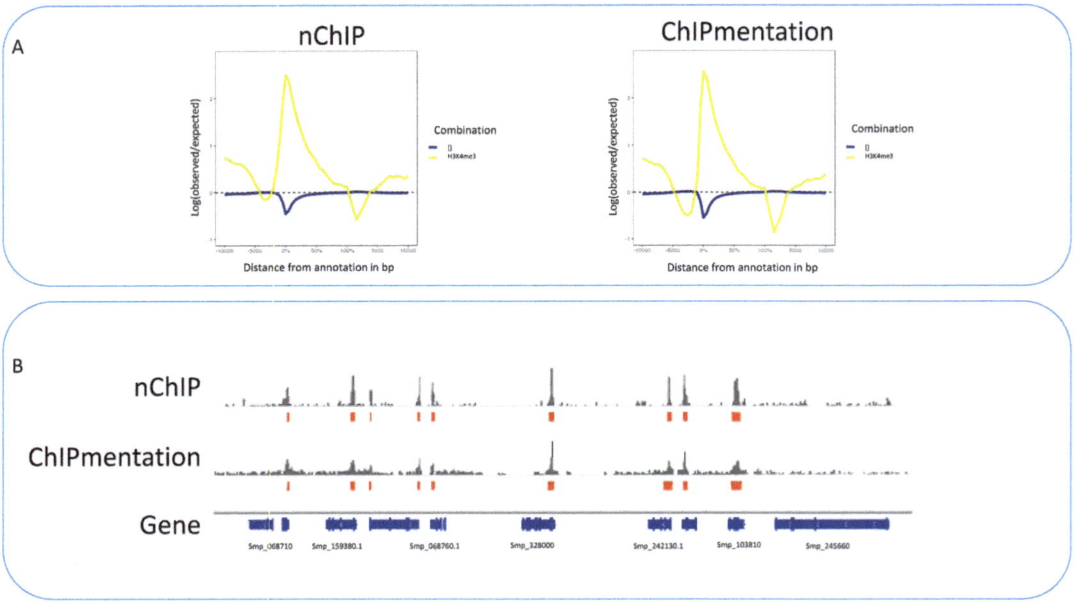

Fig. 3 (**a**) Metagen profile of enrichment of H3K4me3 (yellow line) over genes of *S. mansoni*. *X*-axis over genes in % of length with 0% transcription start site (TSS) and 100% transcription end site, upstream and downstream in base-pair (**b**) genome browser snapshot showing an example of H3K4me3 enrichment (gray peaks) and peak detection (red peak) at TSS

References

1. Kouzarides T (2007) Chromatin modifications and their function. Cell 128:693–705. https://doi.org/10.1016/j.cell.2007.02.005
2. Cosseau C, Wolkenhauer O, Padalino G, Geyer KK, Hoffmann KF, Grunau C (2017) (Epi) genetic inheritance in *Schistosoma mansoni*: a systems approach. Trends Parasitol 33:285–294. https://doi.org/10.1016/j.pt.2016.12.002
3. Fneich S, Dheilly N, Adema C, Rognon A, Reichelt M, Bulla J, Grunau C, Cosseau C (2013) 5-methyl-cytosine and 5-hydroxy-methyl-cytosine in the genome of *Biomphalaria glabrata*, a snail intermediate host of *Schistosoma mansoni*. Parasit Vectors 6:167. https://doi.org/10.1186/1756-3305-6-167
4. Picard MAL, Boissier J, Roquis D, Grunau C, Allienne J-F, Duval D, Toulza E, Arancibia N, Caffrey CR, Long T, Nidelet S, Rohmer M, Cosseau C (2016) Sex-biased Transcriptome of *Schistosoma mansoni*: host-parasite interaction, genetic determinants and epigenetic regulators are associated with sexual differentiation. PLoS Negl Trop Dis 10:e0004930. https://doi.org/10.1371/journal.pntd.0004930
5. Roquis D, Rognon A, Chaparro C, Boissier J, Arancibia N, Cosseau C, Parrinello H, Grunau C (2016) Frequency and mitotic heritability of epimutations in *Schistosoma mansoni*. Mol Ecol 25:1741–1758. https://doi.org/10.1111/mec.13555
6. Roquis D, Taudt A, Padalino G, Hoffmann KF, Holroyd N, Berriman M, Aliaga B, Chaparro C, Grunau C, de Carvalho Augusto R (2018) Histone methylation changes are required for life cycle progression in the human parasite *Schistosoma mansoni*. PLoS Pathog 14:e1007066. https://doi.org/10.1371/journal.ppat.1007066
7. Cosseau C, Grunau C (2011) Native chromatin immunoprecipitation. Methods Mol Biol 791:195–212. https://doi.org/10.1007/978-1-61779-316-5_15
8. Brind'Amour J, Liu S, Hudson M, Chen C, Karimi MM, Lorincz MC (2015) An ultra-low-input native ChIP-seq protocol for genome-wide profiling of rare cell populations.

Nat Commun 6:6033. https://doi.org/10.1038/ncomms7033

9. Feng X, Grossman R, Stein L (2011) PeakRanger: a cloud-enabled peak caller for ChIP-seq data. BMC Bioinformatics 12:139. https://doi.org/10.1186/1471-2105-12-139
10. X, Grossman R, Stein L (2011) PeakRanger: a cloud-enabled peak caller for ChIP-seq data. BMC Bioinformatics 12:139. https://doi.org/10.1186/1471-2105-12-139
11. Thorvaldsdóttir H, Robinson JT, Mesirov JP (2013) Integrative genomics viewer (IGV): high-performance genomics data visualization and exploration. Brief Bioinform 14:178–192. https://doi.org/10.1093/bib/bbs017
12. Taudt A, Nguyen MA, Heinig M, Johannes F, Colome-Tatche M, Chromsta R (2016) Tracking combinatorial chromatin state dynamics in space and time. *bioRxiv*. https://doi.org/10.1101/038612

Chapter 10

Step-by-Step Bioinformatics Analysis of *Schistosoma mansoni* Long Non-coding RNA Sequences

Lucas F. Maciel and Sergio Verjovski-Almeida

Abstract

In the last few years, long non-coding RNAs (lncRNAs) have been widely studied in humans, and their relevance for physiological and pathological conditions has been demonstrated. In parasites, there are only a few works, such as in *Plasmodium falciparum*, where it was shown that an lncRNA regulates the expression of a gene associated with immune system evasion, also indicating the relevance of understanding the role of this class of RNAs in parasites. In *Schistosoma mansoni*, in the last 2 years, there were four published articles related to the annotation of lncRNAs in different life cycle stages using RNA-Seq libraries. In order to make this process of lncRNA identification and annotation more accessible to biologists with no bioinformatics training, considering the growing number of *S. mansoni* RNA-Seq libraries publicly available from different sources, such as ovary tissues from bi-sex and single-sex infections, and the potential of lncRNAs as therapeutic targets, we provide this step-by-step protocol of lncRNA identification and quantification. This guide includes the download of RNA-Seq libraries from a public database and reads processing and mapping against the genome, transcript reconstruction, novel lncRNA identification, transcripts expression level determination, and the identification of differentially expressed lncRNAs.

Key words RNA-Seq, *Schistosoma*, Long non-coding RNAs, Parasitology, Bioinformatics, Transcriptome assembly, Differential gene expression

1 Introduction

Long non-coding RNAs (lncRNAs) are transcripts longer than 200 nucleotides with no coding potential [1]. These transcripts, which have many different mechanisms of action [2], were shown to be important regulators in several vital processes in humans, such as cell cycle [3], pluripotency maintenance [4], immune response [5], and reproduction [6], and their deregulation is associated with cancer [7], Alzheimer disease [8], cardiac diseases [9], and other pathological conditions.

In the malaria parasite *P. falciparum*, several lncRNAs were identified as differentially expressed through the life cycle stages, and one of these transcripts seems to activate the expression of var, a gene related to the virulence and immune system evasion [10]. In

David J. Timson (ed.), *Schistosoma mansoni: Methods and Protocols*, Methods in Molecular Biology, vol. 2151,
https://doi.org/10.1007/978-1-0716-0635-3_10,

S. mansoni, three different reports had identified thousands of lncRNAs [11–13], and some of them were lncRNAs differentially expressed through the life cycle stages, but no work has been done so far to demonstrate their role in this parasite.

The increasing number of publicly available RNA-Seq libraries from different sources, such as the ones provided by Lu et al. [14], that sequenced ovary tissues extracted from females obtained from single and bi-sex infections represents an important source of lncRNAs, so far not explored. In this case, the lncRNAs that one could identify in the Lu et al. [14] dataset could potentially be involved with sexual maturation and reproduction. Identification and study of these lncRNAs are an important step toward understanding the biology of the organism, and lncRNAs may represent future therapeutic targets.

In order to make this process of lncRNA identification and annotation more accessible to biologists with no bioinformatics training, we provide this step-by-step protocol where we used the publicly available RNA-Seq libraries obtained from ovaries isolated from *S. mansoni* female parasites that were recovered from bi-sex (bO) or female single-sex (sO) host infections [14]. This step-by-step protocol has been the basis for the recent analyses of 633 publicly available RNA-Seq libraries, which identified the lncRNAs expressed along the life cycle stages and tissues of *S. mansoni* [15]. The present guide includes the download of RNA-Seq libraries from the public database, processing and reads mapping against the genome, transcript reconstruction, novel lncRNAs identification, transcript expression level determination, and the identification of differentially expressed lncRNAs in ovaries between the sexually mature females (bO) and immature females (sO).

2 Materials

1. For this pipeline, a personal computer or server with access to the Internet is necessary. The computer requirements will vary depending on the number of RNA-Seq libraries. In this guide, we used 62 GB RAM, 16 processor cores, and 1 TB of free disk space (*see* **Note 1**).
2. Ubuntu v. 18.04 (https://www.ubuntu.com/), a free and open-source Linux distribution based on Debian, that can be either installed or run as a virtual machine.
3. The tools used for RNA-seq download, processing, and expression quantification were (*see* **Note 2**):

   SRA-Tools v. 2.9.6 (https://ncbi.github.io/sra-tools/).

   fastp v. 0.19.5 (https://github.com/OpenGene/fastp) [16].

gffread v 0.11.4 (http://ccb.jhu.edu/software/stringtie/gff.shtml).

STAR v. 2.7.0f (https://github.com/alexdobin/STAR) [17].

kallisto v. 0.45.0 (https://pachterlab.github.io/kallisto/) [18].

4. The tools for transcript reconstruction and lncRNAs identification were (*see* **Note 2**):

   Scallop v. 0.10.2 (https://github.com/Kingsford-Group/scallop) [19].

   TACO v. 0.7.3 (https://tacorna.github.io/) [20].

   FEELnc v. 0.1.1 (https://github.com/tderrien/FEELnc) [21].

   CPC2 v. 0.1 (http://cpc2.cbi.pku.edu.cn/) [22].

   bedtools v. 2.27.1 (https://bedtools.readthedocs.io/en/latest/) [23].

5. R v. 3.5.3, a free software designed for statistical computing, and RStudio v. 1.1.419, a free and open-source integrated development environment for R, were used to identify differentially expressed (DEs) transcripts between the experimental groups after RNA-Seq quantification (*see* **Note 2**). The R packages used were:

   tximport v. 1.10.1 [24].

   cowplot v. 0.9.4 (https://github.com/wilkelab/cowplot).

   DESeq2 v. 1.22.2 [25].

   dplyr v. 0.8.1 (https://dplyr.tidyverse.org/).

   magrittr v. 1.5 (https://cran.r-project.org/web/packages/magrittr/vignettes/magrittr.html).

   stringr v. 1.4.0 (https://cran.r-project.org/web/packages/stringr/index.html).

   pheatmap v. 1.0.12 (https://cran.r-project.org/web/packages/pheatmap/index.html).

3 Methods

3.1 RNA-Seq Download from Public Databases and Processing

We recommend the user to familiarize with common commands used in Linux and R before starting this guide (https://maker.pro/linux/tutorial/basic-linux-commands-for-beginners, https://cran.r-project.org/doc/manuals/R-intro.html).

1. The first step is to create a working directory in the user space and to enter it.

```
$ mkdir workdir
$ cd workdir
```

2. Now that we are in the working directory space, we need to create a directory to save all log files that are generated during the analyses so they can be available if we ever need to come back and read them.

```
$ mkdir logs
```

3. Now we need to create a metadata file. This file will contain five columns: (1) full name of the samples, (2) reads length (if they are paired-end reads, the number is the sum of both reads), (3) accession number of the library at SRA, (4) accession number of the sample, (5) group or experimental condition alias. Columns 3 and 4 are important in our case because each of the five samples of ovaries generated by Lu et al. [14], two from single-sex infections and three from bi-sex (or mixed-sex) infections, was separated in each of them into 6 RNAseq libraries and deposited separately. So, the third column will allow us to identify each and all 30 libraries downloaded from the SRA database, and then the information in the fourth one will be used to correctly concatenate all five samples.
4. The metadata need to be downloaded from the SRA database using the project number that is given in the Lu et al. [14] paper, in our case PRJEB1237 (https://www.ncbi.nlm.nih.gov/Traces/study/?acc=PRJEB1237&go=go).
5. We can use any Linux web browser either in our main server or in a local personal computer, where we will download and edit the metadata file, which can be then transferred to the main server.
6. For download, it is necessary to select only the 30 libraries whose names indicate that they come from ovaries and download the files by clicking on the button "RunInfo Table." This downloaded file should contain 30 lines (our 30 libraries of interest), and the SRA file format contains 38 columns; it then must be edited with any text editor in order to retain only the contents of four columns named "Alias" (full name), "AvgSpotLen" (reads length), "Run" (Accession number of the library run that will be downloaded), and "SRA_Sample" (Accession number of the sample); the headers and the data from the other 34 columns are removed, as well as the headers of the four columns that were retained. Data in the fifth column will indicate the experimental condition group that each sample belongs to, and it must be added manually to each line; in our case, this will be **sO** for single-sex infection ovaries or **bO** for bi-sex infection ovaries.
7. After editing the file, we save it as "SRA_files.txt" in our Linux workdir space. The "head" command in Linux allows us to see a snapshot of the first 10 lines of our metadata file that was saved in our working directory as "SRA_files.txt".

```
$ head SRA_files.txt
Schistosoma_mansoni_ovaries__mixed_sex_infection-sc-1878519 200 ERR506075 ERS420092 bO
Schistosoma_mansoni_ovaries__mixed_sex_infection-sc-1878519 200 ERR506099 ERS420092 bO
Schistosoma_mansoni_ovaries__mixed_sex_infection-sc-1878519 200 ERR667053 ERS420092 bO
Schistosoma_mansoni_ovaries__mixed_sex_infection-sc-1878519 200 ERR667077 ERS420092 bO
Schistosoma_mansoni_ovaries__mixed_sex_infection-sc-1878519 200 ERR667101 ERS420092 bO
Schistosoma_mansoni_ovaries__mixed_sex_infection-sc-1878519 200 ERR667125 ERS420092 bO
Schistosoma_mansoni_ovaries__mixed_sex_infection-sc-1878518 200 ERR506074 ERS420091 bO
Schistosoma_mansoni_ovaries__mixed_sex_infection-sc-1878518 200 ERR506098 ERS420091 bO
Schistosoma_mansoni_ovaries__mixed_sex_infection-sc-1878518 200 ERR667052 ERS420091 bO
Schistosoma_mansoni_ovaries__mixed_sex_infection-sc-1878518 200 ERR667076 ERS420091 bO
```

We can see that the first six entries (first six lines) refer to ovaries from mixed-sex (bi-sex, bO) infection set number 1878519 (first column), each with a different ERR run accession number (third column), and all six have a common ERS sample accession number (fourth column). The following four entries in this 10-line snapshot belong to another mixed-sex infection set number 1878518 (first column), again each of the four entries with a different ERR run accession number (third column), and all four have a common ERS sample accession number (fourth column).

8. Now that we have the metadata file, we can start downloading the RNA-Seq library sequence data in the SRA format from the SRA database. For example, if we want to download a hypothetical sample that has the accession number SRR304976 by FTP and save it in a directory named "sra," we would need to use the following commands:

```
$ mkdir sra
$ wget -O sra/SRR304976.sra ftp://ftp-trace.ncbi.
nih.gov/sra/sra-instant/reads/ByRun/sra/SRR/
SRR304/SRR304976/SRR304976.srapi
```

Note that the above ftp://URL points to the SRR304976. sra file using the SRA directory standard structure, which places each SRA file under a directory tree that starts in this case with a directory named "/SRR" (from the first three characters of the file name), then has a directory "/SRR304" (from the first six characters of the file name), and then has a directory "/ SRR304976" (with the full nine characters of the file name), which contains the file of interest, namely, SRR304976.sra.

It is important to keep this SRA directory tree standard structure naming in mind, since we will use it further below, in the automated looping commands, for recurrently downloading all our libraries of interest.

9. In order to automate this process so we do not have to run the same ftp command 30 times (one for each file), we can use the "while read" loop Linux command (https://ccm.net/faq/

1757-how-to-read-a-linux-file-line-by-line). It is important to understand this loop because we are going to use it in almost every step. In this loop, for each line in our metadata file, a given command will be run once. In this command, we consider each column in our metadata file as one variable (in our case five columns, variables v w x y z). As the run accession number (ERR) of each of our 30 libraries is contained in the third column, in order to download them, we use the following command (*see* **Note 3**):

```
$ while read v w x y z; do wget -O sra/${x}.sra ftp://
ftp-trace.ncbi.nih.gov/sra/sra-instant/reads/
ByRun/sra/${x:0:3}/${x:0:6}/${x}/${x}.sra>&
logs/wget_${x}.log; done < SRA_files.txt
```

In this command, the third element from v to z of our loop (the letter x) becomes the third column of the metadata, and the command is run 30 times, once for each accession number, which is contained in one of the 30 lines in the metadata file. The file where the "while read" command retrieves the loop information is specified at the end, in this case by < SRA_files.txt

10. The SRA directories where the files are located, to which the ftp:// function is pointing, are constructed according to the SRA directory tree standard structure naming. Thus, "${x:0:3}" means that for the uppermost directory, we want only the first three letters of the third column, followed by "${x:0:6}" for the next level directory with the first six characters of the file name, while only "${x}" is going to use all nine characters. The ">& logs/wget_${x}.log" gets all the output messages that would be printed in the screen and save them into one file in our log directory.
11. The integrity of the downloaded SRA files can be checked using vdb.

```
$ while read v w x y z; do vdb-validate sra/${x}.sra >& logs/
vdb_${x}.log; done < SRA_files.txt
```

After running the vdb command, we can inspect, for example, the last line of the log file, which will tell if the last downloaded file is consistent.

```
$ tail -n 1 logs/vdb_ERR506075.log
vdb-validate.2.9.6 info: Database 'ERR506075.sra' is consis-
tent
```

12. If all files are consistent, then we must convert the SRA files (which are in a non-human readable binary format) to the human-readable ASCII format, and we must split each SRA sequence run file into two FASTQ files for each record, one

containing the forward read sequences (identified in the SRA records by >name_of_read.1) and one with the reverse read sequences (identified in the SRA records by >name_of_read.2) of the paired-end reads sequencing. The fastq-dump command below does the conversion and the split and saves the output files in a directory of raw reads.

```
$ mkdir raw_reads
$ while read v w x y z; do fastq-dump --split-files --outdir
raw_reads sra/${x}.sra; done < SRA_files.txt
```

The above command should result into 60 files, 2 FASTQ files for each of the 30 records (6 records for each of the 5 different samples). The fastq-dump command adds a suffix "_1.fastq" or "_2.fastq" to the name of each FASTQ file.

3.2 Concatenating the RNA-Seq Libraries

The following steps are necessary because each of our five experimental samples was separated by the authors of the work into six different runs of the same RNA-Seq library, and this section is the one with the most complicated commands; however, it is usually not necessary. If you are using RNA-Seq libraries that have only a single run record per library (usual procedure), you can just move forward to the next section.

1. Now that we have all the 60 FASTQ files (2 FASTQ files for each of the 30 records), we need to concatenate the 30 records into 5 samples. First, we cut columns 4 (sample accession numbers) and 5 (group identifiers) from the metadata, sort them into the 5 unique sample accession numbers along with their group identifiers, and save them into the "samples.txt" file.

```
$ cut -f4,5 SRA_files.txt | sort -u > samples.txt
```

The cat command in this example (with no output file indicated) shows us the contents of the samples.txt file, which has the accession numbers of the five ERS samples along with the experimental condition group identifier of each one (sO or bO).

```
$ cat samples.txt
ERS420088    sO
ERS420089    sO
ERS420090    bO
ERS420091    bO
ERS420092    bO
```

2. Then, we use a while read loop with the grep command (https://www.cyberciti.biz/faq/howto-use-grep-command-in-linux-unix/) to get specifically the accession numbers of the six library runs for each sample.

The cat command (http://www.linfo.org/cat.html) will concatenate the six runs of each sample into one file for each sample and each strand (forward or reverse), and the output is saved into "> raw_reads/concat/${x}_1.fastq" or "> raw_reads/concat/${x}_2.fastq" (see command below).

3. The awk command (https://www.tutorialspoint.com/awk/awk_basic_examples.htm) is used to build the pathway and name to be used by the cat command. In this example, the prefix accession numbers of the library runs present in the third column ("raw_reads/"$3")_will be pasted together with the "_1.fastq" or "_2.fastq" suffix to create the names of the files where the cat command will place the set of concatenated sequences from the six runs into one FASTQ file (forward "_1.fastq" or reverse "_2.fastq").

```
$ mkdir raw_reads/concat
$ while read x y; do grep ${x} SRA_files.txt | cat $(awk '{print "raw_reads/"$3"_1.fastq"}') > raw_reads/concat/${x}_1.fastq; done < samples.txt
$ while read x y; do grep ${x} SRA_files.txt | cat $(awk '{print "raw_reads/"$3"_2.fastq"}') > raw_reads/concat/${x}_2.fastq; done < samples.txt
```

The ls command below will show us the names of the 10 files, 2 files for each of the 5 ERS samples, one of the files with the forward reads ("_1.fastq" suffix) and the other with the reverse reads ("_2.fastq" suffix).

```
$ ls raw_reads/concat/
ERS420088_1.fastq   ERS420089_1.fastq  ERS420090_1.fastq
ERS420091_1.fastq  ERS420092_1.fastq
ERS420088_2.fastq   ERS420089_2.fastq  ERS420090_2.fastq
ERS420091_2.fastq ERS420092_2.fastq
```

3.3 Quality Control and Reads Mapping

1. Now that we have our five samples concatenated, we can remove the adapters and low-quality reads from the RNA-Seq libraries by running the fastp program using 16 processor cores (−w 16) and the syntax provided by the program developer and sending them to the /fastp directory. The fastp directory will then contain the trimmed FASTQ files and the HTML result, with all statistics that can be opened using any browser.

```
$ mkdir fastp
$ while read x y; do fastp -w 16 -h fastp/${x}.html -i raw_reads/concat/${x}_1.fastq -I raw_reads/concat/${x}_2.fastq -o fastp/${x}_1.fastq -O fastp/${x}_2.fastq >& logs/fastp_${x}.log; done < samples.txt
```

2. Once we have the FASTQ files trimmed, we can map the reads against the *S. mansoni* genome in order to assemble the transcripts. The most recent genome assembly (v.7) sequence

FASTA file and the most recent protein-coding gene annotation (v7.1) in GFF3 format are available from the WormBase ParaSite resource [26] at https://parasite.wormbase.org/Schistosoma_mansoni_prjea36577/Info/Index/ and can be downloaded using the following wget commands.

```
$ wget ftp://ftp.ebi.ac.uk/pub/databases/wormbase/parasite/releases/WBPS13/species/schistosoma_mansoni/PRJEA36577/schistosoma_mansoni.PRJEA36577.WBPS13.genomic.fa.gz
$ wget ftp://ftp.ebi.ac.uk/pub/databases/wormbase/parasite/releases/WBPS13/species/schistosoma_mansoni/PRJEA36577/schistosoma_mansoni.PRJEA36577.WBPS13.annotations.gff3.gz
$ gunzip *gz
```

3. We can convert this GFF3 file into GTF ("-T" parameter) and BED ("--bed" parameter) formats that will be needed in the following steps, using gffread (*see* **Note 4**).

```
$ gffread -T schistosoma_mansoni.PRJEA36577.WBPS13.annotations.gff3 -o mansoni.gtf
$ gffread --bed schistosoma_mansoni.PRJEA36577.WBPS13.annotations.gff3 -o mansoni.bed
```

4. We are going to use the STAR program to map all our reads against the genome, but first an index must be built using the genome and the GTF. The parameter "--sjdbOverhang" was set based on reads length (200 − 1 = 199), considering that the read length in this dataset was indicated in the SRA records as being 200.

```
$ mkdir star
$ mkdir star/index
$ STAR --runThreadN 16 --runMode genomeGenerate --genomeDir star/index/ --genomeFastaFiles schistosoma_mansoni.PRJEA36577.WBPS13.genomic.fa --sjdbOverhang 199 --sjdbGTFfile mansoni.gtf
```

5. Then, we map the reads against the genome in a two-pass mode that is recommended for a most sensitive novel splice junction discovery; the basic idea, as described in the STAR's manual, is to run the first time based only on the GTF annotation of splicing events with the usual parameters, then collect all the junctions detected in the first pass that are provided in the file SJ.out.tab, and use them as an additional input of "annotated" junctions for the second pass mapping. All the parameters are set as provided in the manual for the discovery of new transcripts (*see* **Note 5**). The STAR mapping is recursively run over all five samples with the while read loop.

```
$ mkdir star/first
$ mkdir star/second
$ while read x y; do mkdir star/first/${x};done < samples.txt
$ while read x y; do STAR --runThreadN 16 --genomeDir star/
index/ --readFilesIn fastp/${x}_1.fastq fastp/${x}_2.fastq --
outFilterType BySJout --outFilterMultimapNmax 20 --alignSJo-
verhangMin 8 --alignSJDBoverhangMin 1 --alignIntronMin 20 --
outFilterMismatchNoverReadLmax 0.04 --alignIntronMax 1000000
--alignMatesGapMax 1000000 --outFileNamePrefix star/first/
${x}/ --outSAMtype BAM Unsorted --outSAMstrandField intronMo-
tif >& logs/star_first_${x}.log;done < samples.txt
$ while read x y; do mkdir star/second/${x};done < samples.txt
$ while read x y; do STAR --runThreadN 16 --genomeDir star/
index --readFilesIn fastp/${x}_1.fastq fastp/${x}_2.fastq --
outFilterType BySJout --outFilterMultimapNmax 20 --alignSJo-
verhangMin 8 --alignSJDBoverhangMin 1 --alignIntronMin 20 --
outFilterMismatchNoverReadLmax 0.04 --alignIntronMax 1000000
--alignMatesGapMax 1000000 --outFileNamePrefix star/second/
${x}/ --outSAMtype BAM SortedByCoordinate --outSAMstrandField
intronMotif --limitBAMsortRAM 50000000000 --sjdbFileChrStar-
tEnd star/first/*/SJ.out.tab >& logs/star_second_${x}.log;
done < samples.txt
```

6. The read mapping statistics are saved in the "Log.final.out" file. The ls command below will show an example of all the output files generated by STAR second pass for one of the samples, namely, ERS420088.

```
$ ls star/second/ERS420088/
Aligned.sortedByCoord.out.bam Log.final.out Log.out Log.pro-
gress.out SJ.out.tab
```

3.4 Transcriptome Assembly

1. The BAM file ("Aligned.sortedByCoord.out.bam") generated in the second pass for each sample, containing reads mapping coordinates, is used by the scallop program to reconstruct the reads into transcripts. The mapping quality MAPQ threshold was set to 255 because this is the value for uniquely mapping reads, and the parameter "--min_splice_bundary_hits 2" indicates that we need at least two reads confirming a splicing site in order for it to be considered.

```
$ mkdir assembly
$ mkdir assembly/scallop
$ while read x y; do scallop -i star/second/${x}/Aligned.
sortedByCoord.out.bam -o assembly/scallop/${x}.gtf --min_-
transcript_length_base 200 --min_mapping_quality 255 --min_-
splice_bundary_hits 2 --min_num_hits_in_bundle 10 &> logs/
scallop_${x}.log;done < samples.txt
```

2. Scallop will generate a GTF file for each sample; then, a consensus transcriptome for all samples must be constructed. This is performed by the TACO program. We must provide a table manually created, containing the pathway to the GTFs that were generated by scallop; the table should look like as follows:

```
$ head taco.txt
assembly/scallop/ERS420088.gtf
assembly/scallop/ERS420089.gtf
assembly/scallop/ERS420090.gtf
assembly/scallop/ERS420091.gtf
assembly/scallop/ERS420092.gtf
$ taco_run -o assembly/taco/ -p 16 -v --isoform-frac 0.05 --
gtf-expr-attr RPKM --filter-min-length 200 taco.txt >& logs/
taco_run.log
```

3. We have opted to require a minimum length of 200 nucleotides for each transcript (which include long non-coding RNAs) and an option to report transcript isoforms with expression fraction ≧0.05 relative to the highest expressed isoform. For each gene, the highest abundance isoform will be reported with a FRAC of 1.0 (*see* **Note 6**).
4. For the dataset that we have used, TACO has generated a transcriptome composed by 38,224 transcripts; these transcripts are annotated in the BED ("assembly.bed") and GTF format ("assembly.gtf"), which are shown as follows:

```
$ ls assembly/taco/
args.pickle change_points.gtf expr.pos.bedgraph sample_-
stats.txt splice_junctions.bed transfrags.bed
assembly.bed expr.neg.bedgraph loci.txt samples.txt status.
json transfrags.filtered.bed assembly.gtf expr.none.bedgraph
path_graph_stats.txt splice_graph.gtf tmp
```

3.5 LncRNAs Identification

Now that we have all our transcripts, we need to discriminate the lncRNAs from the protein-coding genes.

1. The first step is done by the FEELnc program and uses the GTF genomic coordinates of our transcripts to remove all transcripts in our dataset that have exon-exon overlap at the same strand with the coordinates of all previously annotated protein-coding genes ("-a mansoni.gtf") or to remove unspliced monoexonic transcripts from our dataset, which could represent pre-mRNAs.

```
$ mkdir assembly/lnc_classifier/
$ mkdir assembly/lnc_classifier/feelnc
$ mkdir assembly/lnc_classifier/feelnc/filter
$ FEELnc_filter.pl -i assembly/taco/assembly.gtf -a mansoni.
```

```
gtf -p 16 --monoex=0 -o logs/feelnc_filter.log > assembly/
lnc_classifier/feelnc/filter/Ovaries_filtered.gtf
```

2. Only 5742 transcripts (putative spliced lncRNAs) remained after the above filtering step. Then, we use the FEELnc codpot function to classify these transcripts into (novel) protein-coding genes or lncRNAs; the program uses a random forest machine learning algorithm that is trained here with the previously annotated protein coding genes ("-a mansoni.gtf"). Training and classification are based on intrinsic features of the transcript such as ORF coverage, ORF size, and k-mer frequency.

```
$ FEELnc_codpot.pl -i assembly/lnc_classifier/feelnc/filter/
Ovaries_filtered.gtf -a mansoni.gtf -g /work/verjo102/users/
lucas.maciel/book/schistosoma_mansoni.PRJEA36577.WBPS13.geno-
mic.fa --outdir assembly/lnc_classifier/feelnc/codpot
```

The summary file ("Ovaries_filtered.codpot_RF_summary.txt") contains the number of transcripts in our dataset classified as lncRNAs or mRNAs, and "Ovaries_filtered.codpot_RF_TGROC.png" contains the applied coding probability cut-off (0.35), and the sensitivity and specificity of classification that were achieved, which in our case was equal to 0.962 (Fig. 1).

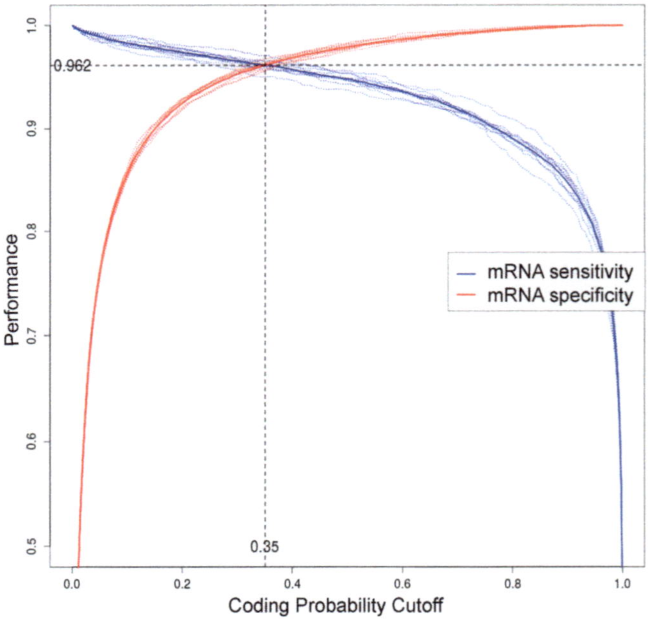

Fig. 1 FEELnc program identifies an optimal coding probability cutoff. Using a random forest machine learning algorithm that was trained with the previously annotated *S. mansoni* protein-coding genes, a coding probability cutoff of 0.35 was identified, which resulted in 0.962 sensitivity (blue) and specificity (red) for discrimination of mRNAs from lncRNAs. This plot was saved in the "assembly/lnc_classifier/feelnc/codpot/Ovaries_filtered.codpot_RF_TGROC.png" file

3. The cat command below will show the contents of the summary file.

```
$ cat assembly/lnc_classifier/feelnc/codpot/Ovaries_filtered.
codpot_RF_summary.txt
# Summary file:
-With_cutoff:    0.3504
-Nb_lncRNAs:     4895
-Nb_mRNAs:       847
```

4. The FEELnc codpot function generates two files: "Ovaries_filtered.codpot.lncRNA.gtf" with 4985 transcripts classified as lncRNAs and "Ovaries_filtered.codpot.mRNAs.gtf" with 847 mRNA transcripts (which would be possible novel *S. mansoni* protein coding genes).

5. The next step is very important, and it was not applied by any of the three works on lncRNAs published until now in *S. mansoni* [11–13]. With the increase in sequencing depth usually obtained in the recent RNAseq datasets, we are now able to capture unprocessed or partially processed pre-mRNAs that are present in the cells [27]. Due to the presence of introns, these pre-mRNAs may be artifactually classified as lncRNAs. This can be seen in Fig. 2, where we show an example of pre-mRNAs that were artifactually classified as lncRNAs in previous works that used an old v5.2 annotation of the transcriptome, and now we see in v7 that there is a protein-coding gene annotated in the locus.

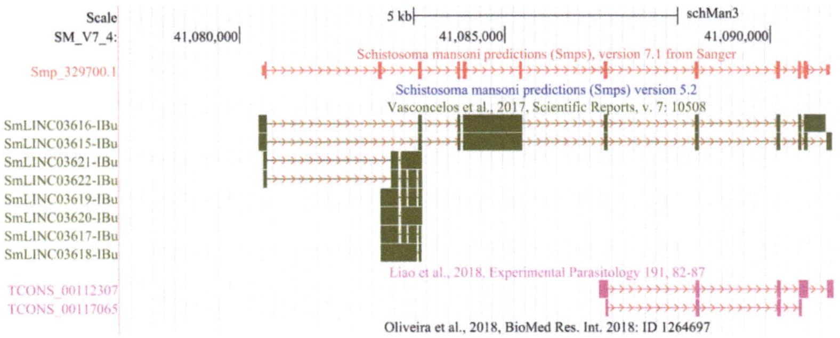

Fig. 2 Previously annotated *S. mansoni* lncRNAs are now recognized as partially processed pre-mRNAs in a genomic locus encoding a **new** protein-coding gene. Snapshot of a *S. mansoni* genome browser image, showing a region spanning 16 kb on chromosome 4 with coordinates SM_V7_4:41,077,745-41,093,812 (top black row). The red track (red first track, top) shows a novel protein-coding gene Smp_329700.1 newly annotated in transcriptome version 7.1 and not present in the old transcriptome version 5.2 (blue track, second track). Partially processed pre-mRNAs with intron retention can be recognized among the eight supposedly intergenic lncRNAs (SmLINC03615-IBu to SmLINC03622-IBu, gray track) that were annotated in the previous work by Vasconcelos et al. [11] and among the two different supposedly intergenic lncRNAs (TCONS_0011xxxx, pink track) that were annotated in the previous work by Liao et al. [13]. No lncRNAs were annotated in this locus by Oliveira et al. [12] (empty black track at the bottom)

So, in order to remove all possible pre-mRNAs in the loci of these novel mRNAs, we filter again the lncRNAs against the 847 transcripts classified as novel mRNAs.

```
$ FEELnc_filter.pl -i assembly/lnc_classifier/feelnc/filter/Ovaries_filtered.codpot.lncRNA.gtf -a assembly/lnc_classifier/feelnc/filter/Ovaries_filtered.codpot.mRNA.gtf -p 16 --monoex=0 -o logs/feelnc_mRNA_filter.log > assembly/lnc_classifier/feelnc/filter/lncRNA_filtered.gtf
```

6. As seen in Fig. 1, FEELnc has specificity and sensitivity of 0.96; so in an attempt to minimize the possible impact of the 0.04 misclassifications, we use another tool to calculate again the protein coding potential of the transcripts that were classified as lncRNAs by FEELnc. We chose the CPC2 tool, which classifies through a support vector machine model trained with human protein-coding mRNAs and human lncRNAs and needs only the *S. mansoni* transcripts in FASTA format as input. We can convert our "lncRNA_filtered.gtf" into a BED format and use it to extract the FASTA file using bedtools.

```
$ gffread --bed assembly/lnc_classifier/feelnc/filter/lncRNA_filtered.gtf -o assembly/lnc_classifier/feelnc/filter/lncRNA_filtered.bed
$ bedtools getfasta -fi schistosoma_mansoni.PRJEA36577.WBPS13.genomic.fa -fo assembly/lnc_classifier/feelnc/filter/lncRNA_filtered.fasta -bed assembly/lnc_classifier/feelnc/filter/lncRNA_filtered.bed -name -split -s
```

7. Now we use this FASTA file as input for CPC2 with default parameters, generating an output table "CPC2.txt" with the classification.

```
$ mkdir assembly/lnc_classifier/cpc2
$ CPC2.py -i assembly/lnc_classifier/feelnc/filter/lncRNA_filtered.fasta -o assembly/lnc_classifier/cpc2/CPC2.txt &> logs/CPC2.log
```

8. With the head command, we can inspect the first 10 lines of the output, and in this case, these first 10 transcripts were also classified by CPC2 as noncoding.

```
$ head assembly/lnc_classifier/cpc2/CPC2.txt
#ID transcript_length peptide_length Fickett_score pI ORF_integrity coding_probability label
TU24683 3249 59 0.25513 4.67706298828 1 0.134893 noncoding
TU24692 1848 60 0.3118 8.08563232422 1 0.0229334 noncoding
TU24685 564 53 0.42092 8.83599853516 1 0.0432804 noncoding
TU24762 5314 108 0.34952 6.02081298828 1 0.344888 noncoding
TU24770 5401 108 0.3287 6.02081298828 1 0.287759 noncoding
TU24841 673 92 0.337 6.05682373047 1 0.154715 noncoding
```

```
TU24756 535 38 0.404 9.21929931641 1 0.0182792 noncoding
TU24867 1237 73 0.37851 9.75677490234 1 0.081108 noncoding
TU24897 5998 132 0.2705 8.89141845703 1 0.19102 noncoding
```

9. Only 60 transcripts were classified by CPC2 as mRNAs. Now we need to filter again our "lncRNA_filtered.gtf" file to remove the 60 mRNAs and other possible pre-mRNAs in their loci. First, we use grep and cut to get the list of transcript names classified as mRNAs and save them into the "coding.txt" file. Then, we use this "coding.txt" as input to grep against the coordinates of all lncRNAs classified by FEELnc ("lncRNA_filtered.gtf") to generate a GTF file containing only the coordinates of these 60 transcripts.

```
$ grep -v noncoding assembly/lnc_classifier/cpc2/CPC2.txt |
cut -f1 | sed 's/$/"/g' > assembly/lnc_classifier/cpc2/coding.
txt
$ grep -f assembly/lnc_classifier/cpc2/coding.txt assembly/
lnc_classifier/feelnc/filter/lncRNA_filtered.gtf > assembly/
lnc_classifier/cpc2/coding.gtf
```

10. Next, we can generate a final GTF file ("final.gtf") using FEELnc to filter out the 60 transcripts classified as mRNAs and any putative lncRNAs that overlap with them.

```
$ FEELnc_filter.pl -i assembly/lnc_classifier/feelnc/filter/
lncRNA_filtered.gtf -a assembly/lnc_classifier/cpc2/coding.
gtf -p 16 --monoex=0 -o logs/feelnc_cpc_filtering.log > as-
sembly/lnc_classifier/final.gtf
```

11. Finally, we classify the remaining bona fide lncRNAs in the "final.gtf" file into intergenic, sense, or antisense lncRNAs using FEELnc classifier, based on the genomic positions of the lncRNAs compared with the positions of protein-coding genes.

```
$ FEELnc_classifier.pl -i assembly/lnc_classifier/final.gtf -a mansoni.gtf -l logs/
feelnc_classifier.log >
assembly/lnc_classifier/lncRNA_classes.txt
$ head assembly/lnc_classifier/lncRNA_classes.txt
isBest lncRNA_gene lncRNA_transcript partnerRNA_gene partnerRNA_transcript direction
type distance
subtype location
1 G2419 TU4311 Smp_319040 Smp_319040.1 antisense genic 0  overlapping intronic
1 G11789 TU20804 Smp_043360 Smp_043360.2 sense intergenic 8542 same_strand downstream
0 G11789 TU20804 Smp_043360 Smp_043360.1 sense intergenic 8542 same_strand downstream
1 G19582 TU34208 Smp_132170 Smp_132170.3 antisense genic 0 nested  intronic
0 G19582 TU34208 Smp_132170 Smp_132170.2  antisense intergenic 6899 convergent downstream
0    G19582 TU34208 Smp_132170    Smp_132170.1  antisense    genic 0    nested intronic
1    G15082 TU26412 Smp_063190    Smp_063190.1  antisense    genic 0    nested intronic
```

```
1   G3596 TU6394 Smp_170340  Smp_170340.1 antisense  genic 0  nested exonic
1   G3746 TU6713 Smp_048230  Smp_048230.1 antisense  genic 0  nested exonic
```

12. Now that we have our final set of 4582 lncRNAs ("final.gtf"), this GTF is converted to BED again ("final.bed"), and the lncRNAs in this BED file are merged with the previously annotated protein-coding genes ("mansoni.bed") with the cat command. After concatenating the protein-coding and lncRNAs BED files, the bedtools sort command sorts the transcripts by their genomic coordinates.

```
$ gffread --bed assembly/lnc_classifier/final.gtf -o assembly/
lnc_classifier/final.bed
$ cat mansoni.bed assembly/lnc_classifier/final.bed > un-
sorted.bed
$ bedtools sort -i unsorted.bed > mansoni_sorted.bed
```

3.6 RNA-Seq Quantification

To identify differentially expressed lncRNAs between ovaries of bi-sex (bO) and single-sex (sO) infections, we now can quantify the expression levels of each of the 4582 lncRNAs and of all previously annotated protein-coding genes using the kallisto tool with recommended default parameters as shown below (*see* **Note 7**).

1. First, kallisto takes a FASTA file of all transcripts to generate an index.

```
$ bedtools getfasta -fi schistosoma_mansoni.PRJEA36577.WBPS13.
genomic.fa -fo mansoni_sorted.fasta -bed mansoni_sorted.bed
-name -split -s
$ mkdir kallisto
$ kallisto index -i kallisto/mansoni_index mansoni_sorted.
fasta &> logs/kallisto_index.log
```

2. Then, kallisto performs a pseudoalignment of reads and counts the abundance of transcripts for each of our five samples. If you do not know if your library is strand specific, you can look for this information in the first 10 lines of the log file logs/scallop_${x}.log that is generated after running Scallop (Subheading 3.4 above). The "–rf-stranded" parameter was used because we know that these RNA-Seq libraries are strand specific.

```
$ while read x y; do kallisto quant -i kallisto/mansoni_index
-o kallisto/${x} --rf-stranded -b 100 fastp/${x}_1.fastq
fastp/${x}_2.fastq &> logs/kallisto_${x}.log; done < sam-
ples.txt
```

This step generates the abundance files with expression levels for each transcript in each sample that can be easily loaded onto R. An example of the output can be inspected for sample ERS420088 further.

```
$ head kallisto/ERS420088/abundance.tsv
target_id        length  eff_length       est_counts       tpm
Smp_329140.1     357     152.189 2.05456 0.739012
Smp_315690.1     965     749.607 0       0
Smp_315690.2     1742    1526.57 0       0
TU2      537     323.024 4       0.677862
Smp_317470.2     1063    847.573 0       0
Smp_317470.3     878     662.732 0       0
Smp_317470.1     1014    798.573 0       0
Smp_318880.1     689     474.131 2.94544 0.34007
Smp_319020.1     1243    1027.57 4       0.21309
```

3.7 Differentially Expressed Transcripts Using DESeq2

Now that we have the expression levels of the lncRNAs and protein-coding genes, we are going to use R to identify the differentially expressed ones.

1. First, we need to know the full path of our working directory to be used with R, which we find out using the command pwd.

```
$ pwd
/work/tutorial/workdir/
```

   This full path, in our case "/work/tutorial/workdir/," will be used in the R environment further.

2. Next, we create a directory to save the results obtained with R, and we run the Rstudio program.

```
$ mkdir R_results
$ rstudio
```

3. From now on, we are going to use the Rstudio environment, so all the syntax used is from R and the commands will be run on Rstudio terminal, and this is why they are indicated in the beginning with ">" instead of "$" that was used until now.
4. The first step is to set a variable (base_dir) as our working directory full path. Then, we use it to load our "sample.txt" table and rename the columns to "ID" and "condition".

```
> base_dir <- "/work/tutorial/workdir/"
> samples <- read.table(paste0(base_dir,"samples.txt"), head-
er = FALSE, stringsAsFactors=FALSE)
> samples <- colnames("ID", "condition")
> samples
         ID condition
1 ERS420088        sO
2 ERS420089        sO
3 ERS420090        bO
4 ERS420091        bO
5 ERS420092        bO
```

5. Now that we have our metadata file loaded, we are going to use it to indicate the pathway of the five abundance files generated by kallisto.

```
> files <- file.path(paste0(base_dir,"kallisto/",samples$ID,
"/abundance.tsv"))
> names(files) <- samples$ID
> files
                                                      ERS420088
"/ work/tutorial/workdir/kallisto/ERS420088/abundance.tsv"
                                                      ERS420089
"/ work/tutorial/workdir/kallisto/ERS420089/abundance.tsv"
                                                      ERS420090
"/ work/tutorial/workdir/kallisto/ERS420090/abundance.tsv"
                                                      ERS420091
"/ work/tutorial/workdir/kallisto/ERS420091/abundance.tsv"
                                                      ERS420092
"/ work/tutorial/workdir/kallisto/ERS420092/abundance.tsv"
```

6. Using tximport package, we will import the abundance files indicated in the variable "files" that we created.

```
> library("tximport")
> txi <- tximport(files, type = "kallisto", txIn = TRUE, txOut
= TRUE)
```

7. We are going to use the package DESeq2 to identify differentially expressed transcripts. In order to do that, we need to load the package and convert our imported files with tximport to the format required by DESeq2 (*see* **Note 8**).

```
> library("DESeq2")
> dds <- DESeqDataSetFromTximport(txi, colData = samples,
design = ~ condition)
```

8. Before starting the differential expression analysis, we can plot the results from a principle component analysis (PCA) to check if the sample similarities are in accordance to their experimental conditions. First, we normalize our data using variance stabilizing transformations (vst) and then use the function plotPCA. It can be seen that the samples were clustered in accordance to their experimental conditions (Fig. 3).

```
> library("cowplot")
> vsd <- vst(dds, blind=FALSE)
> plotPCA(vsd, intgroup="condition")
```

9. Next, as recommended by DESeq2 manual, we perform a filtering step to keep only transcripts that have at least 10 reads total.

```
> keep <- rowSums(counts(dds)) >= 10
> dds <- dds[keep,]
```

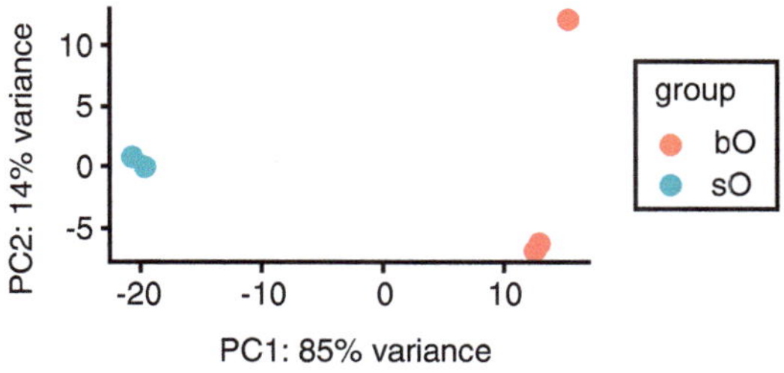

Fig. 3 Clustering of biological replicates assessed by principal component analysis (PCA). A PCA was computed using gene expression data from five different samples and the plotPCA function from DESeq2 package in R. The plot contains data from two biological replicate samples of *S. mansoni* ovaries isolated from immature females (recovered from single-sex host infection, sO) represented by blue circles and from three biological replicate samples of ovaries isolated from mature females (recovered from bi-sex host infection, bO) represented by pink circles. Samples are separated by their first two principal components

10. Then, we set one of our experimental groups as the reference group, in this case we chose the single-sex ovaries (sO).

```
> dds$condition <- relevel(dds$condition, ref = "sO")
```

11. The standard differential expression analysis steps of DESeq2 are wrapped into a single function called "DESeq" and can be easily run.

```
> dds <- DESeq(dds)
```

12. We can use the "results" function to perform an independent filtering, retrieving only the transcripts that have an adjusted p-value below a certain chosen FDR cutoff, which in our case is 0.005 because it was the value chosen by Lu et al. [14] in their analysis of differentially expressed protein-coding genes.

```
> res005 <- results(dds, alpha=0.005)
> summary(res005)
out of 16982 with nonzero total read count
adjusted p-value < 0.005
LFC > 0          : 3915, 23%
LFC < 0 (down)     : 2307, 14%
outliers [1]       : 16, 0.094%
low counts [2]     : 1317, 7.8%
(mean count < 5)
```

13. Based only on the FDR, we have 6222 differentially expressed transcripts, 3915 more highly expressed in bO and 2307 in sO. We can apply a threshold and generate a table that keeps only those transcripts with fold-difference >1.5 between the conditions, which translates into log2 fold-change

≧0.584962501 for upregulated genes or ≦−0.584962501 for downregulated genes.

```
> library(magrittr)
> library(dplyr)
> resSig <- subset(res005, padj < 0.005)
> as.data.frame(resSig) -> export_DE
> rownames(export_DE) -> export_DE$transcript_id
> export_DE %>% filter(log2FoldChange >= 0.584962501 | log2-
FoldChange <= -0.584962501) -> filtered_DE
```

14. As a result of placing a minimum fold-difference cutoff of >1.5, the "filtered_DE" table now contains 5407 differentially expressed genes. Among them, 389 are lncRNAs more highly expressed in bO and 380 lncRNAs more highly expressed in sO, and 4638 are differentially expressed protein-coding genes. This number is higher than the one presented by Lu et al. [14], which was 3600 differentially expressed genes. This happened because we have a different and bigger set of annotated protein-coding genes, since they used version 5.2 and we used version 7.1, and we have the lncRNAs that were not analyzed in their work.
15. The table containing all differentially expressed transcripts can be then exported and saved in the directory that we created in our workdir "R_results."

```
> write.table(filtered_DE, file = paste0(base_dir,"R_results/
DE_transcripts.tsv")
, sep = "\t")
```

16. We can also generate a heatmap plot using all differentially expressed lncRNAs. Stringr is used to get the list of differentially expressed lncRNAs only, because in our case they all start with the "TU" prefix.

```
> library(stringr)
> str_subset(filtered_DE$transcript_id, "TU") -> lnc_list
```

The pheatmap package will be used to generate the heatmap. To do that, data in the "samples" table must be rearranged in the following manner: the contents of the column labeled "ID" in each row is moved to the first column in that row ("rownames(sample_heat)"), where it replaces the number (from 1 to 5) that was used as row name, and then the contents in the column "ID" are deleted. In this way, each sample will be identified in the heatmap by its accession number, not just by a numeral from 1 to 5.

```
> library(pheatmap)
> samples -> sample_heat
```

```
> rownames(sample_heat) <- sample_heat$ID
> sample_heat$ID <- NULL
```

17. We also need the abundance values that were imported using tximport.

```
> txi$abundance -> tpm
```

18. Finally, we provide the TPM values of our differentially expressed lncRNAs only and the modified "sample_heat" table to the pheatmap command that will generate the heatmap (Fig. 4).

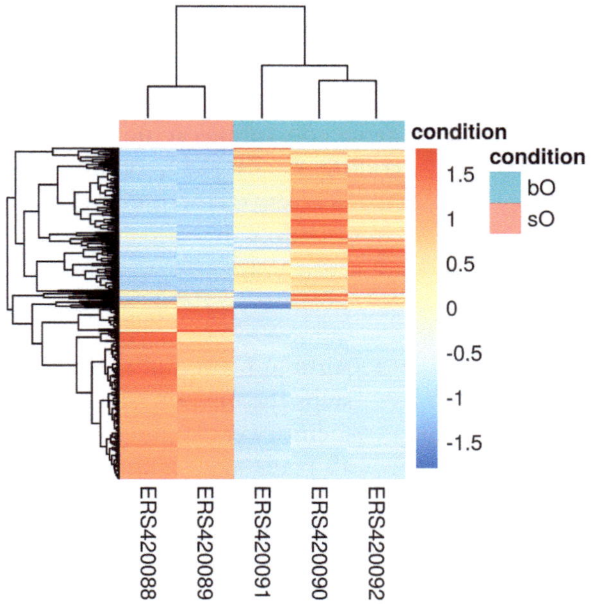

Fig. 4 Gene expression heatmap of differentially expressed lncRNAs detected by RNA-Seq in ovaries of immature and mature female *S. mansoni* parasites. Heatmap of gene expression levels for the 769 lncRNAs (each in one line) detected by our analysis as having a significant differential expression (adjusted *p*-value <0.005; |log2 fold-change| ≧0.584962501) in the ovaries of *S. mansoni* isolated from immature females (recovered from single-sex host infection, sO) compared with the ovaries isolated from mature females (recovered from bi-sex host infection, bO). Expression data from five biological replicate RNA-Seq libraries (one in each column, as indicated at the bottom), being two for the sO experimental condition indicated at the top by the pink color bar (access numbers ERS420088 and ERS420089) and three for the bO condition indicated at the top by the blue color bar (access numbers ERS420090, ERS420091 and ERS420092); data from Lu et al. [14] downloaded from the SRA public database and reanalyzed as described in Subheading 3.7. Unsupervised clustering using the Euclidean distance was performed; expression of each lncRNA gene (each in one line) is shown as the z-score (from −1.5 to 1.5), which is the number of standard deviations above (red) or below (blue) the mean expression value of that gene across all RNA-Seq libraries; the z-score color-scale is shown on the right

```
> pheatmap(tpm[lnc_list,], scale = "row", cluster_rows = T,
show_rownames = F, annotation_col = sample_heat)
```

3.8 Final Remarks

In this chapter, we presented a bioinformatics pipeline that allows users to identify all expressed long non-coding RNAs from RNA-Seq experiments and to compute the differentially expressed ones. In this specific example, using *S. mansoni* ovary samples, we were able to annotate 4582 lncRNAs expressed in ovaries and identify 769 differentially expressed lncRNAs when comparing ovaries isolated from female parasites recovered from bi-sex (bO) or from female single-sex (sO) host infections. We also identified hundreds of potentially novel protein-coding genes that in this chapter were not explored but could be further analyzed. In addition, if other datasets that represent a larger number of different experimental conditions are used, further analyses can be performed, such as gene co-expression network analyses of lncRNAs and protein-coding genes [15], or analyses of the presence of epigenetics marks at the transcription start sites of the lncRNAs [11, 15].

4 Notes

1. Although the full memory available was 62 GB, in almost all steps the memory usage was not higher than 16 GB. The most memory consuming tool used in this guide was STAR, to map all reads against the genome. So, if the user does not have a personal computer or server with 62 GB RAM or more, STAR can be replaced by other mapping tools that require less memory, e.g., Bowtie2 [28] in HISAT [29].
2. The user should be aware that both programs and packages may be updated in the future, and we encourage the user to always try the most recent version, however, realizing that this may cause some software incompatibility.
3. If the user is building a while read loop and wants to check what is going to be run, before running, it is possible to add the "echo" command to the while read loop. With "echo," the while read loop command is only printed in the screen without actually being run, as presented in the example further:

```
$ while read v w x y z; do echo wget -O sra/${x}.sra ftp://ftp-trace.ncbi.
nih.gov/sra/sra-instant/reads/ByRun/sra/${x:0:3}/${x:0:6}/${x}/
${x}.sra;done < SRA_files.txt | head
wget    -O    sra/ERR506075.sraftp://ftp-trace.ncbi.nih.gov/sra/sra-
instant/reads/ByRun/sra/ERR/ERR506/ERR506075/ERR506075.sra
wget    -O    sra/ERR506099.sraftp://ftp-trace.ncbi.nih.gov/sra/sra-
instant/reads/ByRun/sra/ERR/ERR506/ERR506099/ERR506099.sra
wget    -O    sra/ERR667053.sraftp://ftp-trace.ncbi.nih.gov/sra/sra-
instant/reads/ByRun/sra/ERR/ERR667/ERR667053/ERR667053.sra
```

```
wget -O sra/ERR667077.sraftp://ftp-trace.ncbi.nih.gov/sra/sra-instant/reads/ByRun/sra/ERR/ERR667/ERR667077/ERR667077.sra
wget -O sra/ERR667101.sraftp://ftp-trace.ncbi.nih.gov/sra/sra-instant/reads/ByRun/sra/ERR/ERR667/ERR667101/ERR667101.sra
wget -O sra/ERR667125.sra ftp://ftp-trace.ncbi.nih.gov/sra/sra-instant/reads/ByRun/sra/ERR/ERR667/ERR667125/ERR667125.sra
wget -O sra/ERR506074.sraftp://ftp-trace.ncbi.nih.gov/sra/sra-instant/reads/ByRun/sra/ERR/ERR506/ERR506074/ERR506074.sra
wget -O sra/ERR506098.sraftp://ftp-trace.ncbi.nih.gov/sra/sra-instant/reads/ByRun/sra/ERR/ERR506/ERR506098/ERR506098.sra
wget -O sra/ERR667052.sraftp://ftp-trace.ncbi.nih.gov/sra/sra-instant/reads/ByRun/sra/ERR/ERR667/ERR667052/ERR667052.sra
wget -O sra/ERR667076.sraftp://ftp-trace.ncbi.nih.gov/sra/sra-instant/reads/ByRun/sra/ERR/ERR667/ERR667076/ERR667076.sra
```

4. Annotation files that are incorrectly formatted and cannot be rightly recognized by the tools are an often-encountered problem. Then, it is important for the user to know the BED, GTF, and GFF formats (https://genome.ucsc.edu/FAQ/FAQformat.html) in order to be able to recognize this problem and solve it, by manually editing the incorrectly formatted information in the downloaded annotation file.
5. STAR uses a lot of parameters that are very important and impossible to be explained in a few lines. We strongly recommend the user to read the STAR manual (https://physiology.med.cornell.edu/faculty/skrabanek/lab/angsd/lecture_notes/STARmanual.pdf), where everything is explained in details.
6. TACO is an algorithm that reconstructs the consensus transcriptome from a collection of individual assemblies, instead of just merging all transcripts detected in the different assemblies [20]. TACO employs change point detection to break apart complex loci and correctly delineate transcript start and end sites and a dynamic programming approach to assemble transcripts from a network of splicing patterns, using abundance metrics to remove spurious isoforms [20].
7. Kallisto was chosen because it is a fast, simple, and low memory consuming program. However, tools that do not perform pseudoalignment but instead execute reads mapping, for example, RSEM [30] that uses STAR as a mapping tool, are interesting options.
8. In the DESeq2 vignettes, the most basic steps are described in detail. We strongly advise the user to read them point by point to better understand each command and to be able to customize and perform additional plots and analyses (https://bioconductor.org/packages/release/bioc/vignettes/DESeq2/inst/doc/DESeq2.html).

Acknowledgments

Research in our group is funded by Fundação de Amparo a Pesquisa do Estado de São Paulo (FAPESP) grant numbers 2014/03620-2 and 2018/23693-5. LFM received a fellowship from FAPESP number 2018/19591-2 and SVA received a Research Career fellowship award IA from Conselho Nacional de Desenvolvimento Científico e Tecnológico, Brasil.

References

1. Cao H, Wahlestedt C, Kapranov P (2018) Strategies to annotate and characterize long noncoding RNAs: advantages and pitfalls. Trends Genet 34(9):704–721. https://doi.org/10.1016/j.tig.2018.06.002
2. Bhat SA, Ahmad SM, Mumtaz PT, Malik AA, Dar MA, Urwat U, Shah RA, Ganai NA (2016) Long non-coding RNAs: mechanism of action and functional utility. Noncoding RNA Res 1 (1):43–50. https://doi.org/10.1016/j.ncrna.2016.11.002
3. Kitagawa M, Kitagawa K, Kotake Y, Niida H, Ohhata T (2013) Cell cycle regulation by long non-coding RNAs. Cell Mol Life Sci 70 (24):4785–4794. https://doi.org/10.1007/s00018-013-1423-0
4. Rosa A, Ballarino M (2016) Long noncoding RNA regulation of Pluripotency. Stem Cells Int 2016:1797692. https://doi.org/10.1155/2016/1797692
5. Chen YG, Satpathy AT, Chang HY (2017) Gene regulation in the immune system by long noncoding RNAs. Nat Immunol 18:962. https://doi.org/10.1038/ni.3771
6. Golicz AA, Bhalla PL, Singh MB (2018) lncRNAs in plant and animal sexual reproduction. Trends Plant Sci 23(3):195–205. https://doi.org/10.1016/j.tplants.2017.12.009
7. Fang Y, Fullwood MJ (2016) Roles, functions, and mechanisms of long non-coding RNAs in cancer. Genomics Proteomics Bioinformatics 14(1):42–54. https://doi.org/10.1016/j.gpb.2015.09.006
8. Zijian Z (2016) Long non-coding RNAs in Alzheimer's disease. Curr Top Med Chem 16 (5):511–519. https://doi.org/10.2174/1568026615666150813142956
9. Simona G, Antonio SS, Yvan D, Fabio M (2018) Long noncoding RNAs and cardiac disease. Antioxid Redox Signal 29 (9):880–901. https://doi.org/10.1089/ars.2017.7126
10. Jing Q, Cao L, Zhang L, Cheng X, Gilbert N, Dai X, Sun M, Liang S, Jiang L (2018) *Plasmodium falciparum* var gene is activated by its antisense long noncoding RNA. Front Microbiol 9:3117. https://doi.org/10.3389/fmicb.2018.03117
11. Vasconcelos EJR, daSilva LF, Pires DS, Lavezzo GM, Pereira ASA, Amaral MS, Verjovski-Almeida S (2017) The *Schistosoma mansoni* genome encodes thousands of long non-coding RNAs predicted to be functional at different parasite life-cycle stages. Sci Rep 7 (1):10508. https://doi.org/10.1038/s41598-017-10853-6
12. Oliveira VF, Moares LAG, Mota EA, Jannotti-Passos LK, Coelho PMZ, Mattos ACA, Couto FFB, Caffrey BE, Marsico A, Guerra-Sa R (2018) Identification of 170 new long noncoding RNAs in *Schistosoma mansoni*. Biomed Res Int 2018:1264697. https://doi.org/10.1155/2018/1264697
13. Liao Q, Zhang Y, Zhu Y, Chen J, Dong C, Tao Y, He A, Liu J, Wu Z (2018) Identification of long noncoding RNAs in *Schistosoma mansoni* and *Schistosoma japonicum*. Exp Parasitol 191:82–87. https://doi.org/10.1016/j.exppara.2018.07.001
14. Lu Z, Sessler F, Holroyd N, Hahnel S, Quack T, Berriman M, Grevelding CG (2016) Schistosome sex matters: a deep view into gonad-specific and pairing-dependent transcriptomes reveals a complex gender interplay. Sci Rep 6:31150. https://doi.org/10.1038/srep31150. https://www.nature.com/articles/srep31150#supplementary-information
15. Maciel LF, Morales-Vicente DA, Silveira GO, Ribeiro RO, Olberg GGO, Pires DS, Amaral MS, Verjovski-Almeida S (2019) Weighted gene co-expression analyses point to long non-coding RNA hub genes at different *Schistosoma mansoni* life-cycle stages. Front Genet 10:823. https://doi.org/10.3389/fgene.2019.00823
16. Chen S, Zhou Y, Chen Y, Gu J (2018) fastp: an ultra-fast all-in-one FASTQ preprocessor. Bioinformatics 34(17):i884–i890. https://doi.org/10.1093/bioinformatics/bty560

17. Dobin A, Davis CA, Schlesinger F, Drenkow J, Zaleski C, Jha S, Batut P, Chaisson M, Gingeras TR (2013) STAR: ultrafast universal RNA-seq aligner. Bioinformatics 29(1):15–21. https://doi.org/10.1093/bioinformatics/bts635
18. Bray NL, Pimentel H, Melsted P, Pachter L (2016) Near-optimal probabilistic RNA-seq quantification. Nat Biotechnol 34:525. https://doi.org/10.1038/nbt.3519. https://www.nature.com/articles/nbt.3519#supplementary-information
19. Shao M, Kingsford C (2017) Accurate assembly of transcripts through phase-preserving graph decomposition. Nat Biotechnol 35:1167. https://doi.org/10.1038/nbt.4020. https://www.nature.com/articles/nbt.4020#supplementary-information
20. Niknafs YS, Pandian B, Iyer HK, Chinnaiyan AM, Iyer MK (2016) TACO produces robust multisample transcriptome assemblies from RNA-seq. Nat Methods 14:68. https://doi.org/10.1038/nmeth.4078. https://www.nature.com/articles/nmeth.4078#supplementary-information
21. Wucher V, Legeai F, Hédan B, Rizk G, Lagoutte L, Leeb T, Jagannathan V, Cadieu E, David A, Lohi H, Cirera S, Fredholm M, Botherel N, Leegwater PAJ, Le Béguec C, Fieten H, Johnson J, Alföldi J, André C, Lindblad-Toh K, Hitte C, Derrien T (2017) FEELnc: a tool for long non-coding RNA annotation and its application to the dog transcriptome. Nucleic Acids Res 45(8):e57–e57. https://doi.org/10.1093/nar/gkw1306
22. Yang D-C, Kong L, Wei L, Hou M, Kang Y-J, Meng Y-Q, Gao G (2017) CPC2: a fast and accurate coding potential calculator based on sequence intrinsic features. Nucleic Acids Res 45(W1):W12–W16. https://doi.org/10.1093/nar/gkx428
23. Quinlan AR, Hall IM (2010) BEDTools: a flexible suite of utilities for comparing genomic features. Bioinformatics 26(6):841–842. https://doi.org/10.1093/bioinformatics/btq033
24. Soneson C, Love M, Robinson M (2015) Differential analyses for RNA-seq: transcript-level estimates improve gene-level inferences. F1000Res 4:1521. https://doi.org/10.12688/f1000research.7563.1
25. Love MI, Huber W, Anders SJGB (2014) Moderated estimation of fold change and dispersion for RNA-seq data with DESeq2. Genome Biol 15(12):550. https://doi.org/10.1186/s13059-014-0550-8
26. Howe KL, Bolt BJ, Shafie M, Kersey P, Berriman M (2017) WormBase ParaSite—a comprehensive resource for helminth genomics. Mol Biochem Parasitol 215:2–10. https://doi.org/10.1016/j.molbiopara.2016.11.005
27. Clark MB, Amaral PP, Schlesinger FJ, Dinger ME, Taft RJ, Rinn JL, Ponting CP, Stadler PF, Morris KV, Morillon A, Rozowsky JS, Gerstein MB, Wahlestedt C, Hayashizaki Y, Carninci P, Gingeras TR, Mattick JS (2011) The reality of pervasive transcription. PLoS Biol 9(7):e1000625. https://doi.org/10.1371/journal.pbio.1000625
28. Langmead B, Salzberg SL (2012) Fast gapped-read alignment with bowtie 2. Nat Methods 9(4):357–359. https://doi.org/10.1038/nmeth.1923
29. Kim D, Langmead B, Salzberg SL (2015) HISAT: a fast spliced aligner with low memory requirements. Nat Methods 12(4):357–360. https://doi.org/10.1038/nmeth.3317
30. Li B, Dewey CN (2011) RSEM: accurate transcript quantification from RNA-Seq data with or without a reference genome. BMC Bioinformatics 12(1):323. https://doi.org/10.1186/1471-2105-12-323

Chapter 11

Identification of Transposable Elements in *Schistosoma mansoni*

Gisele S. Philippsen and Ricardo DeMarco

Abstract

Transposable elements (TEs) represent a significant portion of eukaryotic genomes and are important players in their dynamics and evolution. Therefore, the description of TEs and the analysis of their distribution in the genomes are important steps to understand their influence in the architecture of genomes. Here we describe the protocol used by us to identify and curate consensus TEs sequences from *S. mansoni*, as well the protocol to map these elements in the *S. mansoni* genome. We expect that these protocols may help researchers interested in studying TEs content in *S. mansoni* or other organisms.

Key words *Schistosoma mansoni*, Genome, Transposable elements, Copies annotation, Bioinformatics analyses

1 Introduction

Bioinformatics analyses aiming to verify the distribution of transposable elements (TEs) insertions are relevant in the understanding of architecture and evolutionary dynamics of genomes. TEs insertions may influence gene expression by introducing new promoters or regulatory sequences, by disrupting existing regulatory structures near the integration locus or by inducing gene silencing [1]. TEs may also contribute with the gene architecture by producing new exons, alternative splice sites, or alternative polyadenylation sites [1–3].

The increasing knowledge about the TEs influence on genomes and genes evolution [4–9] has been obtained with the increasing number of sequenced genomes and developed bioinformatics tools. The availability of *S. mansoni* genome annotation data [10, 11] and TEs consensus sequences described for this species established an appropriate scenario to investigate TEs distribution and influence on genes structure [12]. In this chapter we describe the bioinformatics approach adopted to identify TEs copies in *S. mansoni* genome.

David J. Timson (ed.), *Schistosoma mansoni: Methods and Protocols*, Methods in Molecular Biology, vol. 2151,
https://doi.org/10.1007/978-1-0716-0635-3_11,

2 Materials

The computational approach described in this chapter was performed in a standard computer (processor Intel(R) Core (TM) i7-5500U; RAM memory 8.00 GB) with Ubuntu operational system, employing the following resources:

1. RepeatScout software [13].
2. Standalone BLAST software (*see* **Note 1**).
3. Python with related libraries BioPython and Matplotlib (*see* **Note 2**).
4. R statistical language (*see* **Note 3**).

3 Methods

3.1 Identification and Curation of Representative TEs Sequences from Genomes

A first detection of TEs using a genome sequence can be performed using RepeatScout program [13], which allows for a de novo detection of elements. The sequences provided by RepeatScout can then be annotated using BLASTx searches against appropriated databases containing TEs consensus sequences. Once a group of TEs that are of interest of the researcher is defined, further steps to improve and curate the sequences might be of interest.

One of the problems of the output from RepeatScout and other tools is that they might reflect short or truncated copies of the TEs that usually populate a genome. In our experience, comparison of TE sequences with cDNA sequences might provide information for curation, since sequences from those sources tend to be derived from copies that are transcriptionally active and therefore tend to represent more intact copies of the analyzed element.

1. The first step is to align with BLASTn program the whole TE sequence against a database of transcript reads, such as dbEST or RNASeq libraries, and search for alignments with very high identity (>95%) that are likely to represent transcripts to the same element analyzed (Fig. 1, step 1). Comparison between the query sequences and aligned cDNA may point out for eventual insertions/deletions or point mutations present in the RepeatScout consensus. Mutations that introduce frameshifts and/or stop codons that disrupt the open reading frame are the most obvious choice for curation (Fig. 1, step 2).
2. In addition, data from transcripts can also be used to extend TEs sequences that are incomplete. In that case the extremity to be completed can be used as query against transcript reads databases using BLASTn program in the search for reads that partially aligns the extremity with high identity (>95%) and its unaligned portion extend the extremity sequence (Fig. 1, step

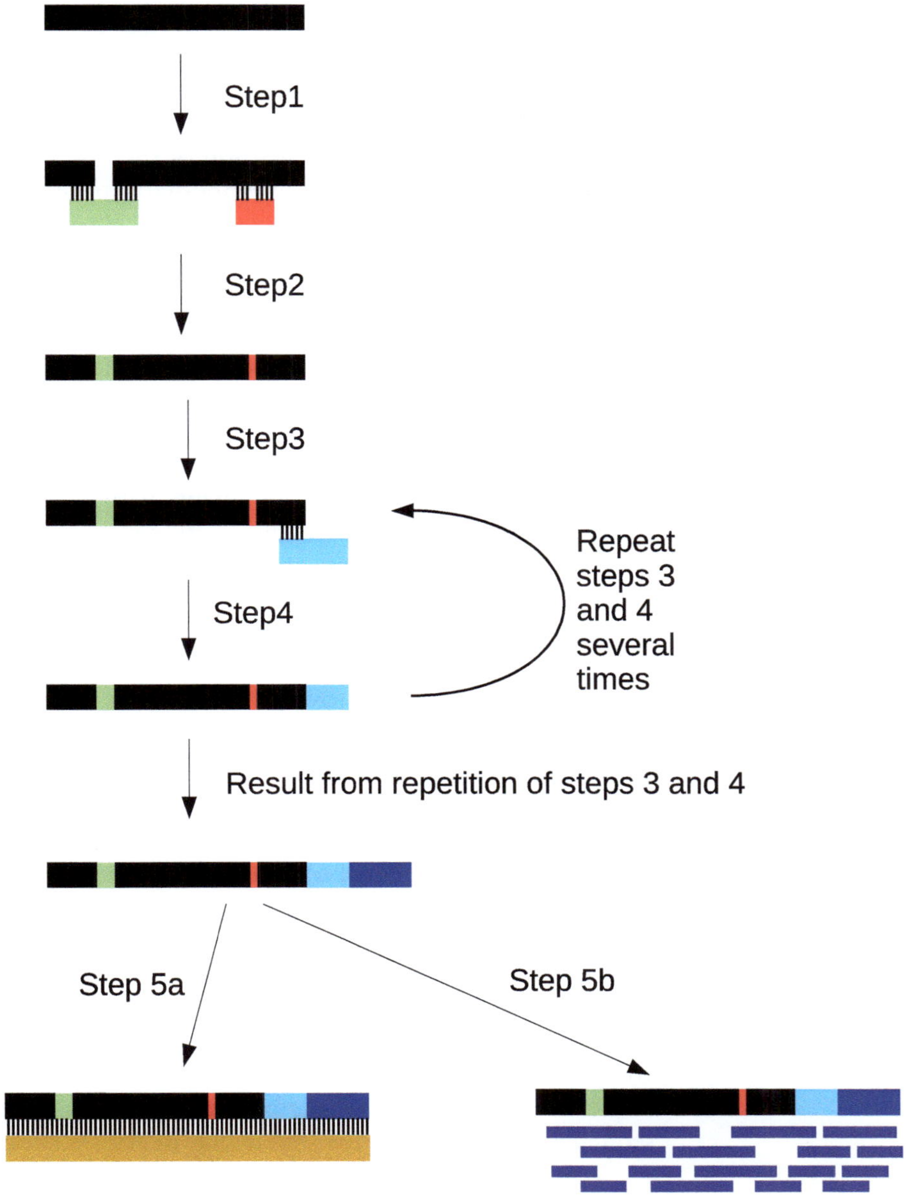

Fig. 1 Schematic representation of curation of a TE sequence. The black bar represents the original TE sequence obtained from a de novo prediction program and colored bars represent sequences obtained from transcript or genome databases. The procedures represented are as follows: (**Step 1**) Alignment of the TE sequence from cDNA sequences. (**Step 2**) Curation of TE sequence based on alignment information. (**Step 3**) Alignment of the TE extremity sequence with cDNA sequence. (**Step 4**) Extension of the TE sequence based on the alignment of its extremity. (**Step 5a**) Confirmation of the reconstructed TE sequence by alignment with a genomic copy. (**Step 5b**) Confirmation of the reconstructed TE sequence by alignment with several cDNA sequences. cDNA sequences are assumed to be partial representation of a full-length transcript as usual for sequence from EST or RNASeq libraries

3). Once such read is found, it can be incorporated into the TE sequence (Fig. 1, step 4). Given the partial nature of ESTs and RNASeq reads, it is expected that incorporation of a single read into TE sequence can only contribute to a modest extension of the TE read, and therefore repetition of such procedure with the newly extended sequence can be desirable to obtain a more complete sequence.

3. Once a complete TE sequence is obtained or further extension is not possible due to the lack of transcript sequences that extend, it may be necessary to verify if the reconstructed portion indeed represent an authentic TE sequence and is not an artifact resulting from incorporation of chimeric reads into the sequence. This could be achieved by looking confirmation to the proposed sequence by aligning it against the genome sequence and examining the alignments with highest scores, which should represent active full-length copies of the TE and align the whole extension of the reconstructed sequence (Fig. 1, step 5a). Alternatively, alignment of the reconstructed sequence with transcript reads may also provide confirmation, as an alignment with several reads from all extension of reconstructed region would provide evidence for its consistency (Fig. 1, step 5b).

The use of transcript data to reconstruct intact TEs sequences is particularly effective in organisms that present high transcriptional activity from TEs, such as *S. mansoni*, in which full-length or near-full-length copies of TEs were obtained using transcript information [14, 15].

3.2 TEs Annotation in the Genome

1. Considering the TEs consensus sequences described for the species, the mapping process starts by aligning TEs sequences into the genome (Fig. 2). In this first stage of the method, the BLASTn search (hits cutoff *e-value* $< 10^{-10}$) of TEs consensus sequences (*queries*) was performed against the genome sequence (*subject*) using the standalone BLAST version.

   The TEs annotation strategy adopted in this method is similar to that defined in the Censor software, where the BLAST output is processed to defragment TEs copies and remove TEs overlaps [16]. Large insertions or deletions may occur along a TE copy after its genome integration, leading to a detection of two alignments instead of one with a large gap [16, 17]. Moreover, TEs insertions in the genome represent regions expected to evolve under a neutral selection and therefore more likely exhibit mutations than regions in which a negative selection of mutations is predominant. Hence, it is expected that these sequences quickly accumulate mutations that may lead to a patchy detection of the element originally inserted. Due to these aspects, a defragmentation process is

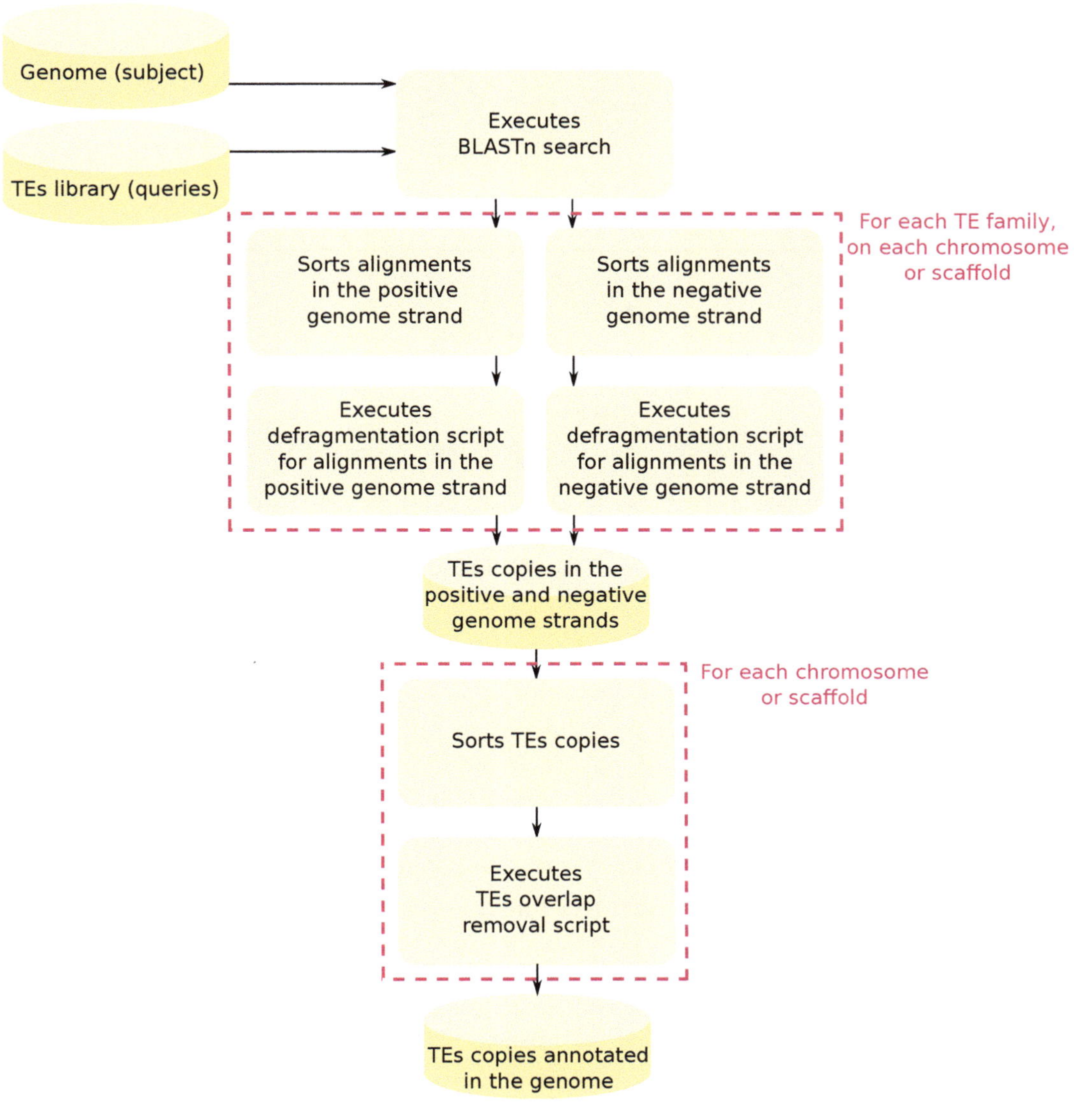

Fig. 2 Stages of the method adopted for the TEs annotation in the genome. In the first stage, the TEs consensus sequences (*queries*) are interrogated against the genome sequence (*subject*) by a BLASTn search. In the second stage, the defragmentation script is executed and, in the third stage, the overlap removal script is executed

necessary to consider the detected fragments as a unique insertion event with biological meaning [17]. This approach is also necessary to avoid the overestimation of the number of TEs copies into the genomes. With this aim, the BLASTn output is processed through script codes written in Python and R languages, as described below.

2. The obtained BLAST alignments are separated according to the genome strand (Fig. 2). For each TE family, in each chromosome or scaffold, the alignments are ordered according to

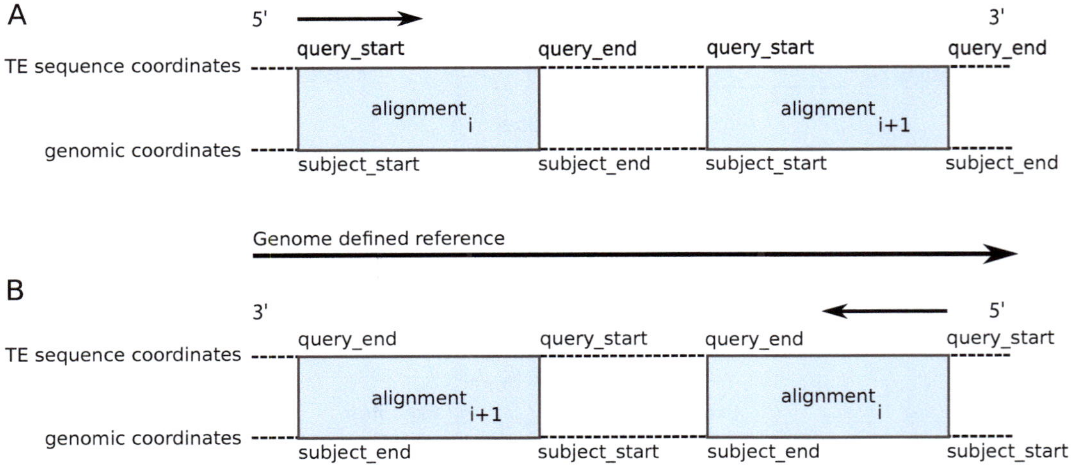

Fig. 3 Schematic representation of successive TE alignments in the genome. Successive alignments of the same TE family are shown in the positive strand (5′–3′) (**a**) and in the negative strand (3′–5′) (**b**) of a specified chromosome or scaffold. If the alignments meet the established criteria, they are considered as a unique insertion event

the DNA strand: the hits in the positive strand (5′–3′) are sorted in ascending order of starting coordinate in the genome (*subject_start* hit parameter), while the hits in the negative strand (3′–5′) are sorted in descending order of genome starting coordinate (Fig. 3). In the case of overlap position in the starting coordinate, the sorter function considers the ascending order of *e-value* hit parameter.

3. After the sorter function execution, the defragmentation step initiates (Fig. 2). In a given chromosome or scaffold strand, two successive alignments of a TE consensus sequence (Fig. 3) are considered as a unique insertion event if the following conditions are satisfied:

   (a) The distance between the end of alignment$_i$ (*subject_end* hit parameter) and the start of alignment$_{i+1}$ (*subject_start* hit parameter) in the genomic coordinates must be lower than 100 base pair (bp). In the case that the alignment$_{i+1}$ overlaps the alignment$_i$ over an extension larger than 50 bp, the alignment$_{i+1}$ replaces the alignment$_i$ if the alignment$_{i+1}$ has lower *e-value* and the alignment$_i$ has not replaced a previous alignment.

   (b) The start of alignment$_{i+1}$ (*query_start* hit parameter) must be higher than the end of the alignment$_i$ (*query_end* hit parameter) in terms of TE consensus sequence coordinates, excepting the 50 bp overlap limit accepted in this method.

At the end of this stage, the alignments belonging to the same TE family, located at the same chromosome, in the same orientation, with a genomic distance lower than 100 bp and showing collinearity regarding TE consensus sequence were considered as a single TE insertion, with a redefined *bit score* parameter (S'). The TEs instances identified are described according to the GFF3 (Generic Feature Format version 3) format, in which the start and end coordinates are established in positive integer coordinates relative to the defined sequence landmark and the DNA strand is specified by a *strand* parameter (*see* **Note 4**).

The *bit score* alignment parameter reported in BLAST searches is the normalized score obtained from the raw alignment score (S) and the k and λ parameters associated with the statistical properties of the scoring system [18], as expressed in Eq. (1). This score normalization approach allows the comparison of alignment scores from distinct searches.

$$S' = (\lambda S - \ln k)/\ln 2 \quad (1)$$

To redefine the *bit score* for the defragmented TE instance, the raw score for each alignment constitutive of the defragmented copy is retrieved—from rearranging the terms in Eq. (1). The sum of these raw scores provides the raw score for the defragmented TE copy, from which the *bit score* calculation is performed.

4. After the defragmentation process, the TEs instances identified in the positive and negative DNA strands are stored in a single data set (Fig. 2). For each chromosome or scaffold, they are sorted in ascending order of starting coordinate in the genome. In the case of overlapping position in the starting coordinate, the sorter function considers the descending order of *bit score* parameter associated with TE copies.
5. The sorted TEs copies constitute the input data for the overlap removal script code (Fig. 2). TEs overlaps may be found due to the expansion of TEs families resulting from mutational events, leading to a scenario in which the distinction limit between them is tenuous. In this step of the method, the overlap identification (Fig. 4) leads to the analysis of the following criterion: if the overlap extension is lower than 50 bp, both TE instances are maintained in the data set; otherwise, insertion$_{i+1}$ replaces insertion$_i$ if insertion$_{i+1}$ has higher *bit score* and insertion$_i$ has not replaced a previous insertion.

After the execution of the method, the TEs copies are annotated in the genome. It is important to highlight the fact that the main goal of the defragmentation procedure is to avoid the overestimation of TEs copies in the genome. In contrast, when the purpose is to analyze the TE content in the coding or regulatory

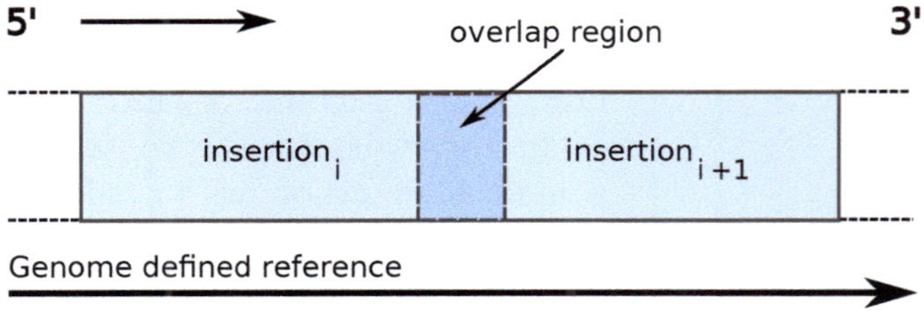

Fig. 4 Schematic representation of TEs overlap in the genome. If the overlap extension exceeds 50 bp, insertion$_{i+1}$ replaces insertion$_i$ if the established criteria defined in the overlap removal script are met

sequences, it is necessary to recover the original alignments that constitute the defragmented TE copy. In this way only the regions actually attributed to the TE are specified, avoiding the inappropriate assignment of intervening genomic sequences as TE derived sequences.

4 Notes

1. One of the most commonly used tools for alignment searches is BLAST (Basic Local Alignment Search Tool) [19, 20]. The standalone version of BLAST is available at www.ncbi.nlm.nih.gov/books/NBK52640.
2. Python (https://www.python.org/) is an appropriate programing language for bioinformatics script codes development [21]. The BioPython library (https://biopython.org/) has useful functions implemented for computational analyses related to molecular biology. In this context, another useful resource available for Python is the plotting library Matplotlib (https://matplotlib.org/).
3. The alignments and insertions ordering steps were performed with the sort function available at R statistical language project (http://www.r-project.org).
4. Generic Feature Format Version 3 (GFF3). Specifications are accessible at http://www.sequenceontology.org/resources/.

Acknowledgments

RDM is recipient of a research productivity fellowship from Conselho Nacional de Desenvolvimento Científico e Tecnológico (CNPq- Project# 304567/2017-5).

References

1. Feschotte C (2008) Transposable elements and the evolution of regulatory networks. Nat Rev Genet 9:397–405. https://doi.org/10.1038/nrg2337
2. Nekrutenko A, Li W-H (2001) Transposable elements are found in a large number of human protein-coding genes. Trends Genet 17:619–621. https://doi.org/10.1016/S0168-9525(01)02445-3
3. Keren H, Lev-Maor G, Ast G (2010) Alternative splicing and evolution: diversification, exon definition and function. Nat Rev Genet 11:345–355. https://doi.org/10.1038/nrg2776
4. Kazazian HH (2004) Mobile elements: drivers of genome evolution. Science 303:1626–1632. https://doi.org/10.1126/science.1089670
5. Janicki M, Rooke R, Yang G (2011) Bioinformatics and genomic analysis of transposable elements in eukaryotic genomes. Chromosom Res 19:787. https://doi.org/10.1007/s10577-011-9230-7
6. Platt RN, Vandewege MW, Ray DA (2018) Mammalian transposable elements and their impacts on genome evolution. Chromosom Res 26:25–43. https://doi.org/10.1007/s10577-017-9570-z
7. Sela N, Kim E, Ast G (2010) The role of transposable elements in the evolution of non-mammalian vertebrates and invertebrates. Genome Biol 11:R59. https://doi.org/10.1186/gb-2010-11-6-r59
8. Philippsen GS, Avaca-Crusca JS, Araujo APU, DeMarco R (2016) Distribution patterns and impact of transposable elements in genes of green algae. Gene 594:151–159. https://doi.org/10.1016/j.gene.2016.09.012
9. Lopes FR, Carazzolle MF, Pereira GAG et al (2008) Transposable elements in *Coffea* (Gentianales: Rubiacea) transcripts and their role in the origin of protein diversity in flowering plants. Mol Gen Genomics 279:385–401. https://doi.org/10.1007/s00438-008-0319-4
10. Protasio AV, Tsai IJ, Babbage A et al (2012) A systematically improved high quality genome and transcriptome of the human blood fluke *Schistosoma mansoni*. PLoS Negl Trop Dis 6:e1455. https://doi.org/10.1371/journal.pntd.0001455
11. Berriman M, Haas BJ, LoVerde PT et al (2009) The genome of the blood fluke *Schistosoma mansoni*. Nature 460:352–358. https://doi.org/10.1038/nature08160
12. Philippsen GS, DeMarco R (2019) Impact of transposable elements in the architecture of genes of the human parasite *Schistosoma mansoni*. Mol Biochem Parasitol 228:27–31. https://doi.org/10.1016/j.molbiopara.2018.12.007
13. Price AL, Jones NC, Pevzner PA (2005) De novo identification of repeat families in large genomes. Bioinformatics 21:i351–i358. https://doi.org/10.1093/bioinformatics/bti1018
14. DeMarco R, Kowaltowski AT, Machado AA et al (2004) Saci-1, -2, and -3 and Perere, four novel Retrotransposons with high transcriptional activities from the human parasite *Schistosoma mansoni*. J Virol 78:2967–2978. https://doi.org/10.1128/JVI.78.6.2967-2978.2004
15. DeMarco R, Machado AA, Bisson-Filho AW, Verjovski-Almeida S (2005) Identification of 18 new transcribed retrotransposons in *Schistosoma mansoni*. Biochem Biophys Res Commun 333:230–240. https://doi.org/10.1016/j.bbrc.2005.05.080
16. Kohany O, Gentles AJ, Hankus L, Jurka J (2006) Annotation, submission and screening of repetitive elements in Repbase: RepbaseSubmitter and censor. BMC Bioinformatics 7:474. https://doi.org/10.1186/1471-2105-7-474
17. Bergman CM, Quesneville H (2007) Discovering and detecting transposable elements in genome sequences. Brief Bioinform 8:382–392. https://doi.org/10.1093/bib/bbm048
18. Fassler J, Cooper P (2011) BLAST glossary. National Center for Biotechnology Information (US), Bethesda, Maryland
19. Teufel A, Krupp M, Weinmann A, Galle PR (2006) Current bioinformatics tools in genomic biomedical research (review). Int J Mol

Med 17:967–973. https://doi.org/10.3892/ijmm.17.6.967
20. Korf I, Yandell M, Bedell J (2003) BLAST. O'Reilly & Associates, Sebastopol
21. Kinser JM (2009) Python for bioinformatics. Jones and Bartlett Publishers, Massachusetts

Chapter 12

Structural and Functional Characterization of *Schistosoma mansoni* Cathepsin B1

Adéla Jílková, Martin Horn, and Michael Mareš

Abstract

Schistosomiasis caused by parasitic blood flukes of the genus *Schistosoma* is a global health problem with over 200 million people infected. *Schistosoma mansoni* cathepsin B1 (SmCB1) is a gut-associated protease critical for digestion of host blood proteins as a source of nutrients. SmCB1 is a validated drug target, and inhibitors of SmCB1 represent promising anti-schistosomals. A comprehensive structural and functional characterization of SmCB1 provides a starting point for the rational design of selective and potent SmCB1 inhibitors. Here, we report optimized protocols for (1) the production of recombinant SmCB1 in the *Pichia pastoris* expression system and its purification, (2) the measurement of SmCB1 activity and inhibition in a kinetic fluorescence assay, and (3) the preparation and crystallization of SmCB1 in complex with a model vinyl sulfone inhibitor, and the determination of its crystal structure.

Key words Parasite, *Schistosoma mansoni*, Protease, Cathepsin B, Expression, Activity assay, Inhibition, Crystal structure

1 Introduction

Schistosomiasis, or bilharzia, is a chronic parasitic disease caused by trematode blood flukes that affects over 200 million people worldwide [1]. *Schistosoma mansoni* is one of five human-infecting schistosomes with geographical distribution in parts of Asia, the Middle East, Africa, and South America. Treatment and control of schistosomiasis rely on a single drug, and concern over drug resistance [1, 2] evokes pressure to identify new schistosomal protein targets and anti-schistosomal drugs [3–5].

Adult schistosomes reside in the cardiovascular system, feeding on blood and producing hundreds of eggs per day. Host blood proteins, which serve as the main source of nutrients, are digested by a network of proteases in the schistosome gut [6, 7]. This chapter focuses on *S. mansoni* cathepsin B1 (SmCB1), which is the most abundant digestive protease [8] critical for parasite survival [9, 10]. SmCB1 has been validated as a chemotherapeutic

David J. Timson (ed.), *Schistosoma mansoni: Methods and Protocols*, Methods in Molecular Biology, vol. 2151, https://doi.org/10.1007/978-1-0716-0635-3_12,

target for cure of schistosomiasis in a murine model using the vinyl sulfone inhibitor K11777 [9]. Inhibition of SmCB1 thus represents an attractive option for anti-schistosomal drug development [10–12].

SmCB1 is synthesized as an inactive zymogen (37 kDa) and is converted to a mature, active enzyme (28 kDa) by proteolytic processing mediated by sulfated polysaccharides or catalyzed by asparaginyl endopeptidase [8, 13]. Mature SmCB1 exhibits both endopeptidase and exopeptidase activities, which are modulated by a flexible occluding loop [10]. Structural studies on SmCB1 have recently provided insights into the natural regulation of this enzyme and the rational design of specific peptidomimetic inhibitors [10, 13].

This chapter focuses on the methods used for the functional and structural characterization of SmCB1 and its interactions with inhibitors. It provides protocols for the cloning and production of recombinant SmCB1 in the *Pichia pastoris* expression system and its purification from expression media. Further, it presents a kinetic activity assay for the high-throughput screening of SmCB1 inhibitors. Finally, it describes crystallization and crystallographic procedures for the structural analysis of SmCB1 in complex with a model inhibitor K11777 that can be applied to other peptidomimetic inhibitors.

2 Materials

2.1 Construction of SmCB1 Expression Vector

1. cDNA from adult *S. mansoni* prepared as described in [14].
2. The forward primer SmCB1.F:
   5′-AGCTACTCGAGAAAAGACATATTTCAGTTAAGAACGAAAAG-3′.
3. The reverse primer SmCB1.R:
   5′-GTAATAGCCGGTCGAATAAACTAAGCGGCCGCACTGC-3′.
4. Proofreading polymerase (e.g., *Pfu* DNA polymerase) and recommended commercial 10× buffer.
5. Deoxynucleotide solution mix (dNTP; 10 mM each dATP, dCTP, dGTP, dTTP).
6. Thermocycler.
7. Equipment and reagents for DNA electrophoresis.
8. DNA gel extraction kit (e.g., QIAquick Gel Extraction Kit, Qiagen).
9. Plasmid pPICZαA (Invitrogen).

10. Restriction enzymes *Xho*I and *Not*I and recommended commercial buffer.
11. PCR clean-up kit (e.g., QIAquick PCR Purification Kit, Qiagen).
12. NanoDrop spectrophotometer.
13. T4 DNA ligase and recommended commercial 10× buffer.
14. Competent *Escherichia coli* cells for the amplification of the recombinant vector (e.g., TOP10, Invitrogen).
15. LB/zeocin plates: 1% (w/v) tryptone, 0.5% (w/v) yeast extract, 0.5% (w/v) NaCl. Adjust the pH with 1 M NaOH to 7.5, add 1.5% (w/v) agar, and autoclave. Add zeocin to a final concentration of 25 μg/mL before pouring plates.
16. LB/zeocin medium: 1% (w/v) tryptone, 0.5% (w/v) yeast extract, 0.5% (w/v) NaCl. Adjust the pH with 1 M NaOH to 7.5, and autoclave. Add zeocin to a final concentration of 25 μg/mL before use.
17. Kit for plasmid DNA isolation (e.g., QIAprep Spin Miniprep Kit, Qiagen).
18. Sequencing primers: 5′AOX1 and 3′AOX1.

2.2 Preparation of Nonglycosylated Mutant of SmCB1

1. Primer pair for Thr168Ala mutagenesis:
   Forward 5′-AGTTCGAAGGAGAATCACGCAGGTTGTG AACCATATC-3′.
   Reverse 5′-GATATGGTTCACAACCTGCGTGATTCTCC TTCGAACT-3′.
2. Primer pair for Thr283Ala mutagenesis:
   Forward 5′-TGGGGTGTGGAAAACAAGGCTCCTTATT GGTTGATTG-3′.
   Reverse 5′-CAATCAACCAATAAGGAGCCTTGTTTTCC ACACCCCA-3′.
3. Proofreading polymerase (e.g., *Pfu* DNA polymerase) and recommended commercial 10× buffer.
4. Deoxynucleotide solution mix (dNTP; 10 mM each dATP, dCTP, dGTP, dTTP).
5. Thermocycler.
6. Restriction enzyme *Dpn*I.

2.3 Transformation of Pichia pastoris by Electroporation

1. Restriction enzyme *Sac*I and recommended commercial 10× buffer.
2. Equipment and reagents for DNA electrophoresis.
3. PCR clean-up kit (e.g., QIAquick PCR Purification Kit, Qiagen).

4. *P. pastoris* strain X-33 (Invitrogen).
5. Electroporation cuvettes.
6. Electroporation device.
7. Sterile 1 M sorbitol.
8. Zeocin (Invitrogen).
9. YPDS/zeocin plates: 1% (w/v) yeast extract, 2% (w/v) peptone, 2% (w/v) dextrose, 2% (w/v) agar, 1 M sorbitol. Autoclave and add zeocin to a final concentration of 100 μg/mL before pouring plates.
10. YPD/zeocin plates: 1% (w/v) yeast extract, 2% (w/v) peptone, 2% (w/v) dextrose, 2% (w/v) agar. Autoclave and add zeocin to a final concentration of 100 μg/mL before pouring plates.
11. Incubator.

2.4 Recombinant Expression of SmCB1

1. Rotary shaking incubator.
2. Centrifuges and centrifuge bottles.
3. BMG medium: 100 mM potassium phosphate, pH 6.0, 1.34% (w/v) yeast nitrogen base, 4×10^{-5}% (w/v) biotin, 1% (v/v) glycerol.
4. BMM medium: BMG medium where glycerol is replaced by 0.5% methanol.

2.5 Purification of SmCB1

1. Lyophilizator.
2. Chromatography column (e.g., 3.2 cm × 60 cm, 470 mL) filled with Sephadex G-25 (GE Healthcare Bio-Sciences).
3. Buffer for gel filtration: 25 mM sodium acetate, pH 5.5, 2.5 mM DTT, 1 mM EDTA.
4. FPLC (e.g., ÄKTAPure, GE Healthcare Bio-Sciences).
5. Mono S 5/50 GL column (GE Healthcare Bio-Sciences).
6. Equilibration buffer for ion-exchange chromatography: 25 mM sodium acetate, pH 5.5, 2.5 mM DTT, 1 mM EDTA.
7. Elution buffer for ion-exchange chromatography: 25 mM sodium acetate, pH 5.5, 2.5 mM DTT, 1 mM EDTA, 1 M NaCl.
8. Equipment and reagents for SDS PAGE.

2.6 Activation of SmCB1

1. Amicon Ultracel-10k centrifugal unit (Millipore).
2. Activation buffer: 25 mM sodium acetate, pH 5.0, 2.5 mM DTT, 1 mM EDTA.
3. Recombinant *S. mansoni* asparaginyl endopeptidase (SmAE) produced using the *P. pastoris* expression system as described previously [15].

4. Storage buffer: 25 mM sodium acetate, pH 5.5, 2.5 mM DTT, 1 mM EDTA.

2.7 Preparation of SmCB1-Inhibitor Complex

1. Argon.
2. Vinyl sulfone inhibitor K11777 (e.g., Adipogen). Prepare 20 mM stock solution in dimethyl sulfoxide (DMSO).
3. Buffer: 5 mM sodium acetate, pH 5.5.

2.8 Kinetic Measurements

1. Activation mix: 100 mM sodium acetate, pH 5.5, 3.5 mM DTT, 0.15% (w/v) PEG 1500.
2. Substrate Z-Phe-Arg-AMC. Prepare 10 mM stock solution in DMSO.
3. Substrate mix: 100 mM sodium acetate, pH 5.5, 0.1 mM Z-Phe-Arg-AMC.
4. Enzyme mix: 0.01–0.05 nM SmCB1 in 100 mM sodium acetate, pH 5.5, 3.5 mM DTT, 0.15% (w/v) PEG 1500.
5. Vinyl sulfone inhibitor K11777 (e.g., Adipogen). Prepare 20 mM stock solution in DMSO.
6. 96-well black flat-bottom microplate.
7. Fluorescence microplate reader.

2.9 Protein Crystallization and Structure Determination

1. Crystallization plate (e.g., EasyXtal 15-well tool, Qiagen).
2. Reservoir solution: 200 mM ammonium acetate, 100 mM sodium citrate, pH 6.2, 30% (w/v) PEG 1500.
3. Microscope.
4. Liquid nitrogen.

3 Methods

3.1 Construction of SmCB1 Expression Vector

This protocol describes the cloning of a coding sequence of the SmCB1 zymogen (without a signal sequence; *see* **Note 1**) into the plasmid pPICZαA, which enables secretion of the expressed SmCB1 protein into the cultivation media.

1. Amplify the SmCB1 zymogen coding sequence (GenBank accession number: AJ506157) by the polymerase chain reaction (PCR) using cDNA from adult *S. mansoni*, the primers SmCB1.F and SmCB1.R (for details, *see* **Note 2**), and a proofreading polymerase according to the manufacturer's protocol.
2. Separate the product of the PCR on a 1% agarose gel. Excise the band of the appropriate size (~1 kbp) and extract and purify the DNA using a DNA gel extraction kit according to the manufacturer's instructions.

3. In two separate tubes, mix the purified DNA (from **step 2**) or 5 μg of the plasmid pPICZαA with 1 μL of *Not*I, 1 μL of *Xho*I, and 5 μL of 10× buffer recommended for the double digest with these restriction enzymes. Adjust the volume to 50 μL using water. Incubate the mixture at 37 °C for at least 30 min. Inactivate the restriction enzymes by heating at 65 °C for 20 min.
4. Purify both the digested DNA and the plasmid pPICZαA using a PCR clean-up kit and measure the concentration of DNA using a NanoDrop spectrophotometer.
5. Mix the purified digested DNA and the plasmid pPICZαA in the molar ratio 5:1, add 2 μL of 10× buffer for T4 DNA ligase and 1 μL of T4 DNA ligase, and fill with water to a total volume of 20 μL. Incubate at 16 °C overnight. After incubation, inactivate the ligase by heating at 65 °C for 20 min.
6. Use 10 μL of the reaction mixture for heat shock transformation of competent *E. coli* cells. Spread the cells on LB/zeocin plates and incubate them at 37 °C overnight.
7. Pick ten colonies from the transformation plate and grow them separately in 5 mL of LB/zeocin medium at 37 °C overnight.
8. Harvest the cells and isolate the plasmid DNA using a kit for plasmid DNA isolation according to the manufacturer's instructions.
9. Confirm the SmCB1 nucleotide sequence by DNA sequencing using vector-specific sequencing 5′AOX1 and 3′AOX1 primers.

3.2 Preparation of Nonglycosylated Mutant of SmCB1

The SmCB1 sequence contains two consensus *N*-glycosylation sites Asn-His-Thr (residues 166–168) and Asn-Lys-Thr (residues 281–283). A nonglycosylated mutant of SmCB1 has been used for structural studies because glycosylation is generally considered problematic for protein crystallization [10, 13]. Here, we describe the disruption of both glycosylation sites using site-directed mutagenesis of Thr to Ala (Thr168Ala and Thr283Ala) by a two-step PCR procedure (*see* **Note 3**).

1. For the PCR mutagenesis reaction, mix 10–50 ng of the plasmid pPICZαA containing SmCB1 (the product of Subheading 3.1), 1.5 μL of forward and reverse primers for Thr168Ala mutagenesis, 1 μL of dNTP, 5 μL of 10× buffer for polymerase, and 1 μL of proofreading polymerase. Add water to a total volume of 50 μL.
2. Perform the PCR reaction with the following amplification conditions: the initial denaturation (95 °C for 2 min), followed by 16 cycles of the denaturation (95 °C for 30 s), the annealing (55 °C for 1 min), the extension (72 °C for 10 min), and the final extension (72 °C for 5 min).

3. To remove the template DNA, add 1 μL of *Dpn*I restriction enzyme and incubate at 37 °C for 2 h.
4. Use the product of **step 3** to transform TOP10 *E. coli* cells and prepare the plasmid DNA as described in Subheading 3.1.
5. Verify the desired mutation in the DNA using DNA sequencing.
6. To obtain the double mutated SmCB1, repeat the whole procedure (**steps 1–5**) with the second primer pair for Thr283Ala mutagenesis.

3.3 Transformation of *Pichia pastoris* by Electroporation

This section describes the transformation of *P. pastoris* cells by the expression vector containing the nonglycosylated SmCB1 mutant (constructed in Subheading 3.2).

3.3.1 Linearization of Plasmid DNA

1. Mix 10 μg of the plasmid pPICZαA containing the nonglycosylated SmCB1 mutant (Subheading 3.2) with 2.5 μL of *Sac*I restriction enzyme and 5 μL of 10× buffer for *Sac*I, and adjust the volume to 50 μL using water. Incubate the mixture at 37 °C for at least 30 min.
2. Check a small aliquot of the digested DNA by agarose gel electrophoresis for complete linearization. Purify the linearized DNA using a PCR clean-up kit and measure the concentration of DNA using a NanoDrop spectrophotometer.

3.3.2 Transformation of *P. pastoris* by Electroporation

1. Prepare the competent cells of *P. pastoris* strain X-33 according to the protocol in the *Pichia* Expression kit manual (Invitrogen).
2. Mix 80 μL of the competent *P. pastoris* cells with 5–10 μg of the linearized DNA (from **step 2** in Subheading 3.3.1) and transfer the mixture into an ice-cold electroporation cuvette with a 0.2 cm gap. Place the cuvette on ice and incubate for 5 min.
3. Pulse the cells with 2 kV charging voltage for 5 ms (or follow the manufacturer's instructions for *Saccharomyces cerevisiae*) and immediately add 1 mL of ice-cold 1 M sorbitol to the cuvette.
4. Transfer the cuvette content to a sterile 15 mL tube and incubate without shaking at 30 °C for 1–2 h.
5. Spread different volumes of the culture (e.g., 25, 100, and 200 μL) on YPDS/zeocin plates and incubate at 30 °C until colonies form (3–5 days).
6. Pick 10 colonies and streak on YPD/zeocin plates to isolate single colonies.
7. Perform a small-scale expression with 5–10 zeocin-resistant *Pichia* transformants to pick the transformant with the highest expression level of the SmCB1 (for details follow the protocol in the *Pichia* Expression kit manual, Invitrogen).

3.4 Recombinant Expression of SmCB1

1. Use the *P. pastoris* transformants showing the highest SmCB1 expression level (obtained in **step** 7 in Subheading 3.3.2) for the large-scale expression of recombinant SmCB1. Inoculate a single colony into 25 mL of BMG medium in a 250 mL flask. Grow at 30 °C with shaking at 230 rpm until OD_{600} reaches 2–6 (for 16–18 h).
2. Use this culture to inoculate 500 mL of BMG medium in a 2 L flask and grow at 30 °C with shaking at 230 rpm until OD_{600} reaches 2–6 (for 16–18 h).
3. Centrifuge the cells in sterile centrifuge bottles at 1500 × *g* at 26 °C for 5 min and discard the supernatant. To induce expression, resuspend the pellet in BMM medium to a final $OD_{600} = 1$ (2–6 L). Divide the culture into 500 mL aliquots and transfer them into new sterile 2 L flasks. Incubate at 30 °C with shaking at 230 rpm.
4. Add 100% methanol to a final concentration of 0.5% (v/v) every 12 h to maintain induction. Daily withdraw an aliquot of the medium and control the pH. In the case of a pH drop, adjust the pH to 6.0 using 25% (v/v) ammonium hydroxide to avoid uncontrolled processing of the SmCB1 zymogen.
5. After 3–5 days, centrifuge the cells at 1500 × *g* at 4 °C for 5 min. Collect the supernatant (containing the secreted recombinant SmCB1), add DTT to a final concentration of 0.5 mM, and store at −80 °C until use.

3.5 Purification of SmCB1

All purification steps should be performed under reducing conditions in the presence of 2.5 mM DTT and 1 mM EDTA to prevent the active site cysteine of SmCB1 from oxidation (*see* **Note 4**).

3.5.1 Purification of SmCB1 by Gel Filtration

1. Concentrate the supernatant containing the recombinant SmCB1 (from **step 5** in Subheading 3.4) at least tenfold using lyophilization.
2. Equilibrate Sephadex G-25 column (3.2 cm × 60 cm, 470 mL) with at least one column volume of 25 mM sodium acetate, pH 5.5, 2.5 mM DTT, 1 mM EDTA at the flow rate of 2 mL/min.
3. Load the concentrated supernatant (from **step 1**) on the column. The volume of the loaded sample should not exceed 1/10 of the column volume. Elute with 25 mM sodium acetate, pH 5.5, 2.5 mM DTT, 1 mM EDTA. Collect 10 mL fractions. Monitor absorbance at 280 nm and conductivity continuously or in collected fractions. Pool the fractions containing protein and store at −80 °C until use.

3.5.2 Purification of SmCB1 Using Ion-Exchange Chromatography

1. Ion-exchange chromatography is performed using an FPLC system. Equilibrate the Mono S 5/50 GL column with 25 mM sodium acetate, pH 5.5, 2.5 mM DTT, 1 mM EDTA at the flow rate of 1 mL/min.
2. Load the protein sample containing pooled fractions after the gel filtration (**step 3** in Subheading 3.5.1) on the column.
3. Elute SmCB1 with a gradient of 0–1 M NaCl in 25 mM sodium acetate, pH 5.5, 2.5 mM DTT, 1 mM EDTA. Collect 2 mL fractions. Monitor absorbance at 280 nm and conductivity continuously. *See* Fig. 1a for a typical ion-exchange chromatography profile.
4. Analyze the fractions using SDS-PAGE (*see* **Note 5**) and pool the fractions containing SmCB1 (*see* **Note 6**).

3.6 Activation of SmCB1

SmCB1 is expressed in *P. pastoris* as an inactive zymogen that has to be converted into the fully active mature form by proteolytic processing. For SmCB1 activation, use asparaginyl endopeptidase from *S. mansoni* (SmAE), which can be produced recombinantly as described previously [15] (*see* **Note 7**).

1. Buffer-exchange the pooled fractions after the Mono S chromatography containing the SmCB1 zymogen (from **step 4** in Subheading 3.5.2) into 25 mM sodium acetate, pH 5.0, 2.5 mM DTT, 1 mM EDTA (*see* **Note 4**), and concentrate to a final concentration of approximately 1–2 mg/mL using an Amicon Ultracel-10k centrifugal unit.
2. Mix the concentrated SmCB1 zymogen with SmAE in the molar ratio ~15:1 and incubate under argon atmosphere at 26 °C for approximately 12 h.
3. Monitor processing to the active mature SmCB1 using a kinetic assay with fluorogenic substrate Z-Phe-Arg-AMC (*see* Subheading 3.8.1) and SDS-PAGE (*see* **Note 5**). Confirm the identity of the final product using N-terminal protein sequencing after electroblotting of the SDS-PAGE gel (*see* **Note 8**).
4. Purify the activated SmCB1 on a Mono S 5/50 GL column (as described in Subheading 3.5.2). Analyze the collected fractions using SDS-PAGE (*see* **Note 5**) and SmCB1 activity assay (described in Subheading 3.8.1). Pool the elution fractions containing SmCB1.
5. Buffer-exchange the pooled fractions into 25 mM sodium acetate, pH 5.5, 2.5 mM DTT, 1 mM EDTA (*see* **Note 4**) and concentrate to 2–5 mg/mL using an Amicon Ultracel-10k centrifugal unit (*see* **Note 9**).

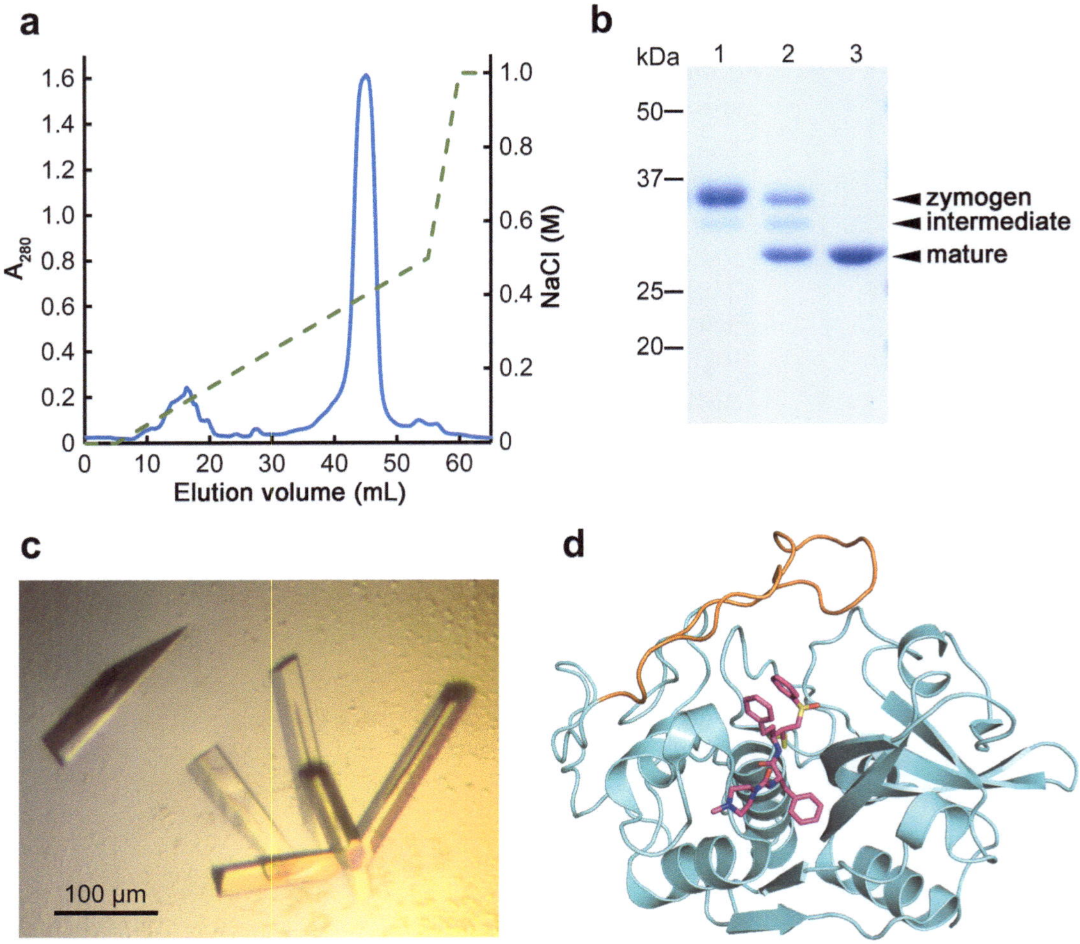

Fig. 1 Purification, activation, and crystallization of SmCB1. (**a**) Ion-exchange chromatography of the SmCB1 zymogen on a Mono S column. (**b**) Activation of the SmCB1 zymogen by asparaginyl endopeptidase analyzed by SDS-PAGE. (*Lane 1*) Nonactivated SmCB1 zymogen. (*Lane 2*) Activation mixture containing three processing forms of SmCB1 (zymogen, intermediate, and mature). (*Lane 3*) Fully activated and purified mature SmCB1. (**c**) Crystals of SmCB1 in complex with a peptidomimetic vinyl sulfone inhibitor. (**d**) Crystal structure of the SmCB1-inhibitor complex (PDB code: 3S3R). SmCB1 is shown in *cyan cartoon* with the occluding loop in *orange*; the inhibitor K11777 is in *magenta sticks*

3.7 Preparation of SmCB1-Inhibitor Complex

In this section, we describe the preparation of the SmCB1 complex with the vinyl sulfone inhibitor K11777, which is a prototype lead compound for the design of SmCB1 inhibitors. This procedure can be used for other peptidomimetic inhibitors of SmCB1.

1. Mix the activated SmCB1 (from **step 5** in Subheading 3.6) with a fivefold molar excess of the inhibitor K11777 in 25 mM sodium acetate, pH 5.5, 2.5 mM DTT, 1 mM EDTA (*see* **Note 4**).
2. Incubate the mixture under argon atmosphere at 26 °C until the inhibition of SmCB1 is complete (approximately 1–3 h). To monitor completeness of SmCB1 inhibition, withdraw an

aliquot of the mixture at different time points and measure SmCB1 activity using a kinetic assay with fluorogenic substrate Z-Phe-Arg-AMC as described in Subheading 3.8.1. As a control, incubate SmCB1 at the same conditions without an inhibitor. Since the concentration of SmCB1 in the mixture is high, the withdrawn samples should be diluted for the activity assay (e.g., 100–1000 times).

3. Buffer-exchange the SmCB1-inhibitor complex into 5 mM sodium acetate, pH 5.5, and concentrate to 2–5 mg/mL using an Amicon Ultracel-10k centrifugal unit (*see* **Note 9**).

3.8 Kinetic Measurements

3.8.1 Activity Assay

SmCB1 activity is measured in a kinetic fluorescence assay. Perform the assay in triplicate in a 96-well microplate (100 μL assay volume) at 37 °C.

1. Pipette 10 μL of the SmCB1 sample with appropriate dilution in 100 mM sodium acetate, pH 5.5 (typically 0.1–0.5 nM, *see* **Note 10**) into the well, and add 70 μL of the activation mix.
2. Incubate at 37 °C for 10 min.
3. Add 20 μL of the substrate mix containing the fluorogenic substrate Z-Phe-Arg-AMC (*see* **Note 11**) and immediately start the measurement in a fluorescence microplate reader. Continuously monitor the kinetics of product release at excitation and emission wavelengths of 360 and 465 nm, respectively.

3.8.2 Inhibition Assay

For inhibition assay, use the purified active SmCB1 (from **step 5** in Subheading 3.6). SmCB1 inhibition is measured in a kinetic fluorescence assay. Perform the assay in triplicate in a 96-well microplate (100 μL assay volume) at 37 °C. The final concentration of DMSO (inhibitor and substrate diluent) should not exceed 2%.

1. Prepare a dilution series of the inhibitor (typically 1 nM–100 μM) in 100 mM sodium acetate, pH 5.5.
2. Pipette 10 μL of the diluted inhibitor into the well and add 70 μL of the enzyme mix (containing 0.01–0.05 nM SmCB1, *see* **Note 10**).
3. Incubate at 37 °C for 15 min.
4. Add 20 μL of the substrate mix containing the fluorogenic substrate Z-Phe-Arg-AMC (*see* **Note 11**) and immediately start the measurement in a fluorescence microplate reader. Continuously monitor the kinetics of product release at excitation and emission wavelengths of 360 and 465 nm, respectively.
5. Determine the IC_{50} values by nonlinear regression (e.g., using GraFit software).

3.9 Protein Crystallization and Structure Determination

This section describes the preparation of SmCB1 crystals by vapor diffusion in a hanging drop and the determination of the SmCB1 crystal structure using molecular replacement. The crystallization conditions are given for the nonglycosylated SmCB1 in complex with the model inhibitor K11777 (prepared in Subheading 3.7).

1. For crystallization, use the protein concentration of the stock solutions between 2 and 5 mg/mL. On the cover slide of the crystallization plate, place a drop of the protein solution and mix it with the reservoir solution in different ratios (e.g., 2:1, 1:1, 1:2). Equilibrate the drops over 1 mL reservoir solution and grow protein crystals at 20 °C. Monitor the drops under a microscope daily until the crystals form (1 to 4 weeks).
2. Flash-cool crystals of diffraction quality (*see* Fig. 1c) by plunging into liquid nitrogen. Collect the diffraction data from the crystals and perform data processing using any of the processing suites available (e.g., XDS, HKL-3000).
3. The structure of the SmCB1-inhibitor complex can be determined by molecular replacement using the structure of the mature SmCB1 (PDB code: 4I07) [13] as the search model as described in [10] (Fig. 1d).

4 Notes

1. The signal sequence for SmCB1 was predicted using the SignalP software.
2. The SmCB1.F forward primer contains (1) the *Xho*I restriction site at the 5′-end, followed by (2) the kexin endopeptidase cleavage site (Lys-Arg), and (3) the complementary sequence for the 5′-end of the coding region of the SmCB1 zymogen. This enables expression of SmCB1 with a native N-terminus. The reverse primer SmCB1.R contains (1) the complementary sequence for the 3′-end of the coding region of SmCB1, (2) the native translation termination codon, and (3) the *Not*I restriction site.
3. Both recombinant SmCB1 forms expressed in *P. pastoris*, the wild-type and the nonglycosylated mutant, exhibit analogous enzymatic activity, activation processing, and inhibition properties [8, 10, 13].
4. As an alternative to DTT, other reducing agents can be used to prevent the active site cysteine of SmCB1 from oxidation (e.g., 2-mercaptoethanol, cysteine or tris(2-carboxyethyl)phosphine, typically in concentration of 5 mM, 20 mM, and 1 mM, respectively). Using the inert atmosphere of argon or nitrogen over buffers and protein samples strengthens the protective effect.

5. Figure 1b shows the transactivation of SmCB1 zymogen by SmAE analyzed by SDS-PAGE. The zymogen (37 kDa) is processed via an intermediate form (32 kDa) to the mature form (28 kDa).
6. The typical yield is around 6 mg of the purified SmCB1 from 1 L of the *P. pastoris* expression media.
7. Asparaginyl endopeptidases from other organisms can be used as an alternative to SmAE.
8. SmAE converts SmCB1 zymogen into the mature form by the cleavage at the Asn69-Val70 site [8, 13]. The N-terminus of the mature SmCB1 starts with the amino acid sequence Val-Glu-Ile-Pro.
9. Prepare small aliquots of SmCB1 or SmCB1-inhibitor complex stock solutions to minimize the freezing/thawing process.
10. The exact concentration of the active SmCB1 in the stock solution can be determined by active site titration. Use an irreversible inhibitor (e.g., CA074, K11777), and perform the inhibition assay as described in Subheading 3.8.2 with the following modifications: use a 100–1000 times higher concentration of SmCB1 and incubate the mixture for 1 h before the measurement of the residual activity.
11. K_m value of SmCB1 determined for the substrate Z-Phe-Arg-AMC is 25 μM [10].

Acknowledgments

This work was supported by the project InterBioMed LO1302 from the Ministry of Education, Youth and Sports of the Czech Republic, grant NV18-05-00345 from the Ministry of Health of the Czech Republic, and the institutional project RVO 61388963.

References

1. Utzinger J, N'goran EK, Caffrey CR, Keiser J (2011) From innovation to application: social-ecological context, diagnostics, drugs and integrated control of schistosomiasis. Acta Trop 120(Suppl 1):S121–S137
2. Caffrey CR (2007) Chemotherapy of schistosomiasis: present and future. Curr Opin Chem Biol 11(4):433–439
3. Caffrey CR, Rohwer A, Oellien F, Marhofer RJ, Braschi S, Oliveira G, McKerrow JH, Selzer PM (2009) A comparative chemogenomics strategy to predict potential drug targets in the metazoan pathogen, *Schistosoma mansoni*. PLoS One 4(2):e4413
4. Abdulla MH, Ruelas DS, Wolff B, Snedecor J, Lim KC, Xu F, Renslo AR, Williams J, McKerrow JH, Caffrey CR (2009) Drug discovery for schistosomiasis: hit and lead compounds identified in a library of known drugs by medium-throughput phenotypic screening. PLoS Negl Trop Dis 3(7):e478
5. Rai G, Sayed AA, Lea WA, Luecke HF, Chakrapani H, Prast-Nielsen S, Jadhav A, Leister W, Shen M, Inglese J, Austin CP, Keefer L, Arner ES, Simeonov A, Maloney DJ, Williams DL, Thomas CJ (2009) Structure mechanism insights and the role of nitric oxide donation guide the development of

oxadiazole-2-oxides as therapeutic agents against schistosomiasis. J Med Chem 52 (20):6474–6483. https://doi.org/10.1021/jm901021k
6. Caffrey CR, McKerrow JH, Salter JP, Sajid M (2004) Blood 'n' guts: an update on schistosome digestive peptidases. Trends Parasitol 20 (5):241–248
7. Delcroix M, Sajid M, Caffrey CR, Lim KC, Dvorak J, Hsieh I, Bahgat M, Dissous C, McKerrow JH (2006) A multienzyme network functions in intestinal protein digestion by a platyhelminth parasite. J Biol Chem 281 (51):39316–39329
8. Sajid M, McKerrow JH, Hansell E, Mathieu MA, Lucas KD, Hsieh I, Greenbaum D, Bogyo M, Salter JP, Lim KC, Franklin C, Kim JH, Caffrey CR (2003) Functional expression and characterization of *Schistosoma mansoni* cathepsin B and its trans-activation by an endogenous asparaginyl endopeptidase. Mol Biochem Parasitol 131(1):65–75
9. Abdulla MH, Lim KC, Sajid M, McKerrow JH, Caffrey CR (2007) Schistosomiasis mansoni: novel chemotherapy using a cysteine protease inhibitor. PLoS Med 4(1):e14
10. Jilkova A, Rezacova P, Lepsik M, Horn M, Vachova J, Fanfrlik J, Brynda J, McKerrow JH, Caffrey CR, Mares M (2011) Structural basis for inhibition of Cathepsin B drug target from the human blood fluke, *Schistosoma mansoni*. J Biol Chem 286(41):35770–35781. https://doi.org/10.1074/jbc.M111.271304
11. Horn M, Jilkova A, Vondrasek J, Maresova L, Caffrey CR, Mares M (2011) Mapping the pro-peptide of the *Schistosoma mansoni* Cathepsin B1 drug target: modulation of inhibition by heparin and design of mimetic inhibitors. ACS Chem Biol 6(6):609–617. https://doi.org/10.1021/cb100411v
12. Fanfrlik J, Brahmkshatriya PS, Rezac J, Jilkova A, Horn M, Mares M, Hobza P, Lepsik M (2013) Quantum mechanics-based scoring rationalizes the irreversible inactivation of parasitic *Schistosoma mansoni* cysteine peptidase by vinyl sulfone inhibitors. J Phys Chem B 117 (48):14973–14982. https://doi.org/10.1021/jp409604n
13. Jilkova A, Horn M, Rezacova P, Maresova L, Fajtova P, Brynda J, Vondrasek J, McKerrow JH, Caffrey CR, Mares M (2014) Activation route of the *Schistosoma mansoni* cathepsin B1 drug target: structural map with a glycosaminoglycan switch. Structure 22 (12):1786–1798. https://doi.org/10.1016/j.str.2014.09.015
14. Stefanic S, Dvorak J, Horn M, Braschi S, Sojka D, Ruelas DS, Suzuki B, Lim KC, Hopkins SD, McKerrow JH, Caffrey CR (2010) RNA interference in *Schistosoma mansoni* schistosomula: selectivity, sensitivity and operation for larger-scale screening. PLoS Negl Trop Dis 4(10):e850. https://doi.org/10.1371/journal.pntd.0000850
15. Caffrey CR, Mathieu MA, Gaffney AM, Salter JP, Sajid M, Lucas KD, Franklin C, Bogyo M, McKerrow JH (2000) Identification of a cDNA encoding an active asparaginyl endopeptidase of *Schistosoma mansoni* and its expression in *Pichia pastoris*. FEBS Lett 466 (2–3):244–248

Chapter 13

Characterization of *Schistosoma mansoni* Dihydrofolate Reductase (DHFR)

Vitor Hugo B. Serrão, Jéssica F. Scortecci, and Humberto D'Muniz Pereira

Abstract

Dihydrofolate reductase (DHFR) is an essential enzyme for nucleotide metabolism used to obtain energy and structural nucleic acids. *Schistosoma mansoni* has all the pathways for pyrimidine biosynthesis, which include the thymidylate cycle and, consequentially, the DHFR enzyme. Here, we describe the characterization of *Schistosoma mansoni* DHFR (*Sm*DHFR) using isothermal titration calorimetry for the enzymatic activity and thermodynamic determination, also the folate analogs inhibition. Moreover, X-ray crystallography was used to determine the enzyme atomic model at 1.95 Å.

Key words Isothermal titration calorimetry, Enzymatic assays, X-ray crystallography, Protein atomic model

1 Introduction

Schistosoma mansoni possesses all pyrimidine pathways, which include the de novo, salvage pathway and thymidylate cycle [1–3]. One of the essential members of the thymidylate cycle is the dihydrofolate reductase (DHFR, E.C. 1.5.1.3) an enzyme commonly found in all the organisms being well described as an anti-parasitic target [4–6]. DHFR enzymatic activity and structures were also completely described for several organisms [7]. Briefly, the DHFR catalyzes the nicotinamide adenine dinucleotide phosphate ($NADP^+$) reduction converting 7,8-dihydrofolate (DHF) to 5,6,7,8-tetrahydrofolate (THF), which is normally used as an intermediate to purines, 5′-thymidylic acid and amino acid synthesis, being crucial for nucleotides metabolism and consequentially for the worm survival [4–7]. Here we demonstrate the enzymatic activity and thermodynamics parameters from *Schistosoma mansoni* DHFR (*Sm*DHFR) enzyme using isothermal titration calorimetry (ITC). This microcalorimetry method is widely used to understand the interaction between macromolecules [8, 9]; however, it is completely capable to determine the enzymatic activities, especially

David J. Timson (ed.), *Schistosoma mansoni: Methods and Protocols*, Methods in Molecular Biology, vol. 2151,
https://doi.org/10.1007/978-1-0716-0635-3_13,

for reactions that are impossible to use spectroscopic methods, as fluorescence or absorbance. This technique is related to measure the heat exchange during the enzymatic reaction and, after the data treatment, identify the substrate consumption or product formation rate in relation to the injected concentration into the experimental cell.

Furthermore, the atomic model was determined by X-ray crystallography. This technique is one of the methods commonly used to solve protein structure in atomic resolution [10, 11]. The protein crystal is composed of many molecules of the protein well packed and ordered [12]. When the X-ray focuses on the crystal, the diffraction occurs at various angles and the intensity is observed in the diffraction pattern. This diffraction pattern reflects the arrangement of the atoms in the crystal, making possible the structure determination using mathematical methods as Bragg's Law and Fourier transform. One problem that arises from X-ray crystallography is called the phase problem [13]. The intensities observed are proportional to squared amplitude, which is a wave function. So the phase is lost in every experiment of X-ray crystallography. There are many ways to solve this problem: MR (molecular replacement), SAD/MAD (single- or multiple-wavelength anomalous diffraction), SIR/MIR (single or multiple isomorphous replacements), and SIRAS/MIRAS (a combination of anomalous dispersion and isomorphous replacement). The most common method is a molecular replacement; 61.3% of solved structures in Protein Data Bank (PDB) [14] were obtained using the MR method. The downside of this method is the requirement of a previously solved homolog protein with at least 30% of identity [15, 16]. The other methods require the use of heavy metal that can be added into the preformed crystal or intrinsic of the protein. Finding the position of this heavy metal leads to phase determination. Other techniques are used for protein structure determination in atomic resolution as nuclear magnetic resonance (NMR) and cryo-electron microscopy (Cryo-EM), but X-ray crystallography is by far the most widely used until now [17, 18].

2 Materials

Prepare all the solutions using ultrapure milli-Q water (deionized water at 18 MΩ-cm at 25 °C) and analytical grade reagents.

2.1 SmDHFR Gene Preparation

1. *Schistosoma mansoni* enriched cDNA library. Store at −20 °C.
2. Oligonucleotides sequences (5′-ACTGTATCATATGATGCG TCTAAATGTTGTTGTGGCAG-3′ (forward) and 5′-ACTG TATCATATGATGCGTCTAAATGTTGTTGTGGCAG-3′ (reverse) primers, Sigma-Aldrich) [100 pmol]. Dilute in mili-Q water to 10 pmol and store at −20 °C.

3. pTZ-57R/T propagation vector (Thermo-Fischer, USA).
4. pET28-a(+) expression vector (Novagen, USA).
5. 10 mM dNTPs.
6. *Taq* DNA polymerase (supplied at 100 U/μL). Store at −20 °C.
7. NdeI/XhoI restriction enzymes. Store at −20 °C.
8. LB-agar plates: 10 g Tryptone, 10 g sodium chloride, and 5 g yeast extract. Add 1 L of milli-Q water and adjust pH to 7.5 using 5 M NaOH. Before autoclave, 20 g/L of agar should be added to prepare the plates. Pour 20 mL in each Petri plate in presence of the specific antibiotic.
9. 30 mg/mL kanamycin dissolved in water. Store at −20 °C.
10. 200 mg/mL ampicillin dissolved in water. Store at −20 °C.
11. TAE buffer: 4.844 g Tris–HCl, 0.372 g EDTA sodium salt dihydrate, and add 1.21 mL of acetic acid. Set the volume to 1 L with mili-Q water.
12. Sybr Safe (Invitrogen).
13. Agarose gel extraction kit and Miniprep kit.
14. T4 DNA ligase.

2.2 Recombinant SmDHFR Expression

1. 2XYT media: 16 g Tryptone, 5 g sodium chloride, and 10 g yeast extract. Add 1 L of mili-Q water and adjust pH to 7.0 using 5 M NaOH.
2. 30 mg/mL kanamycin dissolved in water. Store at −20 °C.
3. 1 M isopropyl β-D-1-thiogalactopyranoside (IPTG) prepared in water. Store at −20 °C.
4. Centrifuges for the harvesting of *E. coli* cells, a standard refrigerated centrifuge capable of a speed of up to 20,000 × *g*.
5. *E. coli* DH5-α and *E. coli* BL21 (λDE3) (Novagen).

2.3 Recombinant SmDHFR Purification and Sample Quality Determination

1. Lysis buffer: 50 mM sodium di-hydrogen phosphate pH adjusted to 7.8 using 5 M NaOH, 200 mM sodium chloride, 10 mM imidazole, and 5 mM 2-mercaptoethanol.
2. Wash buffer: 50 mM sodium di-hydrogen phosphate pH adjusted to 7.8 using 5 M NaOH, 200 mM sodium chloride, 20 mM imidazole, and 5 mM 2-mercaptoethanol.
3. Elution buffer: 50 mM sodium di-hydrogen phosphate pH adjusted to 7.8 using 5 M NaOH, 200 mM sodium chloride, 200 mM imidazole, and 5 mM 2-mercaptoethanol.
4. Dialysis buffer: 20 mM Tris–HCl pH adjusted to 7.8 using 5 M NaOH, 200 mM sodium chloride, and 5 mM 2-mercaptoethanol.

5. 5 mL cobalt-Talon affinity chromatography resin pre-package in 10 mm open gravity flow column. Wash the column with 5 column volumes (CV) of water and 5 CV of lysis buffer before loading the crude extract.
6. 50 mg/mL lysozyme. Store at −20 °C.
7. Cellulose membrane concentrator tubes and filters (Amicon, 10,000 Da MWCO).
8. SDS-PAGE equipment and 15% gels.
9. Nanodrop2000 (Thermo Fischer).
10. Baffled conical flasks.
11. Probe sonicator suitable for lysing *E. coli*.
12. 10 mm open gravity flow column.
13. Thrombin protease.

2.4 Enzymatic Assays Using Isothermal Titration Calorimetry (ITC)

1. Kinetics buffer: 20 mM Tris–HCl pH adjusted to 7.8 using 5 M NaOH, 200 mM sodium chloride supplemented with 1.5 mM potassium chloride.
2. 1 mM nicotinamide adenine dinucleotide phosphate (NADPH) prepared in kinetics buffer.
3. 1 mM dihydrofolic acid (DHF) prepared in kinetics buffer.
4. 0.5 mM aminopterin prepared in kinetics buffer.
5. 1 mM methotrexate (MTX) prepared in kinetics buffer.
6. VP-ITC (MicroCal) (Fig. 1a).
7. Origin 7.0 software integrated into the VP-ITC system.

2.5 Sm*DHFR* Structure Determination by X-Ray Crystallography

1. Crystallization buffer: 100 mM Hepes buffer pH adjusted to 8.0 using 5 M NaOH, 2.0 M ammonium sulphate.
2. Cryo-protector: 20% glycerol diluted in crystallization buffer.
3. Spin-pins and cryo-loops.
4. Unipucks and dry-Dewar for crystals shipment.
5. Honeybee 963 crystallization robot (Genomic Solution).
6. Sitting-drops plastic 96-well plates.
7. ID29 beamline at European Synchrotron Radiation Facility (ESRF, Grenoble – France) or equivalent facility at another synchrotron.
8. XDS package [19].
9. Phaser [20].
10. COOT [21] and PyMOL [22].

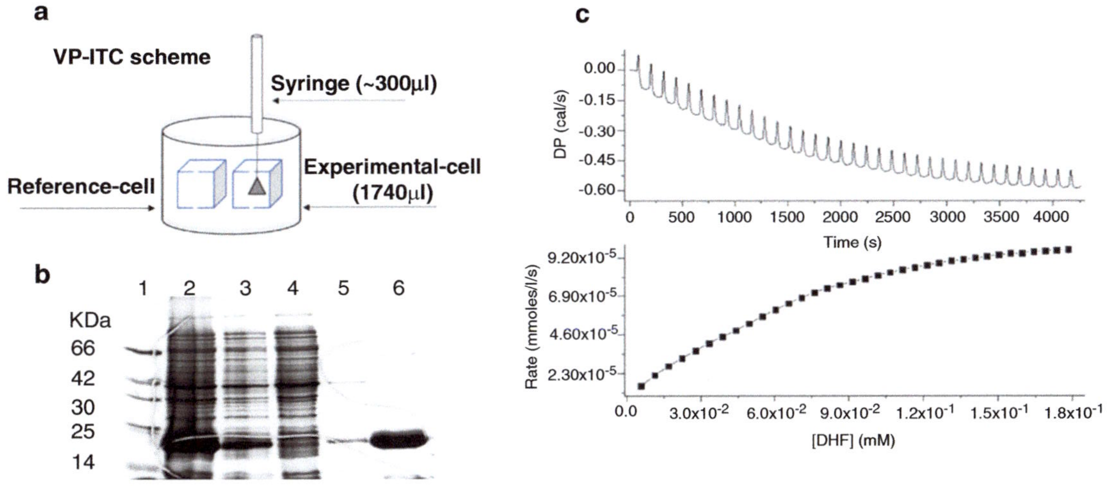

Fig. 1 Schematic isothermal titration calorimetry (ITC) experiments and *Sm*DHFR sample preparation. (**a**) VP-ITC scheme showing the syringe (50 μL < loading volume < 500 μL) and experimental and reference cells (~2 mL volume each) that should be filled with analyte and the reference buffer, respectively. (**b**) Recombinant *Sm*DHFR sample expression and purification evaluated by Coomassie-stained SDS-PAGE 15%. (1) Molecular mass (kDa). (2) Harvested cell pellet after induction. (3) Crude extract. (4) Cobalt-affinity chromatography flow through. (5) Cobalt-affinity chromatography wash and (6) cobalt-affinity chromatography elution. (**c**) Experimental procedure to measure the DHF kinetics using ITC represented by the raw data (top figure) and the derivate analyzed data (bottom figure) using Michaelis-Menten adjust

3 Methods

Carry out all the procedures on ice or at 4 °C unless otherwise specified.

3.1 SmDHFR Gene Preparation and Cloning

1. Prepare the polymerase chain reaction (PCR) adding 1 μL enriched cDNA from *Schistosoma mansoni*, 10 pmol of primers, 0.5 mM dNTPs, and 100 units of *Taq* DNA polymerase in 50 μL reaction volume.
2. Place into the thermocyclers using the sequential steps: initial 5 min at 94 °C for DNA denaturation followed by 35 cycles of 30 s at 94 °C, 45 s at 68 °C, and 45 s at 72 °C. At the end, keep at 4 °C.
3. Prepare 1% agarose gel by weighing 1 g agarose in 100 mL TAE buffer followed by 60 s into the microwave. The melted agarose was supplemented with 0.1% Sybr Safe and loaded in the gel box with the spacers to create the 50 μL wells. Load the resulted PCR into the gel with 1× loading buffer and a molecular marker in base pair references. Apply a constant voltage of 80 V for 45 min.
4. Illuminate the gel in a transilluminator at 470 nm. Cut the ~560 bp band and proceed with the gel extraction commercial kit. Quantify the *Smdhfr* gene concentration by absorption at 260 nm using Nanodrop2000.
5. Mix 10 μg of pTZ-57R/T plasmid with 100 μg of *Smdhfr*. Perform the ligase reaction adding 10 units of T4 DNA ligase at 16 °C for 18 h.
6. Transform the *Smdhfr*- pTZ-57R/T plasmid into *Escherichia coli* DH5-α cells by adding the ligase reaction with 33 μL of chemo-competent cells for 15 min on ice. Transfer the tube to 42 °C for 1 min and keep on ice for 3 min after that. Add 500 μL LB-media and shake 80 rpm at 37 °C during 1 h.
7. Centrifuge the cells and discard the supernatant. Resuspend the cell in 20 μL LB-media and plate in LB-agar plates supplemented with 20 μg/mL ampicillin. Leave overnight incubating at 37 °C.
8. Inoculate a single colony in 5 mL LB media supplemented with 20 μg/mL ampicillin. Shake at 180 rpm overnight at 37 °C.
9. Conduct the commercial miniprep plasmid extraction kit and check the concentration of *Smdhfr*- pTZ-57R/T plasmid as mentioned in **step 4**.
10. Perform the restriction enzyme digestion using 10 units of NdeI/XhoI enzymes in 100 μg of *Smdhfr*- pTZ-57R/T plasmid at 37 °C for 1 h. Check the digestion reaction using agarose gels as described in **steps 3** and **4**.

11. In parallel, repeat **step 10** in pET28a(+) commercial vector.
12. Mix 10 μg of digested pET28a(+) plasmid with 100 μg of digested *Smdhfr* and repeat **step 5**.
13. Transform the *Smdhfr*-pET28a(+) plasmid into *Escherichia coli* BL21(λDE3) cells by adding the ligase reaction with 33 μL of chemo-competent cells for 15 min on ice. Transfer the tube to 42 °C for 1 min and keep on ice for 3 min after that. Add 500 μL LB-media and shake 80 rpm at 37 °C during 1 h. Repeat **step 7** using LB-agar plates supplemented with 30 μg/mL kanamycin.

3.2 Recombinant SmDHFR Expression

1. Inoculate a single *Escherichia coli* BL21 (λDE3) colony in 5 mL LB media supplemented with 30 μg/mL kanamycin. Shake at 180 rpm overnight at 37 °C.
2. Inoculate 10 mL from the overnight culture into 1 L of 2XYT media supplemented with 30 μg/mL kanamycin. Shake at 180 rpm at 37 °C to reach OD_{600nm} of 0.6. Induce the *Smdhfr* gene expression adding 0.1 mM IPTG for 4 h at 37 °C. Recombinant *Sm*DHFR expressed using the *Smdhfr*-pET28a (+) vector, which adds an N-terminal hexahistidine tag with thrombin protease cleavage site.
3. Harvest the cells by centrifugation at 1800 × *g* at 4 °C for 45 min. Discard supernatant and storage the dry cells pellet at −20 °C for future purification procedures.
4. Check the expression levels using SDS-PAGE 15% using 15 μL aliquots prepared with SDS-sample buffer. Apply a constant voltage of 140 V, for 90 min and after Coomassie-stain for 5 min. The gel is revealed using acetic acid 20%.

3.3 Recombinant SmDHFR Purification and Sample Quality Determination

1. Defrost 2 g of cells pellet on ice for 30 min. Add the 5 mL lysis buffer and homogenize by vortexing during 30 s.
2. Add 420 μg/mL lysozyme and keep it on ice for 30 min.
3. Prepare the cell lysate by sonication at 4 °C (30 s pulse on followed by 45 s pulse off).
4. Centrifuge the cell lysate at 20,000 × *g* for 45 min at 4 °C.
5. Filter the crude extract using 0.22 μm cellulose filters, and apply into the polypropylene column previously filled with 5 mL cobalt-Talon resin and equilibrated with 5 CV of lysis buffer.
6. Load the crude extract into the affinity column, and keep the flow using a gravitational column for collecting all the fractions eluted into a 50 mL conical tube.
7. After the crude extract pass through the column, wash the affinity resin with 5 CV using wash buffer and collect all the fractions.

8. Finally, elute the purified sample using 3 CV of elution buffer. Collect all the fractions eluted from the affinity column.
9. Wash the resin with 20 CV of mili-Q water and storage the column at 4 °C with 10 CV ethanol 20%.
10. Dialyze the eluted *Sm*DHFR sample against dialysis buffer in 1 L cylinder overnight at 4 °C in presence of 1 unit of thrombin per 1 mg of purified *Sm*DHFR for the hexahistidine tag cleavage.
11. Concentrate the *Sm*DHFR sample using the cellulose membrane concentrator for 45 min at 4 °C and 1000 × *g*.
12. Check the expression levels using SDS-PAGE 15% using 15 μL aliquots prepared with SDS-sample buffer. Apply 140 V, 400 mA for 90 min and after Coomassie-stain for 5 min. The gel is revealed using acetic acid 20% (Fig. 1b).
13. Determine the sample concentration using a Nanodrop2000 using absorption at 280 nm and extinction coefficient of 32,430 M^{-1} cm^{-1} (1.538% (=1 g/L)).

3.4 Enzymatic Assays Using Single-Injection Isothermal Titration Calorimetry (ITC)

1. Initially, place milli-Q water into the VP-ITC cell and syringe for checking the quality of the cleaning process. The heat exchange must be small using 20 μL single injection during 40 s and spacing of 1200 s (filter at 2 s) at 25 °C with reference power at 5 μCal/s. The initial delay (120 s) and 100 rpm stirring speed with low feedback gain are other parameters that must be settled.
2. Once the experimental cell and syringe are clean, the same procedure should be used to check the quality of the kinetics buffer.
3. Degas all the buffers, samples, and substrates using the MicroCal Malvern ThermoVac2 sample degasser and thermostat station for 25 min at room temperature before starting the experiments.
4. To evaluate the 5,6,7,8-tetrahydrofolate (THF) formation molar enthalpy, load 1740 μL of 500 nM *Sm*DHFR previously diluted in kinetics buffer and in presence of 1 mM NADPH into the VP-ITC experimental cell.
5. Place 300 μL of 1 mM DHF previously diluted in kinetics buffer in presence of 1 mM NADPH into the VP-ITC syringe.
6. Perform a single injection of 20 μL of NADPH/DHF syringe contain monitoring the heat exchange for 1200 s at 25 °C, reference power at 10 μCal/s. The initial delay (300 s) and 100 rpm stirring speed with low feedback gain.
7. Perform the same injection against 1740 μL kinetics buffer and in presence of 1 mM NADPH into the VP-ITC experimental cell.

8. Using Origin 7.0 coupled to the VP-ITC, determine the baseline and perform the subtraction between **steps 6** and 7 curves. The resulted peak has to be integrated, resulting in the apparent molar enthalpy variation (ΔH_{app}) for DHF binding/kinetics.
9. Conduct the same experiment and analysis (**steps 4–8**) using 1740 μL of 500 nM *Sm*DHFR previously diluted in kinetics buffer and in presence of 1 mM DHF into the VP-ITC experimental cell. This result is the apparent molar enthalpy variation (ΔH_{app}) for NADPH binding/kinetics.
10. Plot the graphs and analyze the data using Origin 7.0 (Fig. 1c).

3.5 Enzymatic Assays Using Multiple-Injection Isothermal Titration Calorimetry (ITC)

1. Degas all the buffers, samples, and substrates for 25 min at room temperature before starting the experiments.
2. To evaluate the 5,6,7,8-tetrahydrofolate (THF) formation Michaelian enzymatic profile using multiple injections, place 1740 μL of 5 nM *Sm*DHFR previously diluted in kinetics buffer and in presence of 1 mM NADPH into the VP-ITC experimental cell.
3. Place 300 μL of 1 mM DHF previously diluted in kinetics buffer in presence of 1 mM NADPH into the VP-ITC syringe.
4. Perform multiple injections of 2 μL each of NADPH/DHF syringe contain monitoring the heat exchange for 120 s at 25 °C, reference power at 5 μCal/s. The initial delay (300 s) and 300 rpm stirring speed with low feedback gain.
5. Perform the same injection against 1740 μL kinetics buffer and in presence of 1 mM NADPH into the VP-ITC experimental cell.
6. Using Origin 7.0 coupled to the VP-ITC, determine the baseline and perform the subtraction between **steps 4** and **5**. The resulted apparent molar enthalpy variation (ΔH_{app}) from the single-injection experiment is used to set the enthalpy heat amplitude observed during the experiment. Load the ΔH_{app} value, set the curve origin, and choose the Michaelis-Menten model using a single substrate. The curve is automatically fitted, and the kinetics parameters are available (maximum substrate consumption, V_{max}; Michaelian constant, K_M; turnover, k_{cat}; and the enzymatic efficiency, k_{cat}/K_M).
7. Conduct the same experiment and analysis (**steps 2–6**) using 1740 μL of 5 nM *Sm*DHFR previously diluted in kinetics buffer and in presence of 1 mM DHF into the VP-ITC experimental cell. This results in the kinetics parameters for NADPH kinetics.

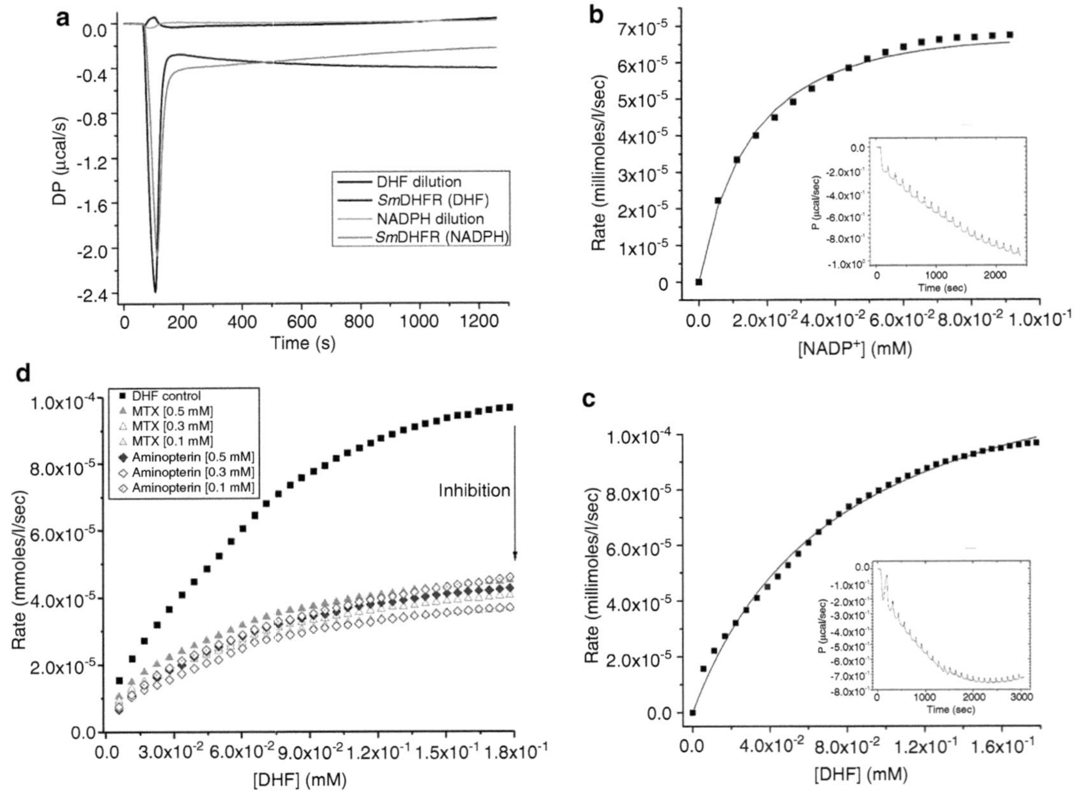

Fig. 2 *Sm*DHFR enzymatic characterization by ITC. (**a**) ΔH_{app} determination by ITC using 20 μL single injections of DHF and NADP$^+$ revealing the exothermic reactions ($\Delta H_{app}^{DHF} = -4.5$ kcal/mol and $\Delta H_{app}^{NADP^+} = -6.6$ kcal/mol, respectively). The dilutions were performed under the same experimental conditions. (**b**, **c**) The Michaelis-Menten kinetics of DHF and NADP$^+$, respectively. (**d**) *Sm*DHFR enzymatic competitive inhibition by folate analogs addition, clearly showing the V_{max} and K_M variation

8. Plot the graphs and analyze the data using Origin 7.0 (Fig. 2a–c). The biological comparison is made based on online databanks for Human Metabolomes (HMDB) and enzymatic parameter (BRENDA).

3.6 Estimation of Sm*DHFR Inhibition by Folate Analogs*

1. Degas all the buffers, sample, substrates, and inhibitors for 25 min at room temperature before starting the experiments.
2. To evaluate the folate analogs inhibition, place 1740 μL of 5 nM *Sm*DHFR previously diluted in kinetics buffer and in presence of 1 mM NADPH into the VP-ITC experimental cell.
3. Place 300 μL of 1 mM DHF previously diluted in kinetics buffer in presence of 1 mM NADPH and 0.5 mM of MTX into the VP-ITC syringe.
4. Perform **steps 4–6** from Subheading 3.5.

5. After the data acquisition is finished and the VP-ITC is clean, redo the **steps 2–5** with 0.3 mM and 0.1 mM MTX, respectively.
6. Using Origin 7.0 coupled to the VP-ITC, conduct the same analysis as previously described in Subheading 3.5, **step 6**. In the end, V_{max} and K_M values are different and are possible to evaluate the *Sm*DHFR inhibition per MTX concentration, allowing the comparison with other folate analog inhibitors and possible targets to be used as chemotherapeutic.
7. Conduct the same experiment and analysis (**steps 1–6**) using 300 μL of 1 mM DHF previously diluted in kinetics buffer in presence of 1 mM NADPH and 0.5 mM of aminopterin into the VP-ITC syringe.
8. Plot the graphs and analyze the data using Origin 7.0 (Fig. 2d). The biological comparison is made based on online databanks for Human Metabolomes (HMDB) and enzymatic parameter (BRENDA).

3.7 *Sm*DHFR Structure Determination by X-Ray Crystallography

1. Place the purified *Sm*DHFR into the concentrator, centrifuge at 1000 × *g*, 4 °C, 15 min, and check the protein concentration using the nanodrop2000. Repeat this process until the protein concentration of 12 mg/mL is reached. The *Sm*DHFR sample quality is checked by SDS-PAGE 15% as described in Subheading 3.3, **step 12**.
2. Remove the *Sm*DHFR sample and transfer into a 1.5 mL tube and keep it on ice.
3. Start to prepare the crystallization plate: in a hanging drop plate, add 2 μL of crystallization buffer at the bottom. In the glass treated with silicon add 2 μL of crystallization buffer in the center. Take 2 μL of the *Sm*DHFR and dispense into the drop of crystallization buffer. Apply the cover at the border of the well and gently invert the glass and place the drop faced to the well. Press the glass against the well gently; otherwise it will break and no vapor diffusion will happen. Store the plates in the incubator at 18 °C and monitor the crystal formation and nucleation.
4. Many crystallization trials must be tested before finding the proper condition for crystallization of a purified sample.

3.8 Prepare the Crystal for Data Collection: Freezing

1. The crystal needs to be cryo-protected to data collection because it occurs at 100 K. In another tube, prepare the cryo-protect buffer: add 8 μL of crystallization buffer and 2 μL of 80% (v/v) glycerol. Place the 10 μL in a siliconized glass coverslip.

2. Open the hanging drop plate by removing the glass containing the crystals. With the cryo-loops, scoop the crystal and transfer to cryo-protect buffer. Immediately, take the crystal of this solution and transfer quickly to liquid nitrogen (LN_2) in the unipuck. Some crystals require different cryo-protector as high sugar concentration, PEG low-MW, or ethylenoglycols. It must be tested. After the crystals are frozen, transfer the unipuck into the dry-ship dewar.

3.9 X-Ray Crystallography Data Collection at a Synchrotron Beamline and Data Processing

1. At the beamline, the unipucks are settled into the storage LN_2 storage dewar for automatic loading in the beamline (ID29-ESRF). The crystal is transferred at 100 K at ID29-ESRF using Pilatus 6 M detector and wavelength = 0.9762 Å.
2. After the crystal's alignment in the beam, the strategy is determined by collecting 2 independent and 90° rotated images. After planning the data collection strategy, the parameters are settled for 0.2° oscillation in 360° range and 0.5 s exposure time at 150 mm detector distance.
3. The dataset is reduced using XDS package [19] and scaled by CCP4 (Aimless). For molecular replacement phasing, it is

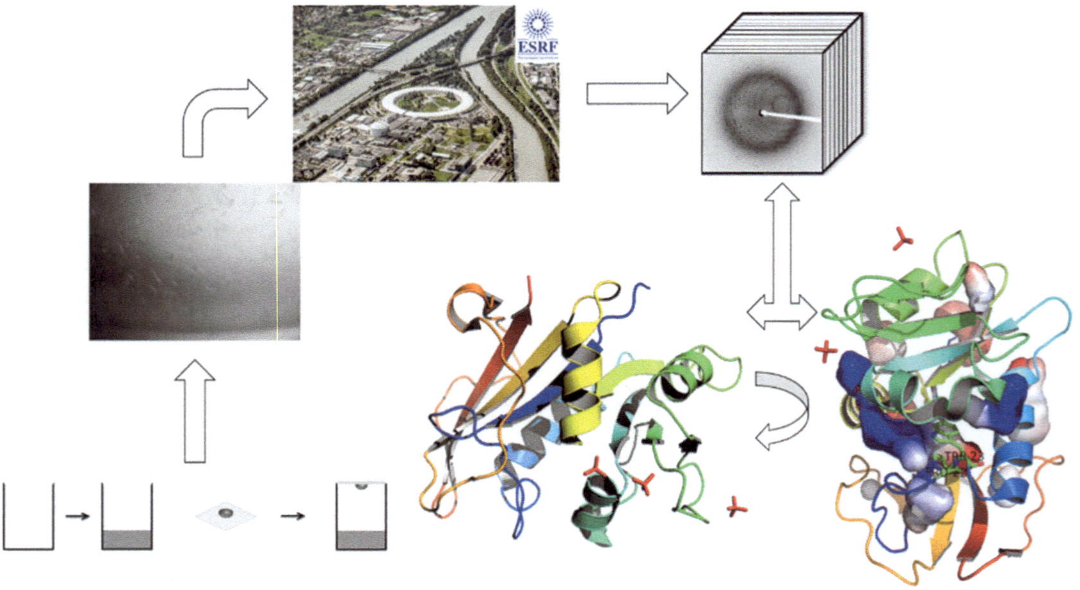

Fig. 3 Crystallographic analysis pipeline: Scheme of hanging drop crystallization plates sealed for vapor diffusion. After the crystal's growth/optimization, the crystals are cryo-protected and shipped for a Synchrotron Radiation Source (like ESRF in Grenoble-France) for high-resolution data collection. With the acquisition of several images, it is possible to conduct the data processing and obtain the crystallographic model, here, showing the secondary structure conformation in alpha-beta-alpha sandwich and highlighting the active pocket and the electrostatic contains that allow *Sm*DHFR to recognize DHF/$NADP^+$ to perform the conversion to THF

possible to use PHASER (provided by PHENIX) [20] and the homolog used was PDB.ID: 8DFR.

4. For the refinement and validation, Coot [21], PHENIX, and Molprobity are commonly used and the final data is generated by PDB EXTRACT.
5. Structural visualization, alignment analysis, and figure generation, PyMOL [22], and Coot are the most common programs used. The PDB file is uploaded to Protein DataBank (PDB) and receives a unique code (PDB.ID: 3VCO) (Fig. 3).

Acknowledgments

We acknowledge the Fundação de Amparo à Pesquisa do Estado de São Paulo (FAPESP) grants 2012/23730-1 (VHBS), 2014/16005-4 and 2016/20977-7 (JFS), 2012/142239 (HMP) also CNPq grant 474402/2013-4 and 140636/2013-7 for the financial support. We also thank the ESRF and beamline scientists that helped us during the data acquisition.

References

1. Senft AW, Crabtree GW (1983) Purine metabolism in the schistosomes: potential targetsfor chemotherapy. Pharmacol Ther 20:341–356
2. Senft AW, Miech RP, Brown PR, Senft DG (1972) Purine metabolism in *Schistosoma mansoni*. Int J Parasitol 2:249–260
3. Senft AW, Senft DG, Miech RP (1973) Pathways of nucleotide metabolism in *Schistosoma mansoni*. disposition of adenosine by whole worms. Biochem Pharmacol 22:437–447
4. Serrão VHB, Romanello L, Cassago A, Torini JR, Cheleski J, deMarco R, Brandão-Neto J, Pereira HM (2017) Structure and kinetics assays of recombinant *Schistosoma mansoni* dihydrofolate reductase. Acta Trop 170:190–195
5. Jaffe JJ, McCormack JJ, Meymarian E (1972) Comparative properties of schistosomal and filarial dihydrofolate reductases. Biochem Pharmacol 21:719–731
6. Jarabak J, Bachur NR (1971) A soluble dihydrofolate reductase from human placenta: purification and properties. Arch Biochem Biophys 142:417–425
7. Liu CT, Hanoian P, French JB, Pringle TH, Hammes-Schiffer S, Benkovic SJ (2013) Functional significance of evolving protein sequence in dihydrofolate reductase from bacteria to humans. Proc Natl Acad Sci U S A 110:10159–10164
8. Todd MT, Gomez J (2001) Enzyme kinetics determined using calorimetry: general assay-forenzymeactivity? Anal Biochem 296:179–187
9. Wiseman T, Williston S, Brandts JF, Lin LN (1989) Rapid measurement of binding constants and heats of binding using a new titration calorimeter. Anal Biochem 179:131–137
10. Dauter Z, Wlodawer A (2016) Progress in protein crystallography. Protein Pept Lett 23 (3):201–210
11. Lattman EE (1994) Protein crystallography for all. Proteins 18(2):103–106
12. McPherson A, Gavira JA (2013) Introduction to protein crystallization. Acta Crystallogr F Struct Biol Commun 70:2–20
13. Taylor G (2003) The phase problem. Acta Crystallogr D Biol Crystallogr 59:1881–1890
14. Burley SK, Berman HM, Kleywegt GJ, Markley JL, Nakamura H, Velankar S (2017) Protein data Bank (PDB): the single global macromolecular structure archive. Methods Mol Biol 1607:627–641
15. Scapin G (2013) Molecular replacement then and now. Acta Crystallogr D Biol Crystallogr 69:2266–2275
16. Abergel C (2013) Molecular replacement: tricks and treats. Acta Crystallogr D Biol Crystallogr 69:2167–2173

17. Pedrini B, Serrano P, Mohanty B, Geralt M, Wüthrich K (2013) NMR-profiles of protein solutions. Biopolymers 99(11):825–831
18. Egelman EH (2016) The current revolution in Cryo-EM. Biophys J 110:1008–1012
19. Kabsch W (2010) XDS. Acta Crystallogr D Biol Crystallogr 66:125–132
20. McCoy AJ, Grosse-Kunstleve RW, Adams PD, Winn MD, Storoni LC, Read RJ (2007) Phaser crystallographic software. J Appl Crystallogr 40:658–674
21. Emsley P, Cowtan K (2004) Coot: model-building tools for molecular graphics. Acta Crystallogr D60:2126–2132
22. The PyMOL Molecular Graphics System, Version 2.0. LLC, Schrödinger

Chapter 14

Characterization of *Schistosoma mansoni* Glutamate-Gated Chloride Channels

Timothy Lynagh

Abstract

Electrophysiology is the standard method for characterizing ion channel function. Two-electrode voltage clamp is a robust and relatively simple version which can be applied to the characterization of glutamate-gated chloride channels from *Schistosoma mansoni*, a potential schistomicidal target. Here, the method is described in detail, with an emphasis on the investigation of *S. mansoni*. GluCls.

Key words GluCl, *Schistosoma mansoni*, Ion channels, Glutamate, Receptor, Pharmacology

1 Introduction

At the synapse and at the neuromuscular junction, neurotransmitter binds to and activates ligand-gated ion channels in the postsynaptic cell membrane. Na^+ and Ca^{2+} ion current through cation-selective channels leads to depolarization and thus excitation of postsynaptic cells, whereas Cl^- current through anion-selective channels generally hyperpolarizes cells or shunts depolarization and is inhibitory [1]. This conversion of neurotransmitter binding to excitation or inhibition is vital to animal function, and ligand-gated ion channels are common targets for pharmacological modulation. As glutamate-gated chloride channels (GluCls) occur in flatworms, such as *Schistosoma mansoni*, and are absent from vertebrates, they constitute a target for potential schistomicidal drugs [2].

The gold standard for functional and pharmacological characterization of specific ion channels is patch clamp electrophysiology, where channels are recombinantly expressed in cultured cells, and membrane currents in response to agonist and/or drug application are recorded [3]. A slightly easier method is two-electrode voltage clamp, where cultured cells are replaced with larger, easier-to-handle oocytes from *Xenopus laevis* (African clawed frog) [4]. With the

David J. Timson (ed.), *Schistosoma mansoni: Methods and Protocols*, Methods in Molecular Biology, vol. 2151,
https://doi.org/10.1007/978-1-0716-0635-3_14, © Springer Science+Business Media, LLC, part of Springer Nature 2020

identification of GluCls and other ion channels from *S. mansoni*, schistomicidal drug discovery could benefit from electrophysiological and pharmacological characterization of these channels. This chapter serves to describe the characterization of GluCls by two-electrode voltage clamp.

2 Materials

Electrophysiology setups can be pieced together with components from various suppliers but are often built around the main components of amplifier and digitizer from a particular supplier. The following outlines the setup commonly used by this author, much of which is part of a convenient two-electrode voltage clamp package, TEV-700 (Warner Instruments), but equivalent components are also available from, e.g., Molecular Devices and NPI Electronic.

2.1 Two-Electrode Voltage Clamp Setup

1. Amplifier (OC-725C, Warner Instruments, includes bath clamp headstage).
2. Digitizer (Axon Digidata 1550B, Molecular Devices).
3. Acquisition software (pClamp 11, Molecular Devices).
4. Computer (*see* **Note 1**).
5. Baseplate (BPM-1, Warner Instruments, but *see* **Note 2**).
6. Stereomicroscope and stand (Z850, Unitron; *see* **Note 3**).
7. Oocyte recording chamber (RC-3Z, Warner Instruments).
8. Perfusion system (VCS-8-Pinch, Warner Instruments).
9. Waste collection system (DWV, Warner Instruments).
10. Left and right micromanipulators (MM-33 L and MM-33R, and MB/B magnetics stands, Warner Instruments).
11. Electrode holders (E45W-F15VH for current cable and ESW-F15V for voltage probe, Warner Instruments).
12. Silver wire.
13. Maneuverable light source.

2.2 Other Equipment

1. Micropipette puller (P-1000, Sutter Instruments).
2. Glass capillaries (Item 504,949 from World Precision Instruments for RNA injection, GC120T-10 from Harvard Apparatus for two-electrode voltage clamp).
3. Tweezers (Dumont Teezers #5, 0.1 × 0.06 mm tips, World Precision Instruments).
4. Microinjector (Nanoliter 2010, World Precision Instruments; this also requires a stereomicroscope and micromanipulators, much like the two-electrode voltage clamp setup).

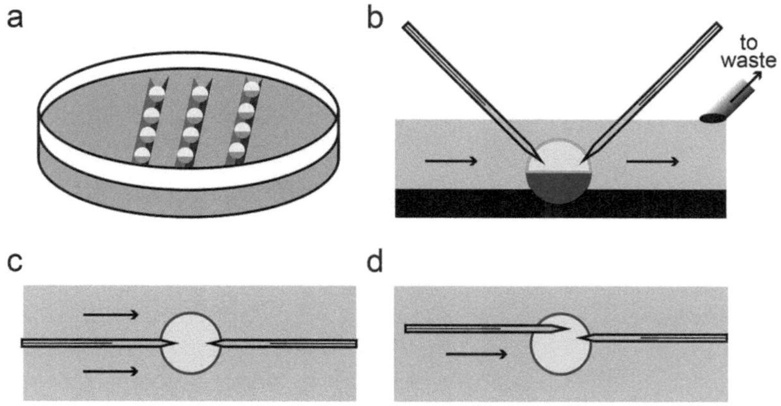

Fig. 1 Oocytes during injection of RNA and during two-electrode voltage clamp experiments. (**a**) A plastic disc (gray) containing linear groves placed inside a 35 mm dish (white) provides a convenient base for injecting and even maneuvering oocytes during injection of RNA. (**b**) Side view of an oocyte during two-electrode voltage clamp experiment (arrows indicate direction of perfusion). (**c**) Good and (**d**) poor impalement of the oocyte, viewed from above, in which uneven perfusion is likely to rip the oocyte from the microelectrodes

5. MicroFil needles (MF28G-5, World Precision Instruments).
6. Parafilm.
7. Disposable lint-free cloth wipes (Kimtech, Kimberly-Clark).
8. Grooved plate or wire grid. This is a simple plastid disc with grooves or a wire mesh/grid for holding oocytes conveniently when microinjecting (Fig. 1a). These can often be built in-house.
9. *X. laevis* stage V–VI oocytes (Ecocyte Bioscience, but *see* **Note 4**).
10. Data analysis software (Prism 8, GraphPad).

2.3 Reagents

1. Mineral oil.
2. T7 or SP6 capped RNA transcription kits (Ambion mMMESSAGE mMACHINE, Thermo Fisher Scientific).
3. RNeasy Mini Kit (Qiagen).

2.4 Solutions

1. OR2: 82.5 mM NaCl, 2 mM KCl, 1 mM $MgCl_2$, 5 mM HEPES, pH 7.4 with NaOH.
2. Leibovitz's L-15 medium (*see* https://www.thermofisher.com/no/en/home/technical-resources/media-formulation.80.html), supplemented with 3 mM L-glutamine, 2.5 mg/ml gentamycin, 15 mM HEPES, pH 7.4 with NaOH.
3. ND96: 96 mM NaCl, 2 mM KCl, 1.8 mM $CaCl_2$, 1 mM $MgCl_2$, 5 mM HEPES, pH 7.4 with NaOH.

4. ND96/gluconate: 96 mM Na gluconate, 2 mM KCl, 1.8 mM $CaCl_2$, 1 mM $MgCl_2$, 5 mM HEPES, pH 7.4 with NaOH.
5. ND96/choline: 96 mM choline Cl, 2 mM KCl, 1.8 mM $CaCl_2$, 1 mM $MgCl_2$, 5 mM HEPES, pH 7.4 with NaOH.

3 Methods

3.1 Injection of RNA into Oocytes

1. GluCl cDNAs should be cloned into a plasmid vector downstream of T7 or SP6 sequence (for recognition by polymerase in mRNA transcription), *X. laevis* β-globin promoter sequence, and Kozak sequence, and upstream of a poly(A) tail and then a restriction site for DNA linearization. Capped mRNAs are transcribed using kits. This author then purifies mRNA with columns in a kit, avoiding chemicals such as phenol/chloroform and LiCl. By eluting with a small volume, one typically recovers 10–20 μl RNA at >1 μg/μl from a single reaction, which can be stored at −20 °C for days or at −80 °C for months or even years.
2. Protocols for pulling micropipettes vary with different filaments in the micropipette puller, and the P-1000 manual or Sutter micropipette cookbook describes how to establish these protocols. For injection of mRNA, a ~10 mm, stringlike taper is desirable, the tip can be broken off with tweezers or fine scissors, achieving a tip small enough to hold RNA, but large enough to expel mRNA when required.
3. The microinjector should be set to the appropriate volume (*see* **Note 5**). Micropipettes are back-filled with mineral oil using a MicroFil needle and syringe, and the micropipette is inserted to the injector. A lint-free cloth wipe should be kept on hand for removing expelled mineral oil, as excessive oil on gloves, etc. becomes troublesome. A 1–2 μl drop of mRNA (*see* **Not e 5**) is pipetted on to a piece of Parafilm or an upside-down lid from a 1.5 ml centrifuge tube. The drop of mRNA is then drawn into the micropipette, taking care to avoid air bubbles. When the entire drop is drawn, ejecting one unit of mRNA should see a drop of mRNA rapidly form at or drop from the end of the micropipette.
4. Transfer ~20 oocytes into a 35 mm Petri dish filled with OR2 and containing the plastic disc with slots or the wire mesh (Fig. 1a).
5. Impale the oocyte as little as possible while definitely reaching the cytoplasm with the micropipette tip, eject RNA, and remove micropipette quickly but carefully to avoid damaging the oocyte. Another unit of RNA can be ejected in the OR2 to remove debris from the micropipette tip. Various laboratories

report impaling the oocyte at different parts (e.g., middle of the animal pole or vegetal pole), but this author has typically injected into the equator. Oocytes can be rolled over with the micropipette, using the left hand to move the dish (and thus the oocyte) and the right hand to move the micropipette.

6. Although micropipettes can be used again immediately for another mRNA/set of oocytes, it is strongly recommended to replace, in order to avoid both clogging of the micropipette tip with debris and contamination of different RNAs.
7. Oocytes are transferred to a dish of Leibovitz's L-15 medium and stored at 18 °C until experiments (*see* **Note 5**). Oocytes may be stored with continuous stirring, which avoids deformation of the oocytes caused by sitting in the same position for hours/days. If after 1 day the media seems dirty, oocytes should be transferred to new Leibovitz's L-15 medium.

3.2 Two-Electrode Voltage Clamp Recordings

1. Silver wire electrodes (these are included with the electrode holders but can be fashioned in-house with silver wire) should be coated in a AgCl layer by cleaning and then immersing (only as far as is necessary for inserting in KCl within the micropipette) in bleach. Bleach should then be rinsed off with deionized water and if necessary carefully dried with a lint-free wipe (dabbing, to avoid wiping off the AgCl layer).
2. The glass micropipettes should be ~1/3- or ~1/5-filled (depending on length of Ag/AgCl electrode) with a MicroFil needle and syringe with 3 M KCl. Care should be taken to ensure all solution is in the end of the micropipette that will impale the oocyte, away from the top of the micropipette, where it can lead to salt buildup in the electrode holders. The micropipettes are then inserted into the micropipette holders, where part of the AgCl-coated electrode is in KCl.
3. The resistance of the microelectrodes can be tested by dipping them into the ND96 that perfuses the recording chamber. The voltage (difference between one electrode and its reference, bath electrode) here should be zero, and by using the membrane test on the amplifier, a current pulse is injected and the calculated resistance is read out, which should be 0.5–1 MΩ for two-electrode voltage clamp.
4. A single oocyte is placed in the recording chamber with continuous perfusion of ND96 solution. The oocyte is impaled with the two microelectrodes, both perpendicular to the oocyte membrane and no deeper than is necessary for obvious electrical connection, visible via the voltage readout on the amplifier (often around −30 mV when inside the oocyte). A scenario perhaps best described as "symmetrical" is essential for the smooth flow of solution around the oocyte (Fig. 1b, c). If

this is not the case, perfusion may be unequal on different sides of the oocyte (which is already the case with the sides of the oocyte facing and facing away from the source of perfusion), and more importantly, the imbalance can tear the oocyte apart (Fig. 1d). Again, various laboratories may differ in the preference of which parts of the oocyte are impaled with the two microelectrodes, but this author prefers inserting both microelectrodes in the less yolky animal pole.

5. With both microelectrodes inside the oocyte and regular ND96 perfusing the oocyte, the amplifier can be set to voltage clamp, and the display for the voltage electrode will show the membrane potential of the oocyte membrane, depending on the user's input, and for resting GluCls, the display for current should read zero. Steady perfusion is essential (*see* **Note 6**). For oocytes expressing Cl^--selective GluCls, the reversal potential (*see* Subheading 3.3, **step 2**) is likely to be around −20 mV. Therefore, and depending on expression levels, oocytes might have to be clamped at −80 mV for currents of "workable" amplitude.

6. For measurement of agonist sensitivity of expressed GluCls, glutamate is applied for ~20 s between ~1 min rest periods and at increasing concentrations (Fig. 2a) by switching between lines on the perfusion system, one containing ND96, another 1 μM agonist in ND96, another 3 μM agonist in ND96, etc. Peak current amplitude can later be plotted against glutamate concentration (Subheading 3.3, **step 1**). For measurement of relative ion permeability (a measure of ion selectivity), the current–voltage (*I*–*V*) relationship in different charge carriers is measured and the reversal potential of glutamate-gated currents in external ND96 can be compared to that in ND96/gluconate and ND96/choline (*see* Subheading 3.3, **step 2** and Fig. 2d). When replacing bath chloride, the bath electrode will need to be connected to the bath indirectly, via a 3M KCl agar bridge, described in [5]. Numerous pharmacological assessments of GluCls can be performed by altering the mixtures of compounds dissolved in ND96 in different tubes of the perfusion system. Two examples are shown in Fig. 2a–c, including a test for agonist effects of selected compounds and a test for sensitivity of glutamate-gated currents to block by an inhibitor. Using site-directed mutagenesis and recording from mutant GluCls, the contribution of specific amino acid residues to GluCl function can be assessed (Subheading 3.3, **step 3**).

7. Voltage clamp is switched off, micropipettes are removed from the oocyte, and the oocyte is disposed of. The same micropipettes can be used for subsequent oocytes/experiments, but over time, debris builds up on the micropipette tip, altering

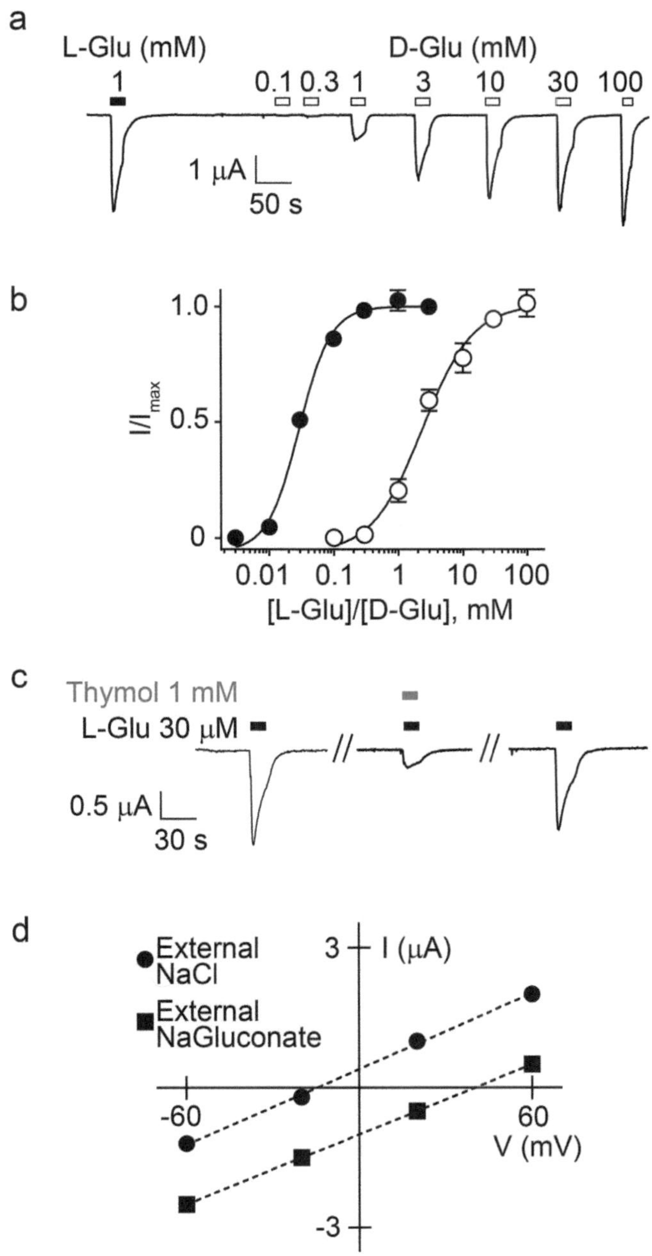

Fig. 2 Characterization of GluCl function with two-electrode voltage clamp. (**a**) Currents activated by application of agonists to oocytes expressing *S. mansoni* GluCl SmGluCl-2 and plot of current amplitude against agonist concentration. (**b**) Currents activated by application of glutamate alone or in the presence of thymol. (**c**) Glutamate-gated currents in the absence and presence of thymol. (**d**) Current amplitude (*I*) plot against membrane potential (*V*) in experiments with regular ND96 (NaCl) and ND96/gluconate

its resistance, and the AgCl layer will be removed from the electrode, often visible by a white, instead of black/brown, electrode. Therefore, glass micropipettes must be replaced and electrodes re-chloridized over the course of a day's recording.

3.3 Data Analysis and Interpretation

1. The sensitivity of GluCls to glutamate, other agonists, and pharmacological modulators is typically quantified with half-maximal effective or inhibitory concentrations, EC_{50} and IC_{50}. These are calculated after plotting current (or inhibition of current) in response to concentration of agonist (or inhibitor) and fitting the data points with nonlinear regression.

$$y = \min + \frac{(\max - \min)}{1 + 10^{(\log\ EC_{50} - x) \times n_H}}$$

where y is the current response, min is the minimum current response, max is the maximum current response, EC_{50} is the half maximal effective agonist concentration, x is the agonist concentration, and n_H is the cooperativity coefficient, or "Hill slope." Minimum current response is likely to be zero, and maximum is often normalized to one for each oocyte, because current amplitude varies across oocytes depending on expression. In the averaged data in Fig. 2b, the current responses (I, on the y-axis) of oocytes expressing SmGluCl-2 from *S. mansoni* to both L-glutamate and D-glutamate are normalized to the maximum current response to L-glutamate. As D-glutamate current amplitude is similar to that of L-glutamate, the former is also considered a full agonist. However, the EC_{50} for L- and D-isomers are ~20 μM and ~1 mM, respectively, suggesting that the apparent affinity of the D-isomer is much lower than that of the L-isomer.

2. As implied by the name, GluCls are selective for Cl^- over cations such as Na^+, K^+, and Ca^{2+}. With two-electrode voltage clamp, this ion selectivity can be measured by substituting external chloride ions for less permeant anions such as gluconate and measuring a change in reversal potential (E_{rev}) of glutamate-gated currents.

   The direction of Cl^- across the membrane through open GluCls depends primarily on two forces. The energy acting on ions due to the membrane potential (voltage) can be described by $-EzF$, where E is the membrane potential, z is the valency of the ion, and F is Faraday's constant. The chemical or diffusion energy acting on ions due to the difference in intracellular and extracellular concentrations can be described by $-RT\ln([\text{ion}]_{out}/[\text{ion}]_{in})$, where R is the universal gas constant and T is temperature in Kelvin. The downward currents (which by convention show movement of positive charge into the cell) in

Fig. 2a and c were recorded at −70 mV and reflect outward flow of Cl^-: despite the high $[ion]_{out}$ (103.6 mM Cl^- in ND96), Cl^- flows out due to the overwhelming electrical force (repulsion from inside the cell). Somewhere around $E = -20$ mV, however, $-EzF = -RT \ln([ion]_{out}/[ion]_{in})$. This is the reversal potential (E_{rev}) of glutamate-gated currents (*x*-intercept in Fig. 2d). Substituting extracellular gluconate for Cl^- (Na gluconate for NaCl) shifts E_{rev} to much more positive potentials, as the chemical energy forcing Cl^- inward is lessened and the electrical energy forcing Cl^- outward dominates (Fig. 2d). In contrast, substituting extracellular choline for Na^+ (choline Cl for NaCl) has no effect on E_{rev} as only impermeant ions have been affected [2].

3. In characterizing the molecular basis of GluCl function, wild-type (WT) glutamate sensitivity (or drug sensitivity, ion selectivity, etc.) is compared to that of mutants containing single-amino acid substitutions. Such experiments can be used to identify drug binding sites by identifying residues whose mutation most significantly affects drug sensitivity (but *see* **Note 7**). Mutations can then be "mapped" onto high-resolution structural (e.g., X-ray crystallographic) data of GluCls, if available, or onto homology models, if not [6].

4 Notes

1. If using an older digitizer, a computer/tower containing PCI or SCSI cards—and importantly, the size to accommodate them—is often necessary. This is not the case with modern digitizers that connect via USB.
2. For electrophysiological recordings from neurons or HEK cells, isolation of the setup from mechanical vibrations and electrical noise is crucial. With larger, more robust oocytes, although a vibration control table (e.g., Cleanbench Laboratory Table from TMC) and a Faraday cage (e.g., FAR01 from Thorlabs, Inc) may be required in some circumstances, a stable laboratory bench will often suffice, especially with a simple, steel base plate described above.
3. The stereomicroscope required for oocytes is generally simple, but when purchasing, a practical working distance (in which microelectrodes must be easily maneuvered) should be considered. This will be listed on the supplier's website, alongside combinations of objectives and eyepieces.
4. Traditionally, *X. laevis* is maintained in an animal facility on site and lobes of ovarian tissue are surgically removed, as described elsewhere [7, 8]. Briefly, the frog is anaesthetized; lobes of

ovarian tissue are removed with forceps via a ~1 cm abdominal incision (away from the midline); the wound is closed with three to five absorbable sutures, first internally, in the body wall, and second externally, in skin; and the frog is left to recover in shallow water ensuring that the head is not submersed. Tissue is then manually separated into clumps of five to ten oocytes with forceps or tweezers, and outer follicular tissue is enzymatically removed by shaking incubation in OR2 containing 1 mg/ml collagenase. Some experimentation with collagenase concentration, shaking speed, incubation time, and temperature (e.g., room temperature vs 37 °C) will ensure optimal defolliculation. Collagenase is removed by several washes in OR2 before storing oocytes at 18 °C until use. Most descriptions of oocyte culture suggest storage in 35 mm Petri dishes. Oocytes may stick to plastic dishes (and subsequently rip), and therefore some trial and error of various plastic dishes, if not glass dishes or agarose coating, may be required.

5. Functional surface expression of GluCls—or any receptor/ channel—differs across clones, batches of oocytes, and laboratories. In this author's hands, the injection of 10 ng mRNA of the *Haemonchus contortus* AVR-14B GluCl yielded ~5 μA currents 1 day after injection, whereas the injection of 10 ng mRNA of *S. mansoni* SmGluCl-2 (from the same plasmid vector and transcription kit) yielded miniscule currents after 1 day, but >2 μA currents after 3 days. High-expressing constructs may require the injection of as little as 0.4 ng mRNA, whereas other constructs may require 40 ng. The injection of mRNA in a volume of 30–40 nl ensures reasonably consistent volume from oocyte to oocyte. The injection of 50 nl and higher is possible but likely to stress the oocyte (visibly). The expression of heteromeric GluCls is achieved by the mixing of different mRNAs shortly before injection [2]. Some channels (e.g., mutants) are constitutively active, leading to a leaky membrane and unhealthy oocytes. In this case, the addition of a channel blocker (e.g., picrotoxin) to the Leibovitz's L-15 medium may prevent the decline in oocyte health.
6. The rate of perfusion and the level of solution in the recording chamber, or the consistency thereof, is crucial for repeatable results. Most laboratories seem to use perfusion rates of 1–2 ml/min. However, equally important to solution entering the recording chamber is solution leaving the chamber: inconsistent vacuum strength will see the level of solution rise and fall. This leads to numerous problems, including changes in the connection with bath electrodes, turbulent flow of perfusion solutions/drugs, and movement of the oocyte, which signals the rapid end to an experiment! Care should be taken to ensure

consistent perfusion and vacuum before starting experiments, e.g., by trimming/adjusting a small piece of tubing where the solution meets the vacuum/waste (Fig. 1b) and by checking tubes/connections in the waste collection system for leaks.

7. Shifts in EC_{50} must be interpreted carefully, as mutations can affect EC_{50} via numerous mechanisms, from ligand recognition to altered channel gating (open–closed channel equilibrium) to desensitization (channel closure in the continued presence of agonist). This is famously discussed by Colquhoun [9], and although the interpretation of mutagenesis/two-electrode voltage clamp data can be aided by homology modeling [6], such questions are best addressed by measuring the kinetics of individual GluCls with more time-consuming patch clamp electrophysiology [10].

References

1. Smart TG, Paoletti P (2012) Synaptic neurotransmitter-gated receptors. Cold Spring Harb Perspect Biol 4(3):a009662. https://doi.org/10.1101/cshperspect.a009662
2. Dufour V, Beech RN, Wever C, Dent JA, Geary TG (2013) Molecular cloning and characterization of novel glutamate-gated chloride channel subunits from *Schistosoma mansoni*. PLoS Pathog 9(8):e1003586. https://doi.org/10.1371/journal.ppat.1003586
3. Molleman A (2003) Patch clamping an introductory guide to patch clamp electrophysiology. Wiley, Hoboken, New Jersey
4. Guan B, Chen X, Zhang X (2013) Two-electrode voltage clamp. In: Gamper N (ed) Ion channels: methods and protocols. Methods in molecular biology 998. Humana press, Totowa, New Jersey
5. Rudokas MW, Varga Z, Schubert AR, Asaro AB, Silva JR (2014) The Xenopus oocyte cut-open vaseline gap voltage-clamp technique with fluorometry. J Vis Exp (85):e51040. https://doi.org/10.3791/51040
6. Lynagh T, Cromer BA, Dufour V, Laube B (2014) Comparative pharmacology of flatworm and roundworm glutamate-gated chloride channels: implications for potential anthelmintics. Int J Parasitol Drugs Drug Resist 4(3):244–255. https://doi.org/10.1016/j.ijpddr.2014.07.004
7. Smith LD, Xu WL, Varnold RL (1991) Oogenesis and oocyte isolation. Methods Cell Biol 36:45–60
8. Wagner CA, Friedrich B, Setiawan I, Lang F, Broer S (2000) The use of *Xenopus laevis* oocytes for the functional characterization of heterologously expressed membrane proteins. Cell Physiol Biochem 10(1–2):1–12. https://doi.org/10.1159/000016341
9. Colquhoun D (1998) Binding, gating, affinity and efficacy: the interpretation of structure-activity relationships for agonists and of the effects of mutating receptors. Br J Pharmacol 125(5):924–947
10. Atif M, Estrada-Mondragon A, Nguyen B, Lynch JW, Keramidas A (2017) Effects of glutamate and ivermectin on single glutamate-gated chloride channels of the parasitic nematode *H. contortus*. PLoS Pathog 13(10):e1006663. https://doi.org/10.1371/journal.ppat.1006663

Chapter 15

A Modified ELISA Method to Evaluate the Interaction of *Schistosoma mansoni* Proteins with Plasminogen

Luis G. V. Fernandes, Rafaela S. Fernandes, Ana L. T. O. Nascimento, and Luciana C. C. Leite

Abstract

An important aspect of host-pathogen interactions is the interference of secreted proteins with the fibrinolytic system. Herein, we describe a modified ELISA method used to evaluate the interaction of a recombinant *Schistosoma mansoni* protein with plasminogen (PLG). Using this protocol, we demonstrated that a secreted protein, recombinant venom allergen-like protein 18 (rSmVAL18) acts as a plasminogen receptor increasing its activation into plasmin in the presence of the urokinase-type plasminogen activator (uPA). PLG binding was determined by immobilizing human PLG in the plate and incubating with the recombinant protein; competitive binding with a lysine analog demonstrated the interaction of the protein lysine residues with PLG Kringle domains. To assess the activation of *S. mansoni* recombinant protein-bound PLG, the amidolytic activity of generated plasmin was measured using the D-Val-Leu-Lys 4-nitroanilide dihydrochloride substrate.

Key words Plasminogen binding, Modified ELISA, Plasmin, *S. mansoni* recombinant proteins, Competitive binding, ACA, Lysine, Kringle domains, uPA

1 Introduction

The identification of host-pathogen interactions is essential to understand the pathogenicity and mechanisms of immune evasion of parasites such as *Schistosoma mansoni*. In this context, the identification of proteins that modify the fibrinolytic system can lead to the discovery of important mechanisms of survival of the parasite in the host and possibly protein targets for immune intervention. Here we describe a modified ELISA protocol [1] used to analyze the binding of a recombinant *S. mansoni* protein to plasminogen [2]. A similar approach has previously been performed to assay the rSmEno-PLG binding [3]. We also describe the method to determine the dose-dependent attachment of the protein with PLG [4] using the same protocol but varying the recombinant protein concentrations. Plasminogen presents Kringle domains that interact

David J. Timson (ed.), *Schistosoma mansoni: Methods and Protocols*, Methods in Molecular Biology, vol. 2151,
https://doi.org/10.1007/978-1-0716-0635-3_15,

with lysine residues in PLG receptors [5]. The protocol for the competitive binding assay was performed adding the recombinant protein to PLG-coated wells together with the lysine analog 6-aminocaproic acid at increasing concentrations (0–20 μM) [2]; this allows demonstration of the participation of lysine residues in the protein-PLG interaction.

The binding of PLG to receptors modifies its conformation, making the molecule more susceptible to tissue (tPA) or urokinase-type plasminogen activator (uPA)-mediated activation, culminating in the generation of plasmin (PLA) [6]. Finally, we report herein the protocol based on the amidolytic activity of generated PLA [7], using the plasmin substrate D-Val-Leu-Lys 4-nitroanilide dihydrochloride, which was performed to evaluate the activation of PLG by *S. mansoni* recombinant proteins.

The ability of schistosomes to interact with plasminogen was first associated with the presence of the enzyme enolase in the tegument of *Schistosoma bovis* adult male worms (SbEno) and *Schistosoma japonicum* schistosomula (SjEno) [8–10]. Additionally, a tegumental enolase (SmEno) able to enhance the activation of PLG was identified in *S. mansoni* 7-day schistosomula and adult parasites [3]. We have recently described that a protein secreted during the infection of the mammalian host, rSmVAL18, a member of the *S. mansoni* venom allergen-like protein family, acts as a PLG-receptor. Specific binding of rSmVAL18 to PLG was demonstrated with a K_D of 7.0 nM (Fig. 1a, b). rSmVAL18 increased PLG conversion into plasmin in the presence of the urokinase-type plasminogen activator (uPA) (Fig. 2) [2].

2 Materials

All solutions must be prepared using ultrapure water and analytical grade reagents. All stock reagents should be stored at room temperature, unless otherwise indicated. Particular attention should be given to waste disposal regulations when disposing waste materials. When preparing diluted solution from concentrated stocks, use a graduated cylinder.

2.1 ELISA for Protein Binding

1. 96-well ELISA plates.
2. Lysine solution at neutral pH: Weigh 0.365 g of L-Lysine monohydrochloride and solubilize it in 90 mL water to prepare a 20 mM lysine solution. Adjust the pH to 7.0 and bring the volume to 100 mL.
3. Plasminogen from human plasma (Sigma). This reagent is provided as a powder, so we recommend adding 1 mL of 20 mM lysine solution to the vial and aliquoting the resulting solution in small volumes (50–100 μL) which can be stored at −20 °C for several months (*see* **Note 1**).

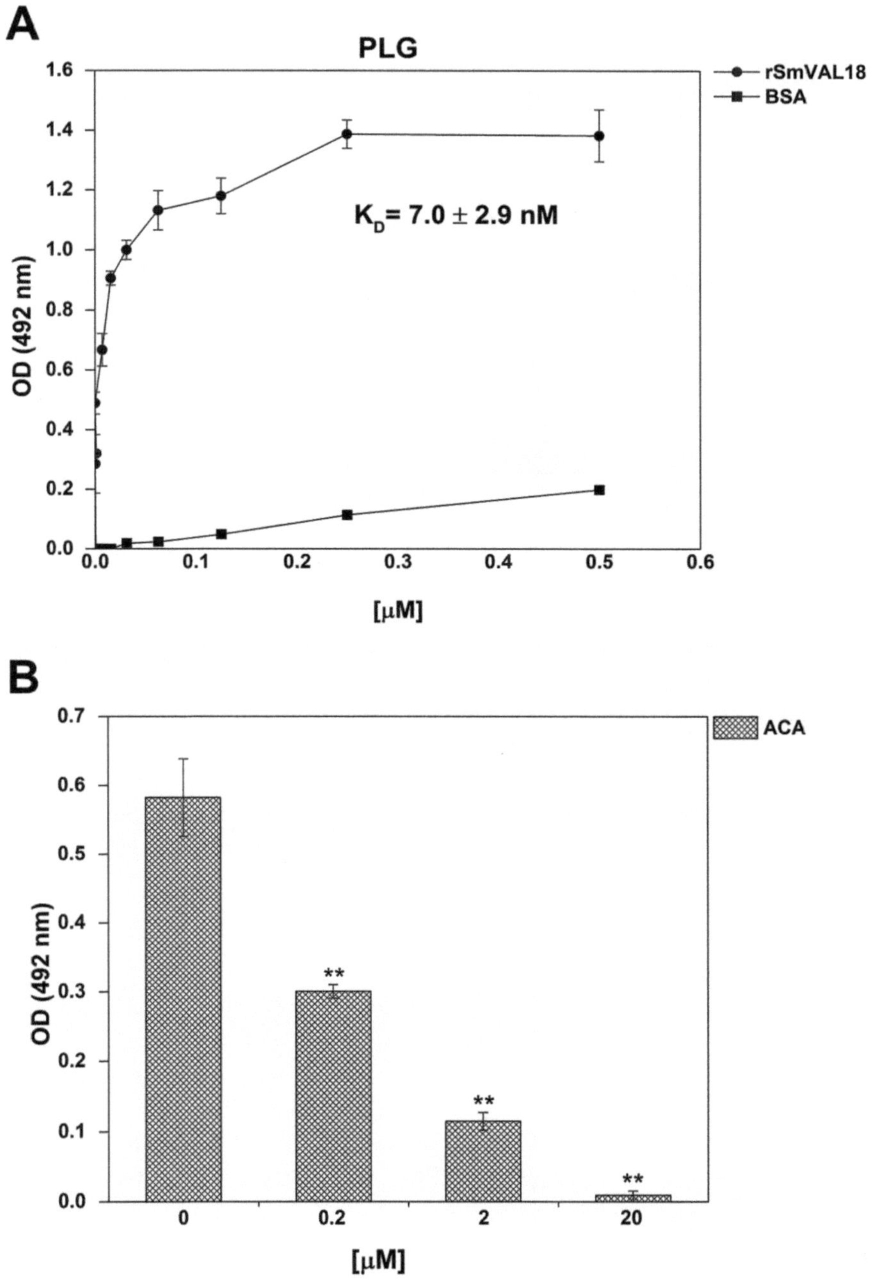

Fig. 1 Dose-dependent attachment of rSmVAL18 to PLG and competitive ACA assay. (**a**) PLG-coated plates received increasing concentrations of rSmVAL18 or the negative control, BSA. Incubation with anti-rSmVAL18 antibodies was followed by incubation with HRP-conjugated anti-IgG secondary antibody. OPD reaction was measured as absorbance values at 492 nm. Each point represents values determined in triplicate and data are expressed as the mean absorbance values with standard errors for each point. (**b**) Binding of rSmVAL18 to PLG was carried out in the presence of the lysine analogue 6-aminocaproic acid (ACA). For statistical analysis, the attachment of the rSmVAL18 in the presence of ACA was compared to its binding to PLG without ACA by the two-tailed *t* test. (Reproduced with minimal modifications from ref. 2 with permission from Elsevier)

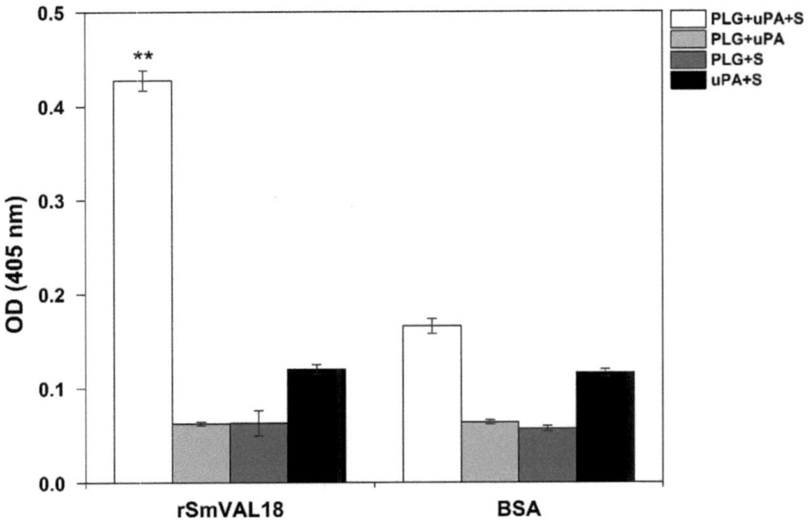

Fig. 2 Plasmin generation as the result of rSmVAL18-PLG binding. The immobilized rSmVAL18 received the following treatment: PLG, uPA, and the specific plasmin substrate S (PLG + uPA + S) or controls lacking one of the three components (PLG + uPA; PLG + S; uPA + S). BSA was used as negative control. Bars represent the mean absorbance values at 405 nm, as a measure of the relative substrate cleavage. Standard deviations are from three replicates for each experimental group and are representative of three independent experiments. Statistically significant difference in comparison to BSA is shown ($**p < 0.01$). (Reproduced from ref. 2 with permission from Elsevier)

4. Recombinant protein, preferably kept in PBS pH 7.4 or Tris–NaCl (150 mM NaCl and 10 mM Tris–HCl, (w/v)) buffer with pH ranging from 6 to 8.
5. Polyclonal antiserum raised against the recombinant protein.
6. Horseradish peroxidase (HRP)-conjugated anti-mouse IgG secondary antibody (*see* **Note 2**).
7. PBS: 137 mM NaCl, 2.7 mM KCl, 10 mM Na_2HPO_4, and 2 mM KH_2PO_4. Prepare a 10X concentrated solution and dilute for use. For a 10× solution: weigh 80.06 g NaCl, 2.01 g KCl, 14.2 g Na_2HPO_4, and 2.72 g KH_2PO_4. Solubilize the reagents in 900 mL water, adjust the pH to 7.4, and complete the volume to 1 L. Autoclave.
8. PBS-Tween 20 0.05% (v/v) (PBS-T). Prepare 1× PBS (100 mL PBS 10× plus 900 mL water) and add 500 μL of Tween 20, which should be drawn slowly with an automatic pipette.
9. Blocking buffer: PBS-T 10% (w/v) dry milk or 1% (w/v) Bovine Serum Albumin (BSA). Weigh 10 g of skimmed dry milk or 1 g of BSA and dilute the powder in 100 mL PBS-T.
10. Reagent reservoir.
11. Incubator.

12. Citrate-phosphate buffer (pH 5.0): Weigh 23.5 g of sodium citrate and 28.47 g sodium phosphate, mix the powder in a 1 L beaker, and add 700 mL milli-Q water. After full solubilization, adjust to pH 5.0. Finally, bring the volume to 800 mL. Autoclave the solution at 121 °C for 20 min.
13. 30% (v/v) hydrogen peroxide solution.
14. *o*-Phenylenediamine dihydrochloride substrate, powder, or tablets.
15. 2 M sulfuric acid solution: Wearing mask and gloves, keep a beaker on ice, and add 178 mL milli-Q water; then, slowly pour 22 mL sulfuric acid and mix the solution slowly with a glass cane. Operate in a fume hood. Keep the solution in an amber flask.
16. Substrate solution for colorimetric detection of HRP activity: Weigh 10 mg OPD and solubilize in 10 mL citrate-phosphate buffer (pH 5.0) and 10 μL of 30% (v/v) hydrogen peroxide. Homogenize well until powder is fully solubilized.
17. Plate sealing tape.
18. ELISA plate reader.

2.2 Dose-Dependent Attachment of the Protein with PLG

1. Recombinant protein in increasing concentrations varying from 0 to 1 μM, preferably kept in PBS pH 7.4 or Tris–NaCl (150 mM NaCl and 10 mM Tris–HCl, (w/v)) buffer with pH ranging from 6 to 8.
2. All materials and solutions described in Subheading 2.1.

2.3 Competitive ACA Binding Assay

1. 1 M 6-aminocaproic acid (ACA—Sigma Aldrich) solution: Weigh 13.12 g ACA and dissolve the powder in 100 mL PBS 1×. Filter the solution with 0.22 μm filter and store at 4 °C for up to 6 months.
2. All material listed in Subheading 2.1.

2.4 PLA Enzymatic Activity

1. Urokinase from human urine: Use lyophilized powder (Sigma). Prepare a 5 ng/μL solution by adding 10 mL PBS to a 50 μg uPA vial. Homogenize, distribute in 1.5 mL tubes and store at −20 °C. When performing the experiment, dilute uPA solution 50× in PBS, making a 0.1 ng/μL solution.
2. Plasmin substrate D-Val-Leu-Lys 4-nitroanilide dihydrochloride (Sigma). Prepare a 0.8 mM solution by adding 11 mL of PBS 1× to 5 mg of substrate. Homogenize vigorously and distribute the solution in 1.5 mL microtubes. Store at −20 °C.
3. All material listed in Subheading 2.1.

3 Methods

Make sure the ELISA plates are fully sealed during 37°C incubation. Washing steps can be performed either manually or by using automated washers.

3.1 ELISA for the PLG-Protein Binding

A graphical summary of this methodology is depicted in Fig. 3.

1. Add 1 μg PLG per well. For each component, we recommend using 5 wells (2 for blank control and 3 for interaction experiment). Save 5 wells for negative control, containing either BSA or milk, depending on the blocking buffer. Negative control proteins will be immobilized during blocking steps.
2. Let immobilization proceed for 3 h at 37 °C or 4 °C for 16 h (overnight incubation). Both coating methods render satisfactory and comparable results.

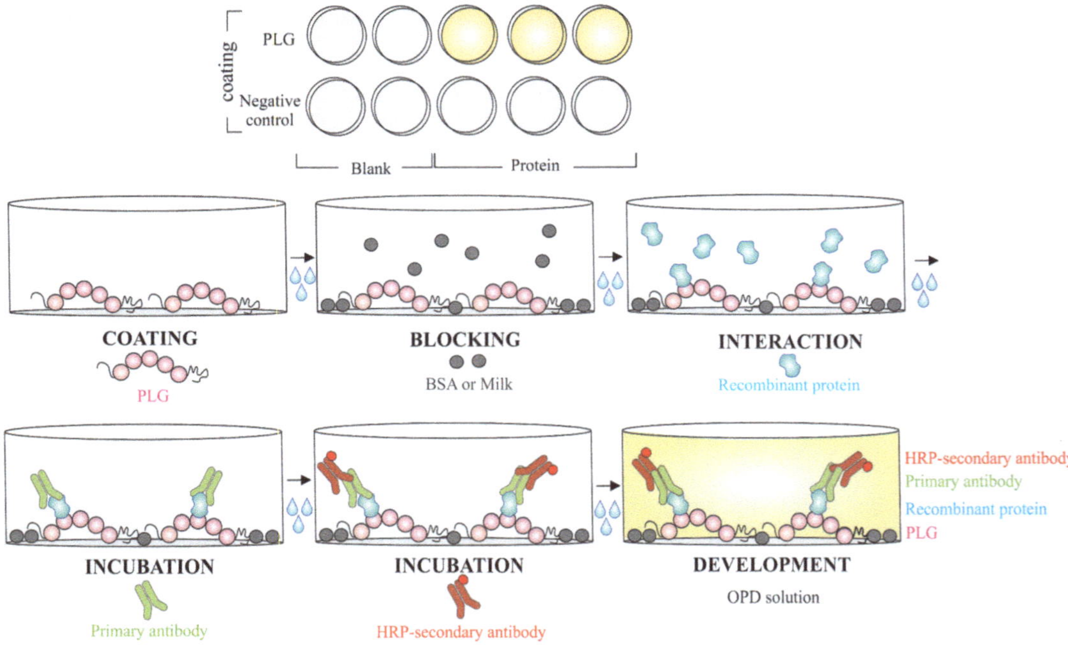

Fig. 3 Schematic ELISA experiment. Wells are coated with PLG (1 μg/well) and negative control protein, normally the same employed in the blocking solution. Then, wells are washed with PBS-T and blocked to prevent nonspecific binding of antigens and antibodies to the well. Recombinant protein is then added to both groups, except to the "blank" wells, used to measure the nonspecific interaction of the antibodies to the coated component. After washing, PLG-bound protein is detected by incubating with primary antibody solution, followed by incubation with HRP-conjugated secondary antibody and development with the OPD plus hydrogen peroxide solution. The values obtained for the blank control is subtracted from the triplicate, and for statistical comparison, discounted values of the PLG group are compared to those from the negative control

3. Wash the wells three times with 200 μL PBS-T. This step ensures that unbound components are washed away.
4. Add 200 μL of blocking buffer and incubate for 2 h at 37 °C.
5. Discard blocking buffer. Washing the plates after this step is optional; normally blockage with PBS-T 10% dried milk causes a little sedimentation in the wells, and in this case, washing 3× with PBS-T is recommended (*see* **Note 3**).
6. Add 100 μL of recombinant protein solution (10 μg/mL) to the wells, totalizing 1 μg of protein per well. If the blocking was performed with 1% (w/v) BSA, we recommend using the same buffer to dilute the protein; if performed with 10% (w/v) milk powder, we suggest using 2% PBS-T milk as diluent (*see* **Note 4**).
7. Let interaction proceed for 2 h at 37 °C.
8. Wash unbound recombinant proteins three times with PBS-T.
9. Add 100 μL/well of anti-recombinant protein antibodies in blocking buffer. Dilution may vary (*see* **Note 5**). Also, when using 6×His-tag containing proteins, HRP-conjugated anti-his antibodies can be employed (*see* **Note 6**).
10. Maintain interaction with primary antibodies for 1 h at 37 °C, and then wash the plates three times.
11. Add secondary antibody diluted in blocking buffer. We used HRP-conjugated anti-mouse IgG (Sigma) at 1:5000 dilution. Incubate for 1 h at 37 °C (*see* **Note 7**).
12. Wash the plates six times to ensure that no unbound HRP-conjugate remains.
13. Add 100 μL substrate solution per well. Let reaction proceeds for 10–15 min. The reaction will produce a yellow-orange product.
14. Stop the reaction by adding 50 μL of 2 M H_2SO_4. In this step, the yellow color turns orange.
15. Read the plate at 492 nm in an ELISA plate reader (*see* **Note 8**).

3.2 Protein-PLG Dose-Dependent Attachment

1. Proceed as described in Subheading 3.1, **steps 1–5**.
2. Add 100 μL of recombinant protein solutions ranging from 0 to 1 μM to the wells (*see* **Note 9**). If the blocking was performed with 1% (w/v) BSA, we recommend using the same buffer to dilute the protein; if performed with 10% (w/v) milk powder, we suggest using 2% PBS-T milk as diluent.
3. Proceed as described in Subheading 3.1, **steps 7–15**.

3.3 Competitive ACA Binding

1. PLG immobilization in ELISA plates and blocking steps are performed as described in Subheading 3.1, **steps 1–5**.
2. Prepare 3 distinct recombinant protein solutions in PBS: 10 μg/mL protein, 10 μg/mL protein plus 2 mM ACA, and 10 μg/mL protein plus 20 mM ACA. Solutions can be prepared in 1.5 mL tubes, and as normally 300 μL of recombinant protein solution is used (100 μL per well, in triplicate), 350 μL is enough for preparing the solutions. For 20 mM ACA, add 7 μL of 1 M ACA solution, and for 2 mM, add 0.7 μL; then add 3.5 μg of recombinant protein and complete to 350 μL with PBS.
3. Transfer 100 μL of the solution to each well and let protein interaction to immobilized PLG proceed for 2 h. When no ACA is added, we consider it 100% of binding (no competition).
4. Perform **steps 8–15** of Subheading 3.1.

3.4 Plasmin Formation

1. Immobilize 1 μg of recombinant protein per well. We recommend preparing a 10 μg/mL solution and adding 100 μL per well. For each protein tested, include 12 wells, to 4 triplicates. This is because one triplicate will constitute the complete system (PLG, uPA, and substrate); the other triplicates will be controls lacking one component at a time. Also, save 12 wells in the pate for the negative control, BSA (*see* **Note 10**). A scheme of well disposition is illustrated in Fig. 4.

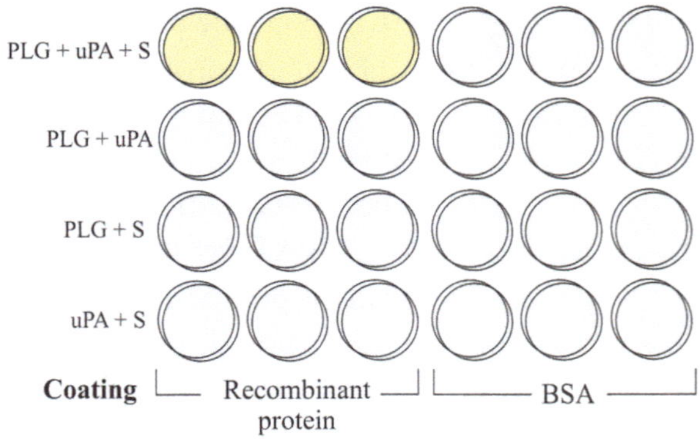

Fig. 4 Experimental design for assessing PLA formation. For each group, there will be 12 wells, referring to 4 treatments in triplicate. The wells are coated with the recombinant protein and the negative control protein and washed with PBS-T. PLG is added to all wells, except for the "uPA + S" treatment. After interaction, wells are washed and 50 μL of uPA and/or substrate (S) solutions are added. Both solutions will be added to the "PLG + uPA + S" (complete system that would render PLA formation) and "uPA + S" (treatment without PLG, no signal). No significant signal is expected in the BSA group, even with the complete system

2. Wash the wells three times with PBS-T and then add 200 μL PBS-T 1% (w/v) BSA.
3. Incubate the plate at 37 °C for 2 h.
4. Discard the blocking solution.
5. Add 100 μL of a 10 μg/mL PLG solution in blocking buffer. In one of the triplicates, *do not* include PLG; instead, add 100 μL of blocking buffer alone.
6. Let the interaction occur for 2 h and then wash the plate three times with PBS-T to eliminate unbound component.
7. Prepare a 0.8 mM PLA substrate solution and a 0.1 ng/ μL uPA solution. In the complete system triplicate (PLG + uPA + S) and in the control lacking PLG (uPA + S), add 50 μL of each solution; in the other 2 controls, add 50 μL of either solution and complete the volume with 50 μL of PBS. The final concentration of substrate will be 0.4 mM and the mass of uPA will be 5 ng per well.
8. Incubate the reaction for 16 h at 37 °C (*see* **Note 11**).
9. Measure absorbance at 405 nm.

4 Notes

1. Final plasminogen concentration can be inferred by protein quantification methods such as Bradford reagent or running an SDS-PAGE, for example.
2. Monoclonal HRP-conjugates anti-His antibodies (Sigma) can also be employed in the case of recombinant proteins expressed with a 6×His-tag epitope.
3. When using 10% (w/v) dried milk for blocking, sometimes three washings with PBS-T are not enough. Check for any sedimentation in the border of the well, since it could mechanically capture some of the subsequent components. If needed, perform additional washing steps until no blocking solids remain.
4. In the "blank" wells, *do not* add recombinant protein. This control accounts for any possible cross-reaction between primary and secondary antibodies against the immobilized component.
5. Before performing binding experiments, we recommend titration of the primary antibody against immobilized recombinant protein. By our experience, we immobilize 1 μg recombinant proteins onto ELISA plates, which are blocked and then incubated with serial dilutions of primary antibody, starting from

1:200. Normally, we use a fixed secondary antibody dilution for detection (1:5000). Based on this titration experiment, we select a dilution in which we observe an OD_{492nm} of around 1 for use in the binding ELISA.

6. In the case of using HRP-conjugated monoclonal anti-His antibodies for protein detection, there is only one step of incubation, not being necessary to add a secondary antibody.
7. If a nonspecific reaction is observed, especially in the blank controls, we recommend using more diluted solutions of the HRP-conjugated antibodies. This can be performed by titration experiments, keeping the concentration of primary antibody and then adding distinct dilutions of the HRP-conjugated secondary, starting from 1:5000.
8. An ideal interaction result would render basal signal in the blank control (around 0.05 OD_{492nm}) and in the negative controls (in which recombinant protein is incubated with proteins contained in the blocking solution). Significant signal is observed when the recombinant protein is set to interact with immobilized plasminogen.
9. If after revelation, saturable interaction is not obtained, we recommend increasing the maximal recombinant protein concentration, since the affinity constant can only be inferred when saturation is achieved. However, it is important to differentiate saturation due to binding and saturation due to maximal reading on the ELISA plates. Some ELISA readers cannot read past 4–5 of OD_{492nm}, and if your saturation is within this range, we recommend performing the experiment with a more diluted antibody.
10. It is not necessary to immobilize BSA in this step; the blocking buffer (1% (w/v) BSA) will be enough.
11. Due to prolonged incubation, make sure that the plates are properly sealed, especially in the corners. It is recommended to use new sealing films as prolonged use can lead to loss of adhesivity.

Acknowledgments

This work was supported by Fundação de Amparo à Pesquisa do Estado de São Paulo (FAPESP, grants 2012/23124-4, 2010/18486-9, 2014/50981-0 and 2017/06731-8), CNPq (grants 301229/2017-1 and 441449/2014-0) and Fundação Butantan.

References

1. Barbosa AS, Abreu PA, Neves FO et al (2006) A newly identified leptospiral adhesin mediates attachment to laminin. Infect Immun 74 (11):6356–6364
2. Fernandes RS, Fernandes LG, de Godoy AS et al (2018) *Schistosoma mansoni* venom allergen-like protein 18 (SmVAL18) is a plasminogen-binding protein secreted during the early stages of mammalian-host infection. Mol Biochem Parasitol 221:23–31
3. Figueiredo BC, Da'dara A, Oliveira SC et al (2015) Schistosomes enhance plasminogen activation: the role of Tegumental Enolase. PLoS Pathog 11(12):e1005335
4. Lin YP, Lee DW, McDonough SP et al (2009) Repeated domains of leptospira immunoglobulin-like proteins interact with elastin and tropoelastin. J Biol Chem 284 (29):19380–19391
5. Lahteenmaki K, Kuusela P, Korhonen TK (2001) Bacterial plasminogen activators and receptors. FEMS Microbiol Rev 25 (5):531–552
6. Lahteenmaki K, Virkola R, Pouttu R et al (1995) Bacterial plasminogen receptors: in vitro evidence for a role in degradation of the mammalian extracellular matrix. Infect Immun 63(9):3659–3664
7. Vieira ML, Vasconcellos SA, Gonçalves AP et al (2009) Plasminogen acquisition and activation at the surface of leptospira species lead to fibronectin degradation. Infect Immun 77 (9):4092–4101
8. Ramajo-Hernandez A, Pérez-Sánchez R, Ramajo-Martín V et al (2007) Schistosoma bovis: plasminogen binding in adults and the identification of plasminogen-binding proteins from the worm tegument. Exp Parasitol 115 (1):83–91
9. de la Torre-Escudero E, Manzano-Román R, Pérez-Sánchez R et al (2010) Cloning and characterization of a plasminogen-binding surface-associated enolase from Schistosoma bovis. Vet Parasitol 173(1–2):76–84
10. Yang J, Qiu C, Xia Y et al (2010) Molecular cloning and functional characterization of Schistosoma japonicum enolase which is highly expressed at the schistosomulum stage. Parasitol Res 107(3):667–677

Chapter 16

Biophysical Analysis of *Schistosoma mansoni* Septins

Ana Eliza Zeraik, Marina Gabriel Fontes, and Ricardo DeMarco

Abstract

Septins are dynamic filament-forming proteins that are recognized as important components of the cytoskeleton and are involved in numerous functions inside the cells, such as cytokinesis, exocytosis, and ciliogenesis and even in defense against pathogenic bacteria. Despite being highly conserved in eukaryotes, there is scarce literature on the role of septins in organisms other than humans and yeast. Therefore, septins from *Schistosoma mansoni* represent an interesting model to study an unexplored branch of this protein family. Here we described standard protocols for recombinant production and initial characterization of septins from *S. mansoni*. Septins are notably difficult to purify, mostly due to their tendency to assemble into filaments. Therefore, specific protocols to stabilize these proteins have been developed. In this chapter, we systematically describe protocols to clone, express, and purify schistosome septins. We also describe the use of circular dichroism to assess the folding and stability of septins and use of chromatography to characterize their oligomeric state, bound guanine nucleotide, and GTP hydrolysis. We expect that these protocols may help researchers involved in the study of schistosome septins as well as assist to establish protocols for septins from other organisms.

Key words Septins, *Schistosoma mansoni*, Oligomerization, Guanine nucleotides, Circular dichroism

1 Introduction

Septins are polymerizable GTP-binding proteins. They were first discovered in the septating bud neck of *Saccharomyces cerevisiae* dividing cells, in which they were essential participants of cytokinesis [1–3]. More recently, these filamentous proteins have been recognized as cytoskeletal components [4]. Besides their role in cell division, septins are involved in several other cellular processes, including ciliogenesis [5], neurogenesis [6], and exocytosis [7], by acting as scaffolds to other proteins and serving as diffusion barriers that help compartmentalize cells [8–10]. In order to perform these functions, septins polymerize into nonpolar complexes and these assemble into higher-order structures such as filaments, bundles, rings, and gauzes.

Septins are highly conserved in eukaryotes; however the number of septin genes varies according to the species [11, 12]. For

David J. Timson (ed.), *Schistosoma mansoni: Methods and Protocols*, Methods in Molecular Biology, vol. 2151,
https://doi.org/10.1007/978-1-0716-0635-3_16,

example, the green algae *Chlamydomonas reinhardtii* has only one septin, while there are thirteen septin genes in humans (*SEPT1* to *12* and *SEPT14*). Our research group has studied septins from several organisms, including the platyhelminth *Schistosoma mansoni*. This flatworm is the main cause of schistosomiasis in America, an acute and chronic tropical disease that affects over 240 million people in the developing world [13]. Four septins were identified in *S. mansoni*, which were named *Sm*SEPT5, *Sm*SEPT10, *Sm*SEPT7.1, and *Sm*SEPT7.2, based on their similarities with human septins [14]. Conserved hallmarks were found in *S. mansoni* septins, such as the tendency to form filaments in vitro [15] and interact with model membranes [16]. Additionally, tissue-specific localization of septins in diverse life stages of *Schistosoma mansoni* suggests diverse functions [14].

Here we describe protocols that were optimized for working with *S. mansoni* septins, and we also expect they can serve as a guide to work with septins from other organisms. These include protocols to clone, express, and purify those proteins and some of the most common biochemical and biophysical experiments used to characterize septins.

2 Materials

Prepare all solutions using ultrapure water or analytical grade reagents.

2.1 Expression and Purification of S. mansoni Septins

1. *E. coli* Rosetta (DE3), *E. coli* BL21 (DE3), or other expression bacteria strain.
2. Lysogenic-Broth (LB) media: weigh 20 g of LB powder and dissolve in 1 L water. Autoclave for 15 min at 121 °C. Allow to cool before adding antibiotics.
3. Kanamycin, ampicillin, chloramphenicol (or other antibiotic compatible with the expression vectors). Stock solutions: 50 mg/mL kanamycin and ampicillin in ultrapure water, filter with 0.2 μm syringe filter and store in aliquots at −20 °C. Prepare 34 mg/mL stock solution of chloramphenicol, dissolve it in ethanol, aliquot, and store at −20 °C.
4. 1 M isopropyl β-D-1-thiogalactopyranoside (IPTG) solution in ultrapure water. Filter with a 0.2 μm syringe filter and store in aliquots at −20 °C.
5. Buffer A: 50 mM Tris–HCl buffer pH 8.0, 800 mM NaCl, 10% (v/v) glycerol, 5 mM β-mercaptoethanol and proteinase inhibitor (Roche—complete, EDTA-free Protease Inhibitor Cocktail Tablets; dissolve one tablet in 50 mL of buffer A).
6. Ni-NTA Agarose or Superflow resin.

7. HisTrap FF column, 5 mL (GE *Healthcare*).
8. Buffer B: 50 mM Tris pH 8.0, 300 mM NaCl, 10% glycerol, 5 mM β-mercaptoethanol.
9. 2 M imidazole solution in ultrapure water; adjust the pH to 8.0.
10. Buffer C: 50 mM Tris–HCl pH 8.0, 500 mM NaCl, 12% (v/v) glycerol, 5 mM β-mercaptoethanol.
11. *Polyacrylamide gels:SDS-PAGE* is used to assess purification (15% acrylamide), and for running cross-linking reactions (10% acrylamide). Alternatively, *precast* SDS-PAGE gradient gel (*4–12%* or 3–8%) can be used as well.
12. AKTA purifier FPLC purification system (GE Healthcare Life Sciences) or equivalent equipment from other suppliers.
13. Superdex 200 10/300 GL (GE Healthcare Life Sciences).
14. Protein-Pak DEAE 5PW Column, 10 μm, 7.5 × 75 mm (Waters) driven by a Waters 2695 chromatography system or equivalent.

2.2 Septin Oligomerization

1. Gel filtration calibration kits: low molecular weight (LMW) and high molecular weight (HMW) (GE Healthcare) or equivalent from other suppliers.
2. 100 mM Ethylene glycol bis[succinimidyl succinate] (EGS) stock solution in DMSO.
3. Buffer D: 50 mM HEPES pH 8.0, 300 mM NaCl, 10% (v/v) glycerol, and 5 mM β-mercaptoethanol.
4. 10 mM GTP and 10 mM GDP stock solutions in ultrapure water.

2.3 GTPase Activity

1. 5 M $HClO_4$ stock solution in ultrapure water (for 50 mL solution, add 14.95 mL of $HClO_4$ mL in 35.05 mL of ultrapure water).
2. 5 M CH_3COOH stock solution in ultrapure water (for 50 mL solution, add 14.3 mL of glacial acetic acid in 35.7 mL of ultrapure water).
3. 1 M KH_2PO_4 stock solution (8.709 g in 50 mL of ultrapure water).
4. 3 M KOH stock solution (8.41 g in 50 mL of ultrapure water).
5. Buffer E: 50 mM Tris–HCl pH 8.0.
6. Buffer F: 50 mM Tris–HCl pH 8.0, 1 M NaCl.

2.4 Assessment of Secondary Structure by Circular Dichroism

1. CD spectrometer—Jasco 815 spectropolarimeter or other equivalent equipment.
2. Quartz cuvette.

3 Methods

3.1 *Schistosoma mansoni* Septin Cloning

1. Extract RNA from *S. mansoni* adult worms, which can be performed with TRIZOL or different commercially available kits (according to the manufacturer's instructions) (*see* **Note 1**).
2. Synthesize cDNA using reverse specific primers (a mix of different reverse primers can be used in a single cDNA reaction) for the full-length septin coding sequence. Alternatively, random primers can be used to synthesize the cDNA, allowing the use of the same template for amplification of diverse cDNA.
3. Perform a standard PCR to amplify the full-length septin coding sequences, using cDNA as template and primers containing restriction sites compatible to the expression vector, in this case, pET28a(+), which expresses the protein fused to a hexahistidine tag at the N terminus. When cloning partial constructs, the full-length DNA coding sequence can be used as a template for the PCR (*see* **Note 2**).

3.2 Expression and Purification of Individual *S. mansoni* Septins (Sm*SEPT5* and Sm*SEPT10*) and Their GTP-Binding Domains

1. After sequencing each insert in the expression vector, transform *E. coli* Rosetta (DE3) bacteria strain with the plasmids.
2. Pick one colony from the plate and inoculate into 5 mL of LB medium containing kanamycin (50 μg/mL) and chloramphenicol (34 μg/mL). After 16 h at 37 °C under 250 rpm, dilute this culture in LB medium with the same antibiotics at 1 mL of culture per 50 mL medium (1:50 dilution) and place it back in the shaker at 37 °C and 250 rpm.
3. After the culture reached OD_{600} (optical density at 600 nm) of 0.6–0.8 (about 2 h), induce the expression with 0.4 mM IPTG. Before induction, put the flasks at 4 °C to cool down for about 20 min. Induction is carried out at 18 °C for 16–18 h at 250 RPM.
4. After expression, harvest the cells by pelleting them at ~6000 × *g* for 30 min at 4 °C. The resulting pellet can be stored in the freezer (−20 °C or −80 °C) for purification at a later time or immediately used. Collect aliquots of the culture before and after induction to check the protein expression using SDS-PAGE.
5. Resuspend the pellet in buffer A (*see* **Note 3**) containing protease inhibitor cocktail.
6. Sonicate the cell suspension on ice (to avoid heating of the sample) using 10 cycles of 25 s "ON" followed by 35 s OFF (25% amplitude). The volume added to the sonicator depends on the probe; if using a microtip, sonicate in aliquots of 10 mL.

7. Separate soluble proteins from the pellet by centrifugation (10,000 × *g* for 30 min at 4 °C). If the supernatant is not clear, centrifuge again or filter before proceeding to the Ni-NTA affinity purification step.

3.2.1 Affinity Chromatography

1. Add 2–3 mL of Ni-NTA resin into a column and wash with 10 volumes of ultrapure water to remove the ethanol containing storage solution.
2. Equilibrate the resin with 10 volumes of buffer A.
3. Incubate the supernatant containing the expressed soluble protein with the Ni-NTA resin for 30 min at 4 °C with slow shaking.
4. Wash the column with 10 volumes of buffer A followed by 10 volumes (~30 mL) of buffer B and a last wash with 100 mL of buffer B containing 10 mM imidazole (*see* **Note 4**).
5. Elute the protein with 5 mL of buffer B containing 500 mM imidazole. Incubation with elution buffer with slow rotation for 5–10 min improves the yield compared to just passing it through the column.
6. Depending on the yield, the protein might need to be concentrated before next purification step (SEC), and septins are prone to aggregation at high concentrations. Therefore, we recommend passivating the membrane of the concentrator with 5% (v/v) Triton-X 100 to decrease the adsorption of septins to the membrane [17, 18]. Wash the concentrator with water, incubate with 5% triton for 2 h at room temperature, and wash again, 3–4 times with water before equilibrating the concentrator membrane with buffer.

3.2.2 Size Exclusion Chromatography (SEC)

The second step of purification is performed on a size exclusion chromatography column, Superdex 200 10/300 GL fitted to an AKTA purifier FPLC system. Always centrifuge the protein resulting from the Ni-NTA elution prior to injecting the supernatant on the Superdex column, to avoid adding precipitated protein into the column. Filter and degas buffers used in the FPLC.

1. Inject 1 mL aliquots into the system, previously equilibrated with buffer B. The recommended flow rate is 0.5 mL/min (decrease the flow rate if the column pressure is too high). Monitor the protein peaks by absorbance at 280 nm and collect 1 mL fractions.
2. Take aliquots of the resulting peaks to evaluate protein content using SDS-PAGE. Evaluation of protein content using SDS-PAGE from other steps of the purification may also be useful to assess where adjustments in the protocol should be made.

3.3 Purification of S. mansoni Septin Heterocomplex

The expression of septin heterocomplexes requires co-transformation of *E. coli* BL21 (DE3) with two plasmids containing different replication origins and antibiotic resistance (*see* **Note 5**). The expression and purification of the heterocomplex was mostly carried out following the same protocol described above for individual septins. Differences in the protocol are described below:

1. Due to the lower yield of the heterocomplex, a higher volume of culture is needed for purification, at least 2 L culture was used in each purification, and for individual septins 1 L was used.
2. Use buffer C for lysis and purification steps.
3. Affinity chromatography was performed using HisTrap FF column fitted to an AKTA purifier FPLC system.
4. Wash the column with 5 volumes of ultrapure water and equilibrate with 5 volumes of buffer C. The recommended flow rate is 5 mL/min for the 5 mL column.
5. Filter the sample with a 0.45 μm syringe filter.
6. Inject the sample at a flow rate of 1 mL/min to ensure maximum binding.
7. Wash with 5 volumes of buffer C containing 50 mM imidazole.
8. Elution is performed with 3 volumes of buffer C with 500 mM imidazole.
9. A second affinity chromatography is recommended to further purify septin complexes. This requires that one of the septins from the complex is expressed fused to a tag other than hexahistidine. We recommend using maltose-binding protein (MBP) as second tag. In this scenario, the elution fraction from nickel column would be applied into an amylose column, washed with 10 volumes of buffer and eluted in buffer C containing 10 mM maltose.
10. Size exclusion chromatography can be performed as described before in a Superdex 200 column, although when purifying hexamers and specially octamers, Superose 6 is more appropriate due to the higher molecular mass of the complexes.

3.4 Assessing the Oligomeric State of S. mansoni Septins

Oligomeric state of septins can be assessed using size exclusion chromatography (SEC) and chemical cross-linking.

Size exclusion chromatography was used to estimate the apparent molecular mass of the proteins through its elution volume.

1. Perform a calibration curve using two gel filtration calibration kits, the low molecular weight (LMW) and the high molecular weight (HMW). The standard proteins used and the recommended concentration were aprotinin (6.5 kDa) at 3 mg/mL,

ribonuclease A (13.7 kDa) at 3 mg/mL, carbonic anhydrase (29 kDa) at 3 mg/mL, ovalbumin (43 kDa) at 4 mg/mL, conalbumin (75 kDa) at 3 mg/mL, aldolase (158 kDa) at 4 mg/mL, and ferritin (440 kDa) at 0.3 mg/mL. Mix two protein standards per run to have a good peak resolution. Dilute the protein standards in the same buffer used to purify the protein (buffer B or C) and inject into the Superdex 200 column; there is no need to collect fractions.

2. Prepare a fresh solution of blue dextran (1.0 mg/mL) in the same buffer and inject separately from the other standards.
3. Use the registered elution volumes to build the calibration curve, which will be used to calculate the apparent molecular mass through the following equation:

$$K_{\text{av}} = (V_{\text{e}} - V_{\text{o}})/(V_{\text{c}} - V_{\text{o}})$$

wherein V_{e} (elution volume) is the volume at the center of the elution peak (where the absorbance is maximum); V_{o} is the column void volume, which is the elution volume observed for blue dextran; and V_{c} is the geometric column volume, which corresponds to 24 mL in Superdex 200 columns.

Plot the K_{av} versus the $\log_{10}$ of the molecular mass for each standard protein used and interpolate the K_{av} value obtained for the unknown sample to obtain its estimated molecular mass (*see* **Note 6**).

3.5 Chemical Cross-Linking

Cross-linking experiments were performed using HEPES buffer, which was chosen because it lacks primary amine groups targeted by the cross-linking reagents used.

1. Add different concentrations (0.5–5 mM) of ethylene glycol bis [succinimidyl succinate] (EGS) in 10 μM of the protein in buffer D (*see* **Note 7**).
2. Incubate the reaction on ice and take aliquots after 1, 2, and 4 h.
3. Stop the reaction by adding Tris–HCl pH 8.0 to a final concentration of 100 mM (for 50 μL reaction, add 5 μL of 1 M Tris stock solution) before adding SDS loading buffer.
4. Boil the samples for 5 min and analyze by 10% SDS-PAGE.

3.6 Evaluation of Nucleotide Content and GTPase Activity

Septins bind to guanine nucleotides and, in some cases, are able to hydrolyze GTP to GDP. However, the roles of nucleotide binding and hydrolysis in septin dynamics have not been fully elucidated [19]. Due to the high affinity of septins to GTP and/or GDP, those nucleotides tend to be co-purified after heterologous expression. Presence of those bound nucleotides may influence several aspects of septin behavior. Therefore, it is crucial to verify the nucleotide content of the purified septins, to account for these bound nucleotides in the design of further experiments. The method described below, adapted from Seckler et al. [20], is used to extract the nucleotide from the protein by denaturing it with perchloric acid and then using anion exchange chromatography to distinguish GTP from GDP.

1. Take an aliquot of the purified protein after Ni-NTA elution and after Superdex elution to assess the nucleotide content (*see* **Note 8**). Prepare the controls under the same conditions and buffers. Controls include the buffer alone and buffer with GTP and GDP added to have a reference of retention time for each nucleotide.
2. To denature the protein, add ice-cold $HClO_4$ (final concentration 0.5 M) and incubate on ice for 10 min.
3. Centrifuge for 10 min at 14,000 × *g* and 4 °C to remove the denatured protein.
4. The supernatant is neutralized and buffered with 0.17 M K_2HPO_4, 0.125 M CH_3COOH, and 0.5 M KOH (final concentrations).
5. Mix the sample, freeze to precipitate $KClO_4$, and store at −20 °C until analysis.
6. Thaw and centrifuge for 10 min at 14,000 × *g* to remove precipitated $KClO_4$.
7. Inject aliquots of 200 μL in the HPLC coupled to an anionic exchange column; in this case we used the column Protein-Pak DEAE 5PW Column, 10 μm, 7.5 × 75 mm (Waters) driven by a Waters 2695 chromatography system.
8. Perform a linear NaCl gradient to elute the nucleotides in buffer E: The column should be previously equilibrated in buffer E. Monitor the nucleotides by absorbance at 253 nm at 1 mL/min flow rate at 25 °C.

To evaluate GTPase activity, simply incubate the protein with threefold excess of GTP and 5 mM $MgCl_2$ (required for septin GTPase activity). Take aliquots at different time points, immediately snap freeze the protein in liquid nitrogen and perform the same procedure described above (**steps 1–8**) for each of the aliquots to evaluate their GTP and GDP content. An increase of the peak associated with GDP concomitantly with a diminution of the peak associated with GTP along the time course is indicative of hydrolysis. A control incubating GTP in the same buffer should be performed to evaluate non-enzymatic hydrolysis of GTP in the absence of septins. Examples of typical GTPase experiments for *S. mansoni* septins can be found in Zeraik et al. [16].

3.7 Influence of Nucleotide Content on Hetero-Oligomerization of Septins

It has been shown that nucleotide binding is essential for oligomerization of heterocomplexes [19, 21, 22]. It is possible to verify whether the presence of nucleotides promoted oligomerization of septins by SEC with the following protocol:

1. Add three times excess of GDP, GTP, or GTPγS (non-hydrolysable GTP analog), and incubate for 2 h to overnight. For example, if 15 μM protein solution is used, add 45 μM GDP/GTP in low salt buffer (25 mM HEPES pH 8.0, 100 mM NaCl, 10% (v/v) glycerol, 5 mM β-mercaptoethanol) containing 5 mM of $MgCl_2$.
2. Centrifuge the sample and inject the supernatant onto the Superdex 200 column.

An example on how the presence of nucleotides affects the oligomerization of septins is shown in Fig. 1. *Sm*SEPT5G is a monomer in solution, (34 kDa; see calibration curve on the inset of Fig. 1a) in the presence or absence of guanine nucleotides. When the GTPase domains of *Sm*SEPT5 and *Sm*SEPT10 were individually purified and combined in the absence of guanine nucleotides, a peak corresponding to the mass of a monomer was still predominant (black curve, Fig. 1b). Addition of GDP or GTP favored the appearance of dimers (Fig. 1b). Therefore, we can infer that the heterocomplex *Sm*SEPT5/*Sm*SEPT10 is formed through the G interface, while the homodimerization of *Sm*SEPT5 is not favored through this interface. This result is in agreement with the supposed assembly of a canonical heterocomplex, in which septins

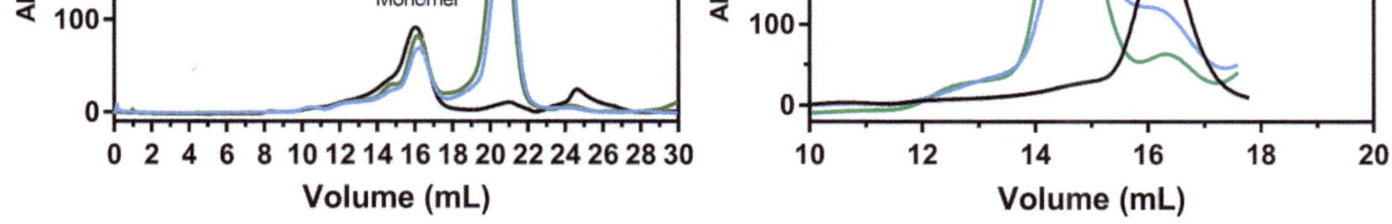

Fig. 1 Typical examples of assays performed to determine the oligomerization state of schistosome septins based on retention time on a size exclusion chromatography. Protein sizes are estimated according to a calibration curve, with an example shown in the inset. Panel (**a**) shows a case where the presence of nucleotides does not induce formation of dimers from a construct of *Sm*SEPT5. In panel (**b**), presence of nucleotide induces the formation of a heterodimer between constructs of *Sm*SEPT5 and *Sm*SEPT10

from the subgroup 2 (*Sm*SEPT5) and subgroup 6 (*Sm*SEPT10) assemble through the GTP-binding domain [22].

3.8 Circular Dichroism (CD)

Circular dichroism is routinely used to assess the secondary structure of proteins. Once a new construct of protein is expressed and purified, use of circular dichroism allows us to quickly obtain information to evaluate if the heterologous protein is correctly folded. Although septin structures display both α-helices and β-sheet, their CD spectra are characteristic of proteins with a high α-helical content, exhibiting two minima, at 208 and 222 nm [23–25].

In using CD to evaluate septins, care should be taken, especially due to the high amount of salt needed to purify septins. Chlorine anions display strong absorbance in wavelengths below 200 nm, rendering spectrum below this wavelength useless to obtain information about protein folding. Therefore, replacement of NaCl by NaF helps to obtain a more complete spectrum [26].

1. Add 200 μL of purified protein (around 3–5 μM) in a 0.1 cm quartz cuvette, and place it on a CD spectrometer at a controlled temperature (20 °C). Other cuvettes can be used, but the concentration must be adjusted accordingly. Cells with shorter pathlengths require higher concentration.
2. Collect the CD spectrum with the following settings: wavelength range 195–260 nm, at 10 °C, scanning speed of 50 nm/min, bandwidth of 1 nm, and a response time of 0.5 s. Eight accumulations to increase the signal to noise is recommended. These parameters can be adjusted according to the sample signal and buffer absorbance (*see* **Note 9**).
3. Measure the spectrum of the buffer alone, which must be subtracted from the protein sample during data processing.

Thermal denaturation can be performed to assess septin stability and whether GTP/GDP binding influences the stability. As septins tend to be unstable in vitro in temperatures above 37 °C, care must be taken in performing thermal denaturation (*see* **Note 10**). Septins are known to aggregate with increasing temperature [24], but the temperature is variable depending on the protein studied; in the case of *S. mansoni* septins, *Sm*SEPT5 is more unstable, aggregating above 30 °C, while *Sm*SEPT10 is stable up to 37 °C [16]. Reading at 222 nm can be recorded with a linear increment of 1 °C/min to monitor α-helical content. Thermal denaturation is followed by a decrease in the 222 nm signal, but also precipitation of protein causes the signal to drop; if any precipitate is observed in the cuvette, the signal cannot be considered unfolded protein but rather due to protein precipitation.

4 Notes

1. Frozen adult worms were used as source of RNA to obtain the cDNA for *Schistosoma mansoni* septins. Other life cycle stages can be used to extract RNA, but be sure to check if the chosen stage expresses the protein of interest. When expressing protein from an organism for the first time, we prefer to have the corresponding cDNA amplified from mRNA to be sure we are working with the correct sequence. Most of *Schistosoma mansoni* sequences in databases are gene predictions, and therefore synthetic genes produced using this information may incorporate prediction errors. If using synthetic genes, look for confirmation of transcript sequence from sequences in databases such as dbEST and SRA to minimize this problem.
2. When designing partial constructs, it is important to use secondary structure prediction programs to preferentially design primers in sequences corresponding to loop regions to increase the chance of obtaining a stable septin construct. Simultaneously designing primers corresponding to different regions adjacent to the portion of interest maximizes the chances of finding a stable construct. A detailed discussion on the choice of septin construct and a protocol for purification of human septins focused on crystallization were recently published by Valadares and Garratt [27].
3. High concentrations of NaCl in the lysis buffer is used to inhibit septin polymerization, which tend to occur spontaneously in vitro under low salt concentrations [28].
4. The concentration of imidazole in the washing solutions should be optimized for each protein. A gradient of imidazole can be performed to evaluate the optimum concentration in the washing buffer that prevents loss of the protein of interest and removes the contaminant proteins. Incubation of the supernatant with 10 mM imidazole minimizes binding of contaminants and a washing step with 40 mM imidazole is recommended. It was observed that *Sm*SEPT5 and *Sm*SEPT10 are loosely bonded to Ni-NTA beads; therefore a relatively low imidazole concentration was chosen.
5. The expression of septin heterocomplexes was performed in *E. coli* BL21 (DE3) because *E. coli* Rosetta strain (used for the individual septin expression) was incompatible with pACYCDuet-1 [29]. The use of bicistronic vectors, which contains two multiple cloning regions, facilitates the co-expression of proteins. For expression of *S. mansoni* septin complex, *Sm*SEPT5 and *Sm*SEPT7.2 were cloned into pETDuet-1 vector, and *Sm*SEPT10 was inserted into pACYCDuet-1 vector. Only *Sm*SEPT7.2 was expressed with a

histidine (His)-tag fusion at the N-terminus, whereas fusion tags were not introduced into the two other septin constructs. This strategy was used to minimize isolation of partial septin complexes or monomers of *Sm*SEPT5 and *Sm*SEPT10 not bound to *Sm*SEPT7.2. The choice of *Sm*SEPT7.2 to harbor the His-tag was based on the lower expression level of this protein, which would result in the maximum yield of a 1:1:1 heterocomplex, but contamination of heterocomplex with monomeric *Sm*SEPT7.2 septin was still observed. More recently, to address this problem, our group have expressed each protein of the complex separately, with proteins of lower yield expressed in a higher volume of media to compensate the lower yield. The bacterial pellets were then pooled together for lysis and purification. Another strategy to improve complete septin complexes is to add two different tags in two septins and perform successive affinity purifications. A His-tag is typically added to one septin and another tag, such as MBP, to a second septin, to be subsequently purified into amylose resin, eliminating, therefore, single tagged septins in the heterocomplex.

6. The protein standards commercially available are largely globular; therefore estimation of molecular mass is less accurate for non-globular proteins. Care must be taken when analyzing septins containing coiled-coil structures, which are elongated and tend to have an overestimated molecular mass in SEC. Analysis with a second technique, such as multiangle light scattering (MALS) or chemical cross-linking, to confirm observations is advisable.
7. The length of the cross-linker space arm does not appear to be a critical factor. Both disuccinimidyl suberate (DSS) and EGS (space arms of 11.4 Å and 16.1 Å, respectively) were successfully used to cross-link septins. Cross-linking reactions to estimate septin oligomeric state can be found in Zeraik et al. [16].
8. Septins are purified in high salt buffer, which is incompatible with the anion exchange chromatography used for separation of nucleotides. Therefore, we dilute the sample using a low salt buffer before the denaturation step to achieve a NaCl concentration of 150 mM. A minimum of 2.5 μM protein is required to have a detectable nucleotide peak.
9. Monitor the high tension (HT) to ensure the signal is below the HT saturation point (around 500–600 V). HT above the saturation point indicates there are no detectable photons reaching the detector, thus indicating unreliable measurements. If the HT is reaching the saturation point, dilute the sample or ignore data from wavelengths corresponding to the saturated region. For more detailed information on CD spectroscopy, we recommend the following Ref. 26, 30, 31.

10. High temperature and concentration promote aggregation of septins. Perform all the purification steps at 4 °C and finish the protocol within the same day. Always keep purified septins on ice when preparing experiments. Avoid concentrating the protein, unless critical for the experiment. If concentration is necessary, passivation of the concentrator's membrane with triton helps septins to not stick to the membranes [17, 18]. After concentration, centrifuge at max speed at 4 °C for 10 min to remove aggregates from the solution, and use the supernatant in the experiments.

Acknowledgments

We acknowledge Andressa Alves Pinto and Ana Paula Ulian de Araujo for helpful discussion and for revising the manuscript and the funding support of FAPESP, grant 2014/15546-1 and fellowship to AEZ (2013/20715-4).

References

1. Hartwell LH (1971) Genetic control of the cell division cycle in yeast. IV. Genes controlling bud emergence and cytokinesis. Exp Cell Res 69:265–276
2. Byers B, Goetsch LE (1976) A highly ordered ring of membrane-associated filaments in budding yeast. J Cell Biol 69:717–721
3. Haarer BK, Pringle JR (1987) Immunofluorescence localization of the *Saccharomyces cerevisiae* CDC12 gene product to the vicinity of the 10-nm filaments in the mother-bud neck. Mol Cell Biol 7:3678–3687
4. Mostowy S, Cossart P (2012) Septins: the fourth component of the cytoskeleton. Nat Rev Mol Cell Biol 13:183–194
5. Kim SK, Shindo A, Park TJ et al (2010) Planar cell polarity acts through septins to control collective cell movement and ciliogenesis. Science 329:1337–1340
6. Tada T, Simonetta A, Batterton M et al (2007) Role of Septin cytoskeleton in spine morphogenesis and dendrite development in neurons. Curr Biol 17:1752–1758
7. Beites CL, Xie H, Bowser R et al (1999) The septin CDCrel-1 binds syntaxin and inhibits exocytosis. Nat Neurosci 2:434–439
8. Kinoshita M (2006) Diversity of septin scaffolds. Curr Opin Cell Biol 18:54–60
9. Field CM, Coughlin M, Doberstein S et al (2005) Characterization of anillin mutants reveals essential roles in septin localization and plasma membrane integrity. Development 132:2849–2860
10. Caudron F, Barral Y (2009) Septins and the lateral compartmentalization of eukaryotic membranes. Dev Cell 16:493–506
11. Cao L, Ding X, Yu W et al (2007) Phylogenetic and evolutionary analysis of the septin protein family in metazoan. FEBS Lett 581:5526–5532
12. Pan F, Malmberg RL, Momany M (2007) Analysis of septins across kingdoms reveals orthology and new motifs. BMC Evol Biol 7:1–17
13. Nahum LA, Mourão MM, Oliveira G (2012) New Frontiers in Schistosoma genomics and Transcriptomics. J Parasitol Res 2012:1–11
14. Zeraik AE, Rinaldi G, Mann VH et al (2013) Septins of Platyhelminths: identification, phylogeny, expression and localization among developmental stages of *Schistosoma mansoni*. PLoS Negl Trop Dis 7:1–14
15. Zeraik AE, Galkin VE, Rinaldi G et al (2014) Reversible paralysis of *Schistosoma mansoni* by forchlorfenuron, a phenylurea cytokinin that affects septins. Int J Parasitol 44:523–531

16. Zeraik AE, Staykova M, Fontes MG et al (2016) Biophysical dissection of schistosome septins: insights into oligomerization and membrane binding. Biochimie 131:96–105
17. Mavrakis M, Tsai FC, Koenderink GH (2016) Purification of recombinant human and drosophila septin hexamers for TIRF assays of actin–septin filament assembly. Methods Cell Biol 136:199–220
18. Bertin A, Nogales E (2016) Preparing recombinant yeast septins and their analysis by electron microscopy. Methods Cell Biol 136:21–34
19. Abbey M, Gaestel M, Menon MB (2019) Septins: active GTPases or just GTP-binding proteins? Cytoskeleton (Hoboken) 76:55–62
20. Seckler R, Wu GM, Timasheff SN (1990) Interactions of tubulin with guanylyl-(-β-γ-methylene)diphosphonate. Formation and assembly of a stoichiometric complex. J Biol Chem 265:7655–7661
21. Sirajuddin M, Farkasovsky M, Zent E et al (2009) GTP-induced conformational changes in septins and implications for function. Proc Natl Acad Sci U S A 106:16592–16597
22. Sirajuddin M, Farkasovsky M, Hauer F et al (2007) Structural insight into filament formation by mammalian septins. Nature 449:311–315
23. Garcia W, Ulian De Araújo AP, De Oliveira Neto M et al (2006) Dissection of a human septin: definition and characterization of distinct domains within human SEPT4. Biochemistry 45:13918–13931
24. Cesar J, Damalio P, Garcia W et al (2012) Self assembly of human septin 2 into amyloid fi laments. Biochimie 94:628–636
25. Sala FA, Valadares NF, Macedo JNA et al (2016) Heterotypic coiled-coil formation is essential for the correct assembly of the Septin Heterofilament. Biophys J 111:2608–2619
26. Miles AJ, Wallace BA (2016) Circular dichroism spectroscopy of membrane proteins. Chem Soc Rev 45:4859–4872
27. Valadares NF, Garratt RC (2016) Septin crystallization for structural analysis. Methods Cell Biol 136:321–338
28. Bertin A, McMurray MA, Grob P et al (2008) *Saccharomyces cerevisiae* septins: Supramolecular organization of heterooligomers and the mechanism of filament assembly. Proc Natl Acad Sci U S A 105:8274–8279
29. Novy R, Drott D, Yaeger K et al (2001) Overcoming the codon bias of *E. coli* for enhanced protein expression. Innovations 512:1–3
30. Greenfield NJ (2007) Using circular dichroism spectra to estimate protein secondary structure. Nat Protoc 1:2876–2890
31. Kelly SM, Jess TJ, Price NC (2005) How to study proteins by circular dichroism. Biochim Biophys Acta 1751:119–139

Chapter 17

Crystallographic Studies of Triosephosphate Isomerase from *Schistosoma mansoni*

Pedro Jimenez-Sandoval, Eduardo Castro-Torres, Corina Diaz-Quezada, and Luis G. Brieba

Abstract

Protein structure determination by X-ray crystallography guides structure-function and rational drug design studies. Helminths cause devastating diseases, including schistosomiasis that affects over one-third of the human population. Trematodes from the genus *Schistosoma* heavily depend on glycolysis; thus enzymes involved in this metabolic pathway are potential drug targets. Here we present a protocol to obtain crystal structures of recombinantly expressed triosephosphate isomerase from *S. mansoni* (SmTPI) that diffracted *in house* to a resolution of 2 Å.

Key words Triosephosphate isomerase, Synthetic gene, Recombinant expression, Protein crystallography

1 Introduction

Protein crystallography depends on the ability to obtain a pure sample that is subject to different set of conditions aimed to create a supersaturated protein solution. This supersaturation is achieved by testing hundreds of conditions like salt, pH, and precipitating agents. Although the process of protein crystallization is based on trial and error [1, 2], modern crystallography follows a pipeline where thousands of conditions are routinely tested [3]. The first step in this pipeline starts with the production of pure protein. Recombinant protein expression allows the purification of proteins from heterologous sources. However, the codon usage of each gene and the intrinsic restriction sites of the gene of interest may have difficulty in cloning into several expression vectors. As an alternative to cloning from cDNA (natural sources), we routinely employ synthetic genes in the form of "gBlocks" from the company IDT (https://www.idtdna.com/) [4–7]. gBlocks are double-stranded DNA fragments that contain a validated DNA sequence

David J. Timson (ed.), *Schistosoma mansoni: Methods and Protocols*, Methods in Molecular Biology, vol. 2151, https://doi.org/10.1007/978-1-0716-0635-3_17,

(in our hands we estimated that at least 95% of the cloned gBlocks exhibit the correct DNA sequence). The use of gBlocks can reduce the cost by up to 50% compared to cloning using synthetic genes. Our cloning strategy consists of directly cloning the gBlocks into cloning a vector like pJET-Blunt (Thermo Fisher), excise the gene from the cloning vector, and subclone it into an expression vector. For triosephosphate isomerases from different sources, including *Schistosoma mansoni*, we routinely clone the desired gene into a modified pET19 vector that replaces the thrombin cleavage site for the one from the human rhinovirus (HRV) type 14 3C protease. Protein expression is carried out in an *Escherichia coli* BL21(*DE3*) strain, and protein purification is achieved by immobilized metal affinity chromatography (IMAC) using a minimal buffer that consists of triethanolamine (TEA) buffer and physiological salt concentrations. Histidine tag cleavage occurs during dialysis and its removal is achieved by protein gel filtration on a Superdex-75 column. Following this protocol, we routinely obtain more than 10 mg of pure triosephosphate isomerase from 1 l of bacterial cell culture. Protein crystallography is executed using a 96-well plate on a Crystal Gryphon robot (Art Robbins Instruments) in the presence and absence of transition state analogs or inhibitors.

2 Materials

2.1 Subcloning into pJET-Blunt and pET-19 Pps

1. MiliQ sterile water and TE Buffer (10 mM Tris, pH 8.0, 1 mM EDTA).
2. 2× reaction buffer for pJET ligation (supplied with the vector).
3. Agarose gel electrophoresis equipment.

2.2 10% SDS-PAGE and Gel Filtration to Monitor SmTPI Expression

1. Luria-Bertani (LB) media: 10 g Tryptone, 10 g NaCl, 5 g yeast extract, 950 ml H_2O.

   Dilute the reagents and adjust the pH to 7 with NaOH. Bring to a final volume of 1 l. Sterilize by autoclaving.
2. 15% SDS-PAGE.

   Assemble a 5 ml resolving gel by mixing:

   2.50 ml 30% acrylamide/*bis*acrylamide.

   1.3 ml 1.5 M Tris (pH 8.8).

   50 μl 10% SDS.

   50 μl 10% ammonium persulfate (make it fresh and store at 4 °C up to a month).

   6 μl TEMED.

   H_2O to 5 ml.

   Assemble a stacking gel (2 ml) by mixing:

   0.33 ml 30% acrylamide/*bis*acrylamide.

0.25 ml 1 M Tris (pH 6.8).

20 μl 10% (w/v) SDS.

20 μl 10% ammonium persulfate.

6 μl TEMED.

H_2O to 3 ml.

2.3 Protein Expression/ Purification

1 ml of Isopropyl β-D-1 thiogalactoside (IPTG) 500 mM.

TEA buffer, (100 mM triethanolamine pH 7.4, 100 mM NaCl).

Buffer A (100 mM TEA buffer, pH 7.4, 500 mM NaCl).

Dialysis buffer (100 mM TEA buffer, pH 7.4, 150 mM NaCl 10 mM DTT, 2 mM EDTA).

PreScission protease (PPS).

Vivaspin® turbo concentration device (Sartorius).

2.4 Protein Crystallography

SmTPI concentrated to 20 mg/ml (0.711 mM) in storage buffer (100 mM TEA buffer, pH 7.4, 150 mM NaCl, 10 mM DTT, 2 mM EDTA).

200 mM 2-phosphoglyceric acid (2-PGA).

Crystallization buffer (0.05 M cesium chloride, 0.1 M MES monohydrate pH 6.5, and 30% v/v Jeffamine M-600).

3 Methods

3.1 Design of the Synthetic SmTPI Gene

The amino acid coding sequences of TPI from and *S. mansoni* (SmTPI) (GenBank: XP_018647623) was synthetically synthesized for its expression in *E. coli* using the *Nde* I and *Bam* HI restriction sites of a modified pET19 vector (pET19pps) courtesy of Drs. Tapan Biswas and Tom Ellenberger. Bacterial expression was optimized using the IDT codon optimization tool (https://www.idtdna.com/CodonOpt) and the SmTPI amino acid sequence. The optimized DNA sequence was modified by manually adding restriction sites for *NdeI* and *BamHI* at the 5′and 3′ends of the gBlock (Fig. 1).

3.2 G-Block Cloning into Cloning and Expression Vectors

3.2.1 pJET gBlock Blunt-End Cloning Protocol (Cloning Vector)

1. Dilute the SmTPI gBlock to a concentration of 50 ng/μl in TE Buffer.
2. Set up the following ligation reaction on ice:

```
catatgatgtcgggttcccggaagttcttcgtgggggaaactggaaaatgaacggctcg
 H  M  M  S  G  S  R  K  F  F  V  G  G  N  W  K  M  N  G  S
cgggacgacaacgacaaattgcttaaacttctgagtgaagctcattttgatgacaacacg
 R  D  D  N  D  K  L  L  K  L  L  S  E  A  H  F  D  D  N  T
gaggtactgatagctccgccgagcgtgttttacatgaaatacgtaagtccctgaaaaag
 E  V  L  I  A  P  P  S  V  F  L  H  E  I  R  K  S  L  K  K
gagattcatgttgccgctcaaaactgctacaaggtgtcgaaaggtgctttcacgggggaa
 E  I  H  V  A  A  Q  N  C  Y  K  V  S  K  G  A  F  T  G  E
atttccccggcaatgattcgtgatataggttgcgattgggtaattcttggccactctgaa
 I  S  P  A  M  I  R  D  I  G  C  D  W  V  I  L  G  H  S  E
cggcgcaacatctttggggaaagcgacgaactgatagcagagaaggtacaacatgcttta
 R  R  N  I  F  G  E  S  D  E  L  I  A  E  K  V  Q  H  A  L
gcggaagggttatcagtgatagcatgtatcggcgagactctttcggagcgcgaatccaat
 A  E  G  L  S  V  I  A  C  I  G  E  T  L  S  E  R  E  S  N
aaaaccgaggaagtttgcgttcggcagttaaaggcaattgctaacaaaatcaaatctgct
 K  T  E  E  V  C  V  R  Q  L  K  A  I  A  N  K  I  K  S  A
gatgaatggaaacgtgttgttgtggcttatgagccagtctgggcaatcggaactggcaag
 D  E  W  K  R  V  V  V  A  Y  E  P  V  W  A  I  G  T  G  K
gtggcaacgccccaacaggctcaagaagtccataattttcttcgtaaatggtttaaaacg
 V  A  T  P  Q  Q  A  Q  E  V  H  N  F  L  R  K  W  F  K  T
aacgcgccaaacggagtggatgaaaagatacggataatctacggagggagcgtaacggcg
 N  A  P  N  G  V  D  E  K  I  R  I  I  Y  G  G  S  V  T  A
gcaaattgtaaagaattagcccagcagcacgatgtcgatggctttcttgtgggtggtgcc
 A  N  C  K  E  L  A  Q  Q  H  D  V  D  G  F  L  V  G  G  A
agcttaaagccggaatttactgagatttgtaaagcccgccaaagataaggatcc
 S  L  K  P  E  F  T  E  I  C  K  A  R  Q  R  -  G  S
```

Fig. 1 Optimized gBlock for SmTPI. *NdeI* and *BamHI* restriction sites are in italics and the stop codon is underlined

2× reaction buffer	5 μl
pJET cloning vector	0.5 μl
G-block (50 ng/ul)	1 μl
Water	2.5 μl
T4 DNA ligase	1 μl

3. Mix by pipetting and incubate at room temperature (21 °C) for 30 min.
4. Transform using 1 μl of ligation reaction in a Top10 of DH5α *E. coli* strain.
5. Screen positive transforms by PCR colony.
6. Obtain plasmid DNA by the boiling lysis method or commercial kits [8].

3.2.2 pET19pps Cloning Protocol (Expression Vector)

1. Set up the following restriction reaction and incubate for 4 h at 37 °C:

10× reaction buffer	5 μl
pJET-gBlock vector (100 ng/ul)	10 μl
Nde I restriction enzyme	1 μl
BamH I restriction enzyme	1 μl
Water	32 μl
Total	50 μl

2. Set up a preparative agarose gel electrophoresis and excise the 750 bp band corresponding to the SmTPI gene.
3. Perform a ligation reaction as in Subheading 3.2.1 on ice using plasmid pET19pps previously digested with *Nde* I and *BamH* I restriction enzymes.
4. Mix by pipetting and incubate at room temperature (21 °C) for 30 min.
5. Transform using 1 μl of ligation reaction into the Top10 of DH5α *E. coli* strain and screen positive transforms by colony PCR.
6. Obtain plasmid DNA by the boiling lysis method and send at least three colonies to DNA sequencing. We routinely use the T7 promoter primer. The reads are long enough to cover the entire gene.

3.3 SmTPI Purification

1. A positive SmTPI clone should be transformed onto the *E. coli* BL21(*DE3*) strain and plated in a LB-agar plate supplemented with 100 μg/ml of ampicillin.
2. A single ampicillin-resistant colony should be used to inoculate 50 ml of Luria-Bertani (LB) media, supplemented with 100 μg/ml of ampicillin and shaken for 8–12 h at 37 °C.
3. This pre-inoculum was used to inoculate 1 l of LB media. The culture should be incubated at 37 °C and shaken at 250 RPM until an $OD_{600} = 0.6$ is reached.
4. SmTPI expression is achieved by adding isopropyl β-D-1 thiogalactoside (IPTG) to a final concentration of 0.5 mM. The culture was shaken for 18 h at 250 RPM at 16 °C, and the bacterial biomass was harvested by centrifugation at 5000 × *g* at 4 °C for 15 min and stored at −80 °C until used.

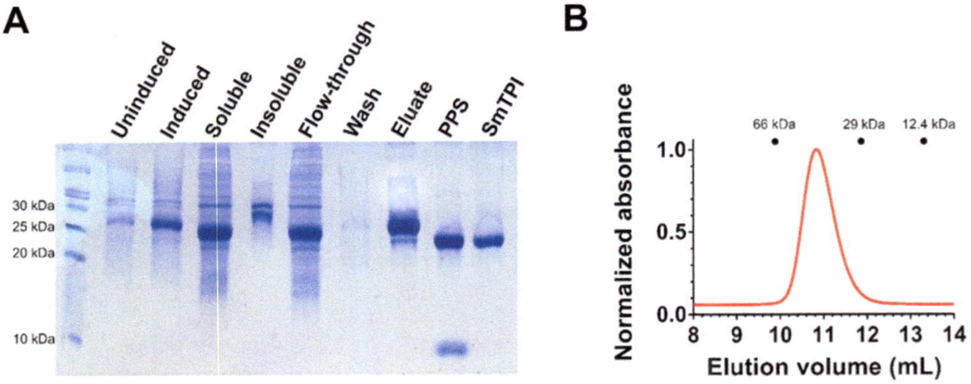

Fig. 2 (**a**) 12% SDS-PAGE and Coomassie-blue stained showing the IMAC and gel filtration purification steps of SmTPI. Lane 1, molecular weight marker; lane 2, uninduced cell culture; lane 3, induced protein fraction; lane 4, soluble fraction; lane 5, insoluble fraction; lane 6, flow-through; lane 7, wash; lane 8, protein eluate; lane 9, SmTPI treated with PPS; lane 10, recombinant SmTPI after gel filtration. (**b**) Chromatogram of SmTPI loaded onto a Superdex 75 column. The dimeric nature of SmTPI is determined by a regression analysis of the elution profile of the molecular-weight standards; albumin (66 kDa), carbonic anhydrase (29 kDa), cytochrome C (12.4 kDa), and aprotinin (6.5 kDa)

5. For bacterial lysis, the bacterial pellet should be resuspended in 20 ml of ice-cold lysis buffer TEA buffer. The resuspended cell culture should be sonicated on ice with 3 pulses of 30 s each.
6. The bacterial lysate should be clarified by centrifugation at 35,000 × *g* for 30 min at 4 °C. The supernatant should be passed through a 5 ml HisTrap™ fast flow column previously equilibrated with buffer A. The column is then washed with 10 volumes of buffer A. Elution of SmTPI is executed with a linear gradient of imidazole from 0 to 0.5 M in running buffer A.
7. SmTPI elutes at 200–250 mM of imidazole. The eluate should be immediately dialyzed in dialysis buffer at 4 °C for 12 h. 0.1 mg of PreScission protease (PPS) was added directly into the dialysis bag.
8. Separation from the histidine tag is performed by running the protein sample onto a Superdex 75 size exclusion column.
9. This protocol produces more than 10 mg of pure protein, per liter of cell culture, which assembles as a dimer on a gel filtration column (Fig. 2).

3.4 Crystallography

3.4.1 Crystallization

1. SmTPI should be concentrated to 20 mg/ml using a Vivaspin® turbo concentration device (Sartorius) and incubated in the presence and absence of 10 mM of 2-phosphoglyceric acid (2-PGA) for 30 min on ice in protein storage buffer.

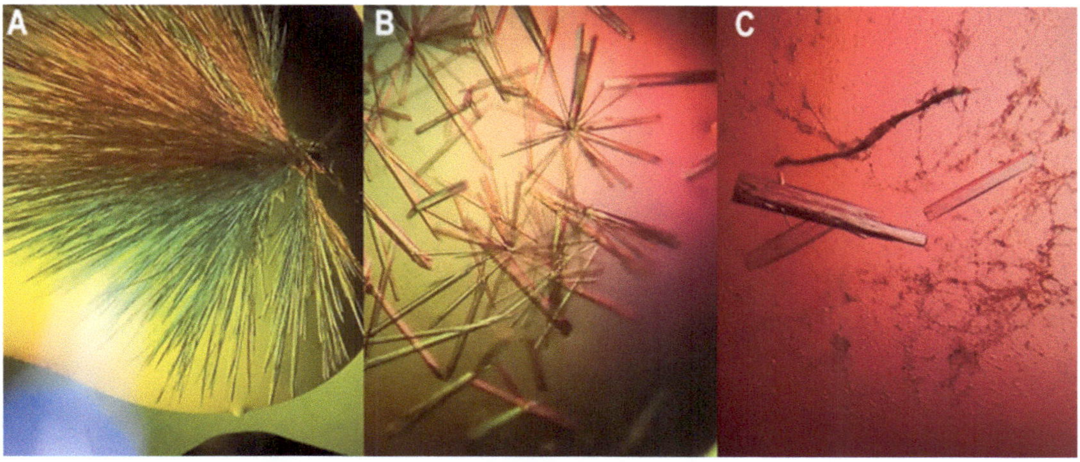

Fig. 3 Protein crystallization and optimization of SmTPI. (**a**) Initial crystal needles. (**b** and **c**) crystals obtained by seeding

2. Crystallization trails should employ the sitting drop method using 0.3 μl of protein-inhibitor complex and 0.3 μl of the reservoir solution on a 96-well plate format. Protein drops were incubated at 21 °C. Thin SmTPI crystal needles appear overnight in a solution containing 0.05 M cesium chloride, 0.1 M MES monohydrate pH 6.5, and 30% w/v Jeffamine M-600 (Fig. 3a).
3. Crystallization trails should also be employed using the hanging drop method and seeds from the sitting drop experiments. A four-by-four matrix using cesium chloride from 0.04 to 0.05 M and 24 to 28% w/v Jeffamine M-600 should be assayed. Individual crystals appear in all conditions (Fig. 3b, c).

3.4.2 Data Collection

1. A single-rod crystal of approximately 1200 × 80 × 80 μm dimensions should be mounted on a Rigaku Micromax 002+ sealed tube X-ray generator equipped with a RAXIS IV++ detector and frozen directly on the cryostream. A diffraction image was taken using an exposition time of 10 min and 0.5° oscillation produced usable spots to a resolution of 2 Å (Fig. 4).

Acknowledgments

We thank "Programa de Salud de la Fundación Miguel Alemán A. C (México)" for funding. We thank Daniela Camacho and Lucia Leyva for their help during crystallogenesis and Misraim Gurrola for his interest in epidemiology that initiated this study.

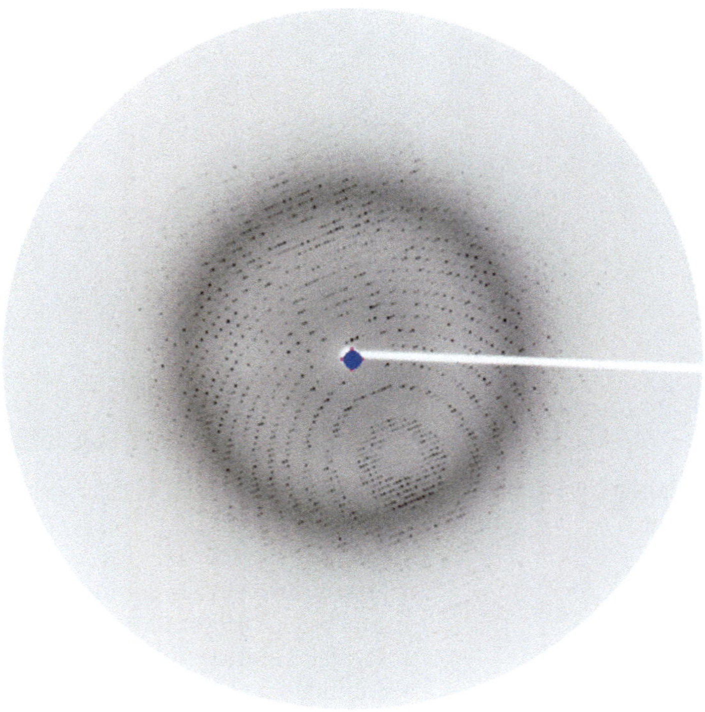

Fig. 4 Diffraction image of SmTPI obtained by 0.5° oscillation and a time of exposition of 10 min on a sealed tube X-ray generator

References

1. McPherson A, Cudney B (2014) Optimization of crystallization conditions for biological macromolecules. Acta Crystallogr F Struct Biol Commun 70:1445–1467
2. McPherson A, Malkin AJ, Kuznetsov YG (1995) The science of macromolecular crystallization. Structure 3:759–768
3. Chayen NE (2009) High-throughput protein crystallization. Adv Protein Chem Struct Biol 77:1–22
4. Trasvina-Arenas CH, Lopez-Castillo LM, Sanchez-Sandoval E, Brieba LG (2016) Dispensability of the [4Fe-4S] cluster in novel homologues of adenine glycosylase MutY. FEBS J 283:521–540
5. Castro-Torres E, Jimenez-Sandoval P, Fernandez-de Gortari E, Lopez-Castillo M, Baruch-Torres N, Lopez-Hidalgo M, Peralta-Castro A, Diaz-Quezada C, Sotelo-Mundo RR, Benitez-Cardoza CG, Espinoza-Fonseca LM, Ochoa-Leyva A, Brieba LG (2018) Structural basis for the limited response to oxidative and Thiol-conjugating agents by triosephosphate Isomerase from the photosynthetic bacteria synechocystis. Front Mol Biosci 5:103
6. Castro-Torres E, Jimenez-Sandoval P, Romero-Romero S, Fuentes-Pascacio A, Lopez-Castillo LM, Diaz-Quezada C, Fernandez-Velasco DA, Torres-Larios A, Brieba LG (2019) Structural basis for the modulation of plant cytosolic triosephosphate isomerase activity by mimicry of redox-based modifications. Plant J 99 (5):950–964
7. Lopez-Zavala AA, Carrasco-Miranda JS, Ramirez-Aguirre CD, Lopez-Hidalgo M, Benitez-Cardoza CG, Ochoa-Leyva A, Cardona-Felix CS, Diaz-Quezada C, Rudino-Pinera E, Sotelo-Mundo RR, Brieba LG (2016) Structural insights from a novel invertebrate triosephosphate isomerase from *Litopenaeus vannamei*. Biochim Biophys Acta 1864:1696–1706
8. Zhu K, Jin H, He Z, Zhu Q, Wang B (2006) A continuous method for the large-scale extraction of plasmid DNA by modified boiling lysis. Nat Protoc 1:3088–3093

Chapter 18

Luminescence-Based, Low- and Medium-Throughput Assays for Drug Screening in *Schistosoma mansoni* Larval Stage

Alessandra Guidi, Roberto Gimmelli, Alberto Bresciani, and Giovina Ruberti

Abstract

Schistosomiasis is one of the major parasitic diseases with more than 200 million people infected worldwide every year. Praziquantel is the drug of choice against the schistosomiasis although the use of a single drug to treat such a large amount of infected people appears particularly worrisome. For this reason, the search of new schistosomicidal compounds is viewed as an urgent goal and a number of screening campaigns have been carried out in the past years. The larval stage of *Schistosoma* (schistosomula) has been widely used in order to identify new compounds against the parasite. Here we describe detailed practical procedures for a luminescence-based assay proven to be highly effective for the selection of schistosomicidal compounds on small and medium-high scale. The assay is based on the quantitation of the parasite ATP, a good indicator of metabolically active cells, as measure of schistosomula viability. This assay is fast and reproducible, and it is suitable either for manual or for semiautomated screenings.

Key words *Schistosoma mansoni*, Drug screening, Luminescence, Viability assay, ATP quantitation

1 Introduction

Trematode flatworms (also referred to as fluke) of the genus *Schistosoma* are the causing agents of schistosomiasis. Among human parasitic diseases, schistosomiasis ranks second, only behind malaria, in terms of prevalence and socioeconomic impact worldwide. It affects the health and well-being of over 200 million people throughout the world. Praziquantel (PZQ), the drug of choice to treat people infected with the three main species of the parasite (*S. mansoni*, *S. haematobium*, and *S. japonicum*), is poorly effective on larval and juvenile developmental stages. Moreover, the use of a single drug can be viewed as a worrisome situation and PZQ resistance needs to be routinely monitored [1]. Therefore, the

David J. Timson (ed.), *Schistosoma mansoni: Methods and Protocols*, Methods in Molecular Biology, vol. 2151,
https://doi.org/10.1007/978-1-0716-0635-3_18,

search of novel schistosomicidal compounds appears critical, and in the past years several efforts have been made toward this aim.

The screening of compound libraries using the larval stage of *S. mansoni* (schistosomula) is a methodology widely used to rapidly select promising hits for follows up in juvenile and adult stages of schistosome. Newly transformed schistosomula (NTS) represent the most suitable stage for small, medium- and high-throughput assays as they can be prepared in large number by mechanical transformation of cercariae. In several drug discovery campaigns a number of different approaches for compound screening have emerged. In particular, some of the latest methodologies have used fluorescent dyes [2], viable colorant [3], or phenotyping screening [4]. Overall the desirable characteristics to perform a good screening are rapidity and reproducibility. We here describe detailed protocols of a luminescent method for drug screening setup which was validated in our laboratories using ATP quantitation as good indicator of parasite live/dead status. This methodology is simple, rapid, and suitable to give objective and quantitated values of parasites viability. Moreover, a relatively small amount of NTS are needed for the assay and it can be applied to medium-large-scale screening campaigns using 384-well microtiter plates as well as small libraries screening using 96-well microtiter plates. This method allowed us to screen 40,000 small molecules from both commercial and non-commercial sources leading to the identification of novel active compounds [5, 6].

2 Materials

2.1 Mechanical Cercariae Transformation and Newly Transformed Schistosomula (NTS) Preparation

1. *S. mansoni* cercariae-infected snails (4–5 weeks after infection).
2. Conditioned water (tap water passed through activated-charcoal filters for at least 48 h).
3. Pyrex crystallizing dishes with a capacity of 100 mL (70 mm × 40 mm).
4. Neon lamp for cercariae shedding.
5. Corex conical glass centrifuge tubes with screw cap (50 mL).
6. Ice bucket with lid.
7. 5× Dulbecco's Modified Eagle Medium (DMEM) high glucose: 69.9 g of DMEM high glucose powder in 50 mL of sterile dechlorinated water. Add 24.66 mL of 7.5% (w/v) sodium bicarbonate solution, 12 mL of antibiotic-antimycotic reagent (100,000 U/mL penicillin, 10,000 μg/mL streptomycin, 25 μg/mL amphotericin B) and water up to 100 mL final volume. Filter into sterile vacuum bottle-top 0.45 μm filters and store at 4 °C.
8. Bench vortex mixer.

9. 10 mL syringes and 22G syringe needles.
10. DMEM high glucose without phenol red: 4500 mg/L glucose, 1 mM Hepes pH 6.98–7.30, 2 mM L-glutamine, antibiotic–antimycotic reagent.
11. Fetal bovine serum (FBS).
12. Hank's balanced salt solution (HBSS) containing calcium and magnesium without phenol red.
13. Percoll cell culture tested, pH 8.5–9.5 (25 °C), density 1129 g/mL.
14. Percoll working solution: mix 9 mL of percoll with 1 mL of HBSS. Use this solution to make 70% percoll gradient with DMEM (not supplemented with serum).
15. Tissue culture polystyrene 6-well plates.

2.2 Luminescent-Viability Assays

1. Dimethyl sulfoxide (DMSO) tissue culture grade.
2. Gambogic acid (GA) 50 mM, stock concentration.
3. Compounds stock solution in DMSO (vehicle).
4. DMEM high glucose w/o phenol red.
5. Flat-bottom 96- and 384-well black tissue culture-treated plates for compounds screening.
6. Acoustic droplet ejection for compound transfer to 384-well assay plates (e.g., ATS-100, EDC Biosystems).
7. CellTiterGlo (CTG) reagent (Promega).
8. Luminescence plate reader device (e.g., Envision PerkinElmer or Varioskan Lux, Thermo Fisher Scientific).

2.3 Additional Material and Instruments

1. Pyrex glass beaker (20, 50,100 mL).
2. Pasteur glass pipettes.
3. Serological pipettes (5–25 mL).
4. Manual cell counter.
5. Pipettes and filter tips (100, 200, 1000 μL).
6. Refrigerated centrifuge with swing rotor and buckets.
7. Stereomicroscope.

3 Methods

A Puerto Rican strain of *S. mansoni* can be maintained by passage through the intermediate snail host *Biomphalaria glabrata* and ICR (CD-1) outbred female mice as definitive host. Mice can be routinely infected by percutaneous exposure to *S. mansoni* cercariae via the tail [7]. Infected livers can be recovered 7–8 weeks after the

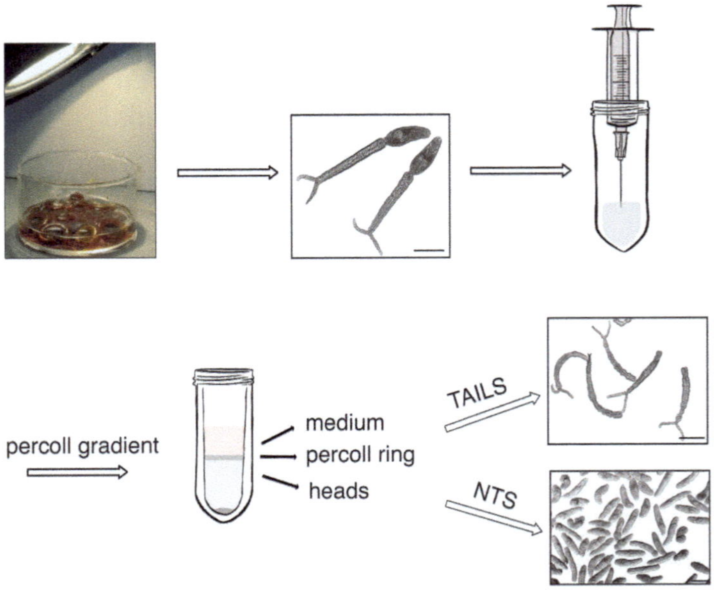

Fig. 1 Schematic procedure of NTS preparation from *S. mansoni* cercariae. Scale-bars: 100 μm

infection and processed to obtain large amount of eggs to infect the snail host. All studies must be authorized according to the ethical and safety rules and guidelines for the use of animals in biomedical research provided by the relevant national and international laws and directives.

3.1 NTS Preparation by Mechanical Transformation of Cercariae

NTS should be prepared using an optimized version [5] of the protocol of Brink et al. [8], previously described by Protasio et al. [9] (Fig. 1) (*see* **Note 1**).

1. Place 20–25 infected snails (4–5 weeks after miracidia infection) into a 100 mL pyrex crystallizing dish filled with 10 mL of conditioned water (*see* **Note 2**). In our experience each snail will release between 1000 and 1500 cercariae (*see* **Note 3**).
2. Leave them under the light for 60 min at 27 °C for cercariae shedding.
3. Use a 10 mL serological pipette to collect the cercariae suspension from all the dishes and transfer them into a 100 mL beaker.
4. Add 10 mL of conditioned water and repeat **steps 2** and **3**.
5. Let the suspension settle down 5 min at room temperature so that all the snails's contaminants will precipitate.
6. Transfer the suspension without the precipitated contaminants into a clean 100 mL beaker (*see* **Note 4**).

7. Assess the number of cercariae by counting 3 × 50 μL of the suspension with Bouin's solution under a bright-field stereomicroscope.
8. Carefully pipette 40 mL of the cercariae suspension into the corex glass tubes and put them on ice, in the dark, for 30 min. This step will immobilize and sediment the cercariae.
9. Remove the supernatant without disturbing the cercarial pellet.
10. Resuspend the pellet in cold DMEM 5× to a final volume of maximum 10 mL (*see* **Note 5**).
11. Vortex for 30 s.
12. Use a 10 mL syringe with 22G needle and vigorously push the liquid back and forth 10–12 times. At this point at least 90% of the tails should be separated from the head (NTS).
13. Leave the suspension for 5 min at room temperature to allow the sedimentation of the heads and carefully remove and discard the supernatant containing the tails (*see* **Note 6**).
14. Prepare fresh 70% (v/v) percoll solution (*see* **Note 7**).
15. Resuspend the pellet in 10 mL of DMEM without FBS and mix well pipetting up and down. When the NTS are all in suspension, add 2–3 mL of 70% (v/v) percoll gradient solution placing the serological pipette at the bottom of the tube and slowly releasing the solution.
16. Centrifuge at 600 × *g* for 10 min at 4 °C. Three phases will appear in the tube: the upper phase (DMEM only), the percoll ring (containing tails), and the bottom phase (containing the pellet of the heads).
17. Remove the upper phase and the percoll ring without disturbing the sedimented heads.
18. Resuspend the heads in 30 mL of DMEM and assess the number of NTS by counting 3 × 50 μL of the suspension with the Bouin's solution using a dish with 2 mm square grid pattern under a bright-field stereomicroscope. In our experience the efficiency of NTS recovery from the initial cercariae shedding is about 90%.
19. Wash the NTS twice with 30 mL of DMEM by centrifuging at 600 x g 5 min at 4 °C (*see* **Note 8**).
20. Add to the NTS the appropriate volume of DMEM supplemented with 10% FBS in order to have the maximum concentration of 2000 parasites/mL.
21. Leave parasites in the incubator (37 °C 5% CO_2) for 24 h before using them in viability assays.

3.2 Luminescent ATP-Dependent NTS Viability Assay

It has been previously demonstrated that ATP quantitation reflects the overall metabolic state of the NTS and it represents a good indicator of their viability [5]. Since the ATP quantitation shows strong correlation with the parasite numbers, it is very important to use a number of NTS in which this correlation is linear [5]. Drug screening with NTS can be performed for small or medium-large number of compounds, manually or semiautomatically, and in 96 or 384 wells flat-bottom black tissue culture-treated plates respectively. The optimal number of NTS to be used is 150–200 NTS/96 wells and 100 NTS/384 wells [5]. Nevertheless higher relative luminescence units (RLU) can be obtained increasing the number of NTS, within the linear range, without changing the readout of compound's LD_{50} [5].

3.2.1 Low-Scale Viability Assay Using 96-Well Plates

1. Use room temperature culture medium and work under sterile conditions.
2. Initial screening can be performed at the final drug concentration of 10 or 50 μM (*see* **Note 9**).
3. Collect all NTS and adjust the concentration to 200–250 parasites/100 μL: pipette 75 μL of the suspension in each well (with the exception of three wells, the blanks) so that each well will contain approximately 150–200 NTS.
4. Dilute the stock solution of compounds (50–100 mM) in supplemented DMEM medium w/o phenol red to a final concentration of 40–200 μM (4× of the final assay concentration).
5. For each compound perform five replicates.
6. Add 25 μL of the 4× compound solution to the replicate wells.
7. Repeat **steps 3–6** for all the compounds.
8. For control samples use the same concentration of vehicle (DMSO) instead of the compound and perform at least six replicates (*see* **Note 10**).
9. Prepare positive control wells as in **steps 3–6** using GA.
10. Add supplemented medium only for the blank measurement in the three wells (without NTS).
11. Compounds can also be tested at different concentrations for dose-response curve.
12. For each compound, repeat **steps 3–6** making 1:2 serial dilutions in order to obtain the following final concentrations 100, 50, 25, 12.5, 6.25, 3.13, 1.60, and 0.78 μM.
13. Prepare positive, negative, and blank controls as in **steps 8–10**.
14. Incubate the tissue culture plate at 37 °C 5% CO_2 for 24 or 72 h.

15. Thaw the CellTiterGlo (CTG) buffer and equilibrate it together with the lyophilized CTG substrate at room temperature 30 min prior use.
16. Prepare the CTG reagent following the manufacturer's instructions.
17. Remove the tissue culture plate from the incubator and leave it 10 min at room temperature.
18. Add 50 μL of CTG reagent to each well, cover the plate with aluminum foil, and mix for 5 min with gentle shaking. Incubate at room temperature for 25 min.
19. Read the plate using a luminescent plate reader device (e.g., Varioskan Lux) (*see* **Note 11**).
20. The viability of parasites in each well is quantified as RLU (relative luminescent units) and raw data are normalized between DMSO (100% viability) and 50 μM GA (0%) viability (Fig. 2).
21. The threshold for screenings can be set to 60% decrease in parasites viability at 10–50 μM compound concentration.
22. Dose-response curve data can be analyzed vs DMSO in order to calculate the LD_{50} value for each compound tested (e.g., GA in Fig. 2).

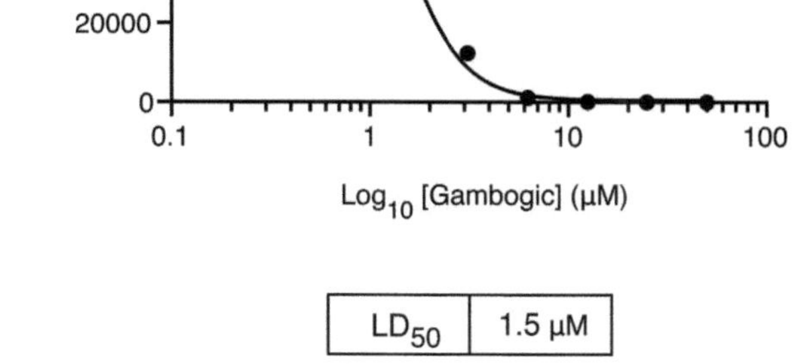

Fig. 2 ATP luminescence signal of gambogic acid (GA) dose-response curve. NTS (150/well) were incubated with serial dilutions of GA (0.78–50 μM) and ATP luminescence signal measured. RLU (raw data) are reported on the *y*-axis. The graph shows the mean values and standard deviations. The calculated LD_{50} is indicated

3.2.2 Medium-Throughput Assay Using 384-Well Plates

1. Prepare 10 mM DMSO solution of the compounds to be tested including GA as positive control.
2. Transfer compounds (10 mM solution), DMSO alone (vehicle), and GA to the bottom of 384-well, black tissue culture-treated plates using the acoustic droplet ejection technology (e.g., ATS-100, EDC Biosystems) in order to achieve the desired concentration in the final volume of 30 μL (*see* **Note 9**).
3. Prepare a suspension of 3.333 NTS/mL in complete DMEM.
4. Add 30 μL of the NTS suspension to the assay plates by a suitable dispenser (e.g., multidrop, Thermo Fisher).
5. Incubate the plate for 24 h at 37 °C and 5% CO_2.
6. Add 30 μL of CTG to each well by a suitable dispenser (e.g., multidrop, Thermo Fisher).
7. Incubate at room temperature for 30 min.
8. Read luminescence by a suitable detector (e.g., Envision, PerkinElmer).

4 Notes

1. Cercariae represent the infective stage of schistosomiasis. Always wear protective equipment such as laboratory coat, two pairs of gloves, and long-sleeved pair over the top while handling cercariae.
2. When shedding the snails, do not overcrowd the dishes and distribute 20–25 snails into each dish in 10 mL conditioned water.
3. The number of cercariae shed from the snails is dependent on how often they have been used to produce the cercariae, and the production gradually drops over time and after each shedding. Seven to eight weeks after the first use, snails are no longer shedding cercariae efficiently.
4. Since the NTS preparation is not performed under sterile conditions, it is important to work as clean as possible to reduce the risk of contamination of the NTS culture.
5. During NTS preparation, keep all media on ice and use cold solutions.
6. Before removing all the supernatants, pipette a small volume into a Petri dish and check under the stereomicroscope that heads are no longer in the suspension.
7. Always prepare fresh 70% (v/v) percoll solution; do not store it.
8. Perform all the washes in sterile conditions.

9. 16 or 320 compounds can be tested simultaneously using a single drug concentration into a 96- or 384-well plate, respectively.
10. To minimize the variability of the readout, increase the number of control samples (DMSO).
11. Before the readout plates can be subjected to 5 s shaking at 300 rpm.

Acknowledgments

A.B. and G.R. are supported by CNR (National Research Council)-CNCCS (Collezione Nazionale di Composti Chimici e Centro di screening) "Rare, Neglected and Poverty Related Diseases—Schistodiscovery Project" (DSB.AD011.001.003).

References

1. McManus DP, Dunne DW, Sacko M et al (2018) Schistosomiasis. Nat Rev Dis Primers 4(1):13. https://doi.org/10.1038/s41572-018-0013-8
2. Peak E, Chalmers IW, Hoffmann KF (2010) Development and validation of a quantitative, high-throughput, fluorescent-based bioassay to detect schistosoma viability. PLoS Negl Trop Dis 4(7):e759. https://doi.org/10.1371/journal.pntd.0000759
3. Mansour NR, Bickle QD (2010) Comparison of microscopy and Alamar blue reduction in a larval based assay for schistosome drug screening. PLoS Negl Trop Dis 4(8):e795. https://doi.org/10.1371/journal.pntd.0000795
4. Paveley RA, Mansour NR, Hallyburton I et al (2012) Whole organism high-content screening by label-free, image-based Bayesian classification for parasitic diseases. PLoS Negl Trop Dis 6(7): e1762. https://doi.org/10.1371/journal.pntd.0001762
5. Lalli C, Guidi A, Gennari N et al (2015) Development and validation of a luminescence-based, medium-throughput assay for drug screening in *Schistosoma mansoni*. PLoS Negl Trop Dis 9(1): e0003484. https://doi.org/10.1371/journal.pntd.0003484
6. Guidi A, Lalli C, Gimmelli R et al (2017) Discovery by organism based high-throughput screening of new multi-stage compounds affecting *Schistosoma mansoni* viability, egg formation and production. PLoS Negl Trop Dis 11(10): e0005994. https://doi.org/10.1371/journal.pntd.0005994
7. Tucker MS, Karunaratne LB, Lewis FA et al (2013) Schistosomiasis. Curr Protoc Immunol 103:Unit 19.1. https://doi.org/10.1002/0471142735.im1901s103
8. Brink LH, McLaren DJ, Smithers SR (1977) *Schistosoma mansoni*: a comparative study of artificially transformed schistosomula and schistosomula recovered after cercarial penetration of isolated skin. Parasitology 74(1):73–86
9. Protasio AV, Dunne DW, Berriman M (2013) Comparative study of transcriptome profiles of mechanical- and skin-transformed *Schistosoma mansoni* schistosomula. PLoS Negl Trop Dis 7 (3):e2091. https://doi.org/10.1371/journal.pntd.0002091

Chapter 19

Testing of *Schistosoma mansoni* Vaccine Targets

Weidong Zhang, Adebayo Molehin, Parth Patel, Eunjee Kim, Adam Peña, and Afzal A. Siddiqui

Abstract

Vaccination has greatly reduced the burden of human diseases caused by infectious pathogens. Systematic development of vaccine targets requires established protocols to assess immunogenicity and efficacy of such vaccine candidates. Using a leading schistosomiasis vaccine candidate, Sm-p80, as an example, we describe standardized approaches for testing the immunogenicity and efficacy of *Schistosoma mansoni* vaccine targets. Unlike other parasite systems in which sterile immunity is required, the goal of *S. mansoni* vaccine targets is overall reduction in morbidity. Methods related to the parasitological parameters described in this chapter allow for the testing of the prophylactic (reduction in adult worm burden), anti-pathology (liver and intestine egg retention), and transmission blocking (fecal egg expulsion and egg hatching rates) efficacies for the vaccine target. The RNA sequencing approaches provide basis for identification of molecular signatures predictive of desirable outcomes for schistosomiasis vaccines.

Key words *Schistosoma mansoni*, Schistosomiasis, Vaccine testing, Sm-p80, Animal models, Egg burden, Immunology, Vaccine efficacy

1 Introduction

The helminth parasite *Schistosoma* is the causative agent of the human disease known as schistosomiasis, and it represents the second most devastating infectious parasitic disease worldwide. The disease is endemic in 79 countries, and current estimates are that over 240 million people are infected with one or more of the three major species of schistosomes (*Schistosoma mansoni*, *S. haematobium*, and *S. japonicum*) with 800 million people at risk of contracting the infection. In addition, the number of annual deaths associated with schistosomiasis is 280,000 with the disability adjusted life years (DALY) of 3.8 million [1]. Furthermore, active schistosomiasis transmission is now being reported in geographical areas previously known to be schistosomiasis-free, for example, Corsica, France [2].

David J. Timson (ed.), *Schistosoma mansoni: Methods and Protocols*, Methods in Molecular Biology, vol. 2151, https://doi.org/10.1007/978-1-0716-0635-3_19,

Over the last four decades, the control efforts against schistosomiasis have largely been predicated on mass drug administration (MDA) with praziquantel (PZQ). Despite the effectiveness of PZQ at killing established schistosome parasites in their infected human hosts, the intensity and prevalence of schistosomiasis has remained unchanged due to sustained reinfection rates, low PZQ coverage, and emergence of drug-resistant parasites [3]. Due to many problems associated with the MDA and water, sanitation, and hygiene (WaSH) programs, experts in the field now agree that global reduction of schistosomiasis burden and possible disease elimination is only feasible through the deployment of an effective vaccine in concert with current control programs [4].

Historically, vaccines have been shown to be the most cost-effective interventions against human diseases caused by infectious pathogens [3]. To date, there is no licensed schistosomiasis vaccine for human use because no candidate antigen has succeeded in performing effectively in human efficacy trials [5]. Several preclinical studies have shown that the large subunit of *S. mansoni* calcium activated neutral protease calpain, termed Sm-p80, offers robust prophylactic, therapeutic, anti-pathology, and transmission-blocking efficacy against *S. mansoni* in mice and nonhuman primate, baboons, models of infection, and disease. Phase I/II human clinical trials of Sm-p80 vaccine are scheduled to begin in the USA and Africa in 2019.

In this chapter, we describe protocols and useful guidelines for assessing the efficacy of Sm-p80 vaccine antigen in animal models. Conventional approaches to evaluate the efficacy of schistosomiasis vaccine candidates in preclinical studies employ rodent models; however, due to our extensive expertise in the nonhuman primate models, we have included both mouse and baboon models of infection and disease. Here we outline procedures for the parasitological parameter for efficacy determination (worm and egg burden, egg hatching rate, circulating anodic antigen (CAA), and circulating cathodic antigen (CCA) tests and immune responses (Th1, Th2, Th17 and humoral response). Although the methodology described in this chapter is based on Sm-p80 vaccine efficacy testing, the protocols described here, it can also be applied to other schistosomiasis vaccine candidate antigens.

2 Materials

2.1 Establish Animal Model

1. Animal: 3–4-week-old female C57/BL6 mice. Male and female nonhuman primates, olive baboons (*Papio anubis*) (*see* **Note 1**).
2. Schistosome vaccine: SchistoShield® vaccine (Sm-p80 plus glucopyranosyl lipid adjuvant formulated in a stable emulsion,

GLA-SE) is a TLR4-based recombinant Sm-p80 vaccine (PAI Life Sciences, Seattle, WA, USA).

3. Anesthetics: Isoflurane for mouse model, ketamine (15 mg/kg) and acepromazine (0.5 mg/kg) for baboon model.
4. EZ-109 Multi-Animal Breather.
5. EZ-155 Stand-Alone Vaporizer.
6. 1 cc syringe.
7. 5 mm lancet.
8. Microcentrifuge tubes.
9. Winged blood collection set safety lock needle with Luer adaptor.
10. Vacutainer heparin tubes and serum separation tubes.
11. Alcohol pads.
12. Gauze.
13. Razor.
14. Clinical digital thermometer.
15. Tube rotator (Benchmark Scientific Roto-Mini™ Plus R2024 Tube Rotator with Adjustable Speed).
16. *Schistosoma mansoni*-infected *Biomphalaria glabrata* snails (provided by Biomedical Science Research Institute, BRI, Rockville, Maryland).
17. Iodine solution: Weigh 2 g potassium iodide and 1 g iodine into a 50 mL conical tube. Fill the tube with 50 mL distilled water and mix well. Aliquot the solution to microcentrifuge tubes and store at room temperature.
18. 12 × 75 mm glass tubes.
19. Dissecting microscope.
20. Stainless steel rings.
21. Euthanasia: As per the recommendations of the Panel on Euthanasia of the American Veterinary Medical association and IACUC-approved carbon dioxide (for mouse model): SomnaSol or Euthasol (for baboon model).
22. Perfusion equipment: Peristaltic perfusion pump; pump head; foot switch for use with driver; silicone pump tubing fitted into 1 cc syringe with a 26-gauge needle.
23. Perfusion buffer: 0.85% (w/v) sodium chloride, 0.75% (w/v) sodium citrate. Weigh 7.5 g trisodium citrate dihydrate and 8.5 g sodium chloride, transfer to a 1 L graduated cylinder or a glass beaker, and prepare a 1 L solution. Store up to 7 days at 4 °C.
24. 150 mm plastic Petri dish with grid.

25. Sharp dissecting scissors and micro-dissecting fine forceps (4 in., full curve, 0.5 mm).
26. Heparin.
27. Appropriate size of catheter.
28. Vacuum aspiration system.
29. Surgery tools: suture, bone snips and scalpel, scissors, tweezers, etc.
30. Glass cannula connected with transparent tubing.
31. Metal and glass pans.
32. Pilsner glasses and beakers.
33. 1.2%(w/v) NaCl: Weigh 12 g NaCl and transfer 1 L beaker. Fill the beaker with purified water till 1 L. After complete dissolution, store it at 4 °C.

2.2 Parasitological Parameters

2.2.1 Worm Burden Determinations

1. Pilsner glass.
2. 150 mm Petri dish.
3. Fine-tip forceps.
4. Perfusion buffer: *see* Subheading 2.1, **item 23**.

2.2.2 Egg Burden Determination in Tissues

1. 4%(w/v) KOH: Weigh 40 g KOH and transfer to 1 L beaker. Fill the beaker with purified water till 1 L. After complete dissolution, store it at room temperature. 4% KOH should be freshly made before use.
2. Centrifuge and shaker.
3. 1.2%(w/v) NaCl: *see* Subheading 2.1, **item 33**.
4. Microscope.
5. Razor blades.
6. Scale.

2.2.3 Egg Hatching in Tissues

1. Scale.
2. Scalpel blades.
3. Collagenase B.
4. Antibiotics: penicillin and streptomycin.
5. Phosphate buffer saline (PBS): 137 mM NaCl, 10 mM Na_2HPO_4, 1.8 mM KH_2PO_4, 2.7 mM KCl, and a pH of 7.4. (Weigh 8 g of NaCl, 0.2 g of KCl, 1.44 g of Na_2HPO_4, and 0.24 g of KH_2PO_4. Place all the chemicals to a 1 L beaker with 800 mL distilled water. Adjust the pH to 7.4 with HCl and add distilled water to a total volume of 1 L. Autoclave the reagent and store it at 4 °C.)
6. Stainless steel sieves (Gilson sieve No. 40, 70, 60, 80, 100, 140, 325; mesh openings of 425 μm, 250 μm,180 μm, 150 μm, 106 μm, and 45 μm).

7. Percoll.
8. 0.25 M sucrose: Weigh 85.58 g sucrose and dissolve it in distilled water. Make final volume to 1 L.
9. PBS containing 1 mM ethylenediaminetetraacetic acid (EDTA), 1 mM ethylene glycol-bis(β-aminoethyl ether)-*N*,*N*,*N*′,*N*′-tetraacetic acid (EGTA): Weigh 0.29 g EDTA and 0.38 g EGTA. Completely dissolve them in 1 L PBS with strong magnetic stirring rod. Store it at 4 °C.
10. 1.2% (w/v) NaCl: *see* Subheading 2.1, **item 33**.
11. Waring blender, with variable speed control (Eberbach, model: 8580 semi-micro container and 2-speed Waring power base unit).
12. Plastic spray bottle.
13. Glass Petri dishes with flat bottoms (Corning Incorporated, 100 × 20 mm).
14. Pasteur pipets.
15. Gauze.
16. Inverted microscope.

2.2.4 Egg Burden Determination in Feces

1. Scale.
2. Gauze.
3. Centrifuge.
4. Kimwipe.
5. Purified water.
6. 10% Formalin: Measure 270 mL 37% formaldehyde using a cylinder and add water to 1 L.
7. Ethyl acetate.
8. 1.2% NaCl: *see* Subheading 2.1, **item 33**.

2.2.5 Egg Hatching in Feces

1. Scale.
2. Gauze.
3. 1.2% NaCl: *see* Subheading 2.1, **item 33**.
4. Stainless steel sieves (Gilson sieve No. 40, 70, 60, 80, 100, 140, 325; mesh openings of 425 μm, 250 μm,180 μm, 150 μm, 106 μm, and 45 μm).
5. Plastic spray bottle.
6. Pasteur pipets.
7. Inverted microscope.
8. Glass Petri dish.

2.2.6 Circulating Anodic Antigen Assay (CAA)

1. Stock AWA-TCA (adult worm antigen): 100,000 pg/mL, stored in −20 °C freezer.
2. Normal baboon urine/serum; negative for CAA.
3. 4% trichloroacetic acid (TCA) (provided by Leiden University Medical Center): 2 mL 100% TCA dissolved 48 mL milliQ water.
4. Urine/serum samples from baboon, standards provided by Leiden University Medical Center.
5. CAA UCP dry conjugate tubes: Stored in closed aluminum bags containing 2 × 96 wells and silica dry packs at room temperature (dry place). Check carefully if the aluminum bag is reclosed well after opening.
6. CAA UCP strips: Stored at room temperature (dry place) in white containers with silica dry packs.
7. Lyophilized HSLF buffer: Supplied in aluminum bags with silica dry packs. Stored at room temperature in closed aluminum bags in a dry place. After solving the buffer, it can be stored at −20 °C.

2.2.7 Point-of-Care CAA Assay

1. POC-CAA kit (Rapidmedical Diagnostic Company).

2.3 Humoral Immune Responses

2.3.1 ELISA for Immunoglobulin Endpoint Titers

1. 96-well, Immulon flat-bottom microtiter plates.
2. 0.22 μm filter units.
3. Costar Cluster tubes.
4. Calibrated μL pipets with clean or sterile tips capable of accurately delivering 5–1000 μL.
5. Calibrated multi-channel pipet and tips capable of accurately delivering 25–200 μL.
6. ELISA Wash Buffer (EWB): Mix 50 mL 20× wash solution concentrate (2.74 M NaCl, 200 mM Na_2HPO_4, 36 mM KH_2PO_4, 54 mM KCl, and 10 mL Tween 20 and a pH of 7.4) with 950 mL reagent quality water (*see* **Note 2**).
7. ELISA Coating Buffer (ECB): Mix 25 mL 10× coating solution concentrate (150 mM Na_2CO_3, 348.89 mM $NaHCO_3$ and a pH of 9.6) with 225 mL reagent quality water (*see* **Note 2**).
8. ELISA Blocking Buffer (EBB): Mix 50 mL 10% BSA diluent/blocking solution (1.37 M NaCl, 100 mM Na_2HPO_4, 18 mM KH_2PO_4, 27 mM KCl, 1.5 mM BSA, and a pH of 7.4) with 450 mL reagent quality water. Add 0.25 mL Tween 20 to the BSA diluent/blocking solution and mix well (*see* **Note 2**).

9. Diluted ELISA Blocking Buffer (DEBB): Dilute EBB 1:10 in EWB for dilution of samples and secondary antibody conjugate (*see* **Note 2**).
10. Polypropylene conical tubes.
11. Plate sealers.
12. Vortex mixer.
13. Plate washer.
14. Microcentrifuge rack.
15. Incubator (20–25 °C).
16. Calibrated microplate reader (SpectraMax series).
17. Analysis Software (SoftMax Pro).

2.3.2 ELISPOT (Kit Was Purchased from U-CyTech, The Netherlands)

1. Sterile distilled water.
2. 70% (v/v) ethanol.
3. Sterile and pyrogen-free PBS (phosphate-buffered saline).
4. Wash buffer.
5. ELISPOT plate with lid.
6. Blocking buffer: For one ELISPOT plate, dilute 2 mL of blocking stock solution B (10×) in 18 mL PBS-I.
7. Culture medium and stimulus.
8. Dilution buffer: For one ELISPOT plate, dilute 1.5 mL dilution buffer B (10×) in 13.5 mL PBS-I.
9. GABA solution (φ-labeled anti-biotin antibodies).
10. Activator I/II Solution: For one ELISPOT plate, mix gently but thoroughly 1.5 mL of Activator I with 1.5 mL Activator II.
11. Light microscope.
12. Dissecting microscope or ImmunoSpot image analyzer.

2.4 Cellular Immune Responses

2.4.1 Quantitative Real Time-Polymerase Chain Reaction (qRT-PCR) for Cytokine Expression

1. Frozen cells (*see* **Note 3**).
2. 37 °C water bath.
3. Complete media: RPMI 1640 media with 10% FBS, 10 μg/mL gentamicin, 100 μg/mL streptomycin, 100 μg/mL penicillin G.
4. 6-well plate.
5. rSm-p80.
6. Centrifuge.
7. RNA isolation kit (GenElute™ mammalian total RNA miniprep kit, MilliporeSigma, St. Louis, MO).
8. SYBR® Premix Ex Taq™ (Tli RNase H Plus) (Takara).

9. Thermo Scientific™ Maxima™ First Strand cDNA Synthesis Kit for RT-qPCR (Thermo Fisher Scientific).
10. StepOne™ plus Real time PCR machine.
11. Software: DataAssist™ (Thermo Fisher Scientific) and GraphPad prism (Graphpad.com).

2.4.2 FACS (See ***Note 3****) (Using Cells from Baboon Model as Illustrated Example)*

1. Phorbol 12-myristate 13-acetate, PMA 1 mg (Sigma). Storage solution: PMA dissolved in DMSO for concentration 100 μg/mL and keep at −20 °C for up to 6 months. Working solution: 1:1000 dilution in RPMI 1640 medium; the concentration should be 100 ng/mL.
2. Ionomycin (Sigma). Storage solution: Ionomycin dissolved in DMSO for concentration of 1 mg/mL and keep at −20 °C. Working solution: 1:1000 dilution in PRMI 1640 medium; the concentration should be 1 μg/mL.
3. BD Cytofix/Cytoperm™ Plus Fixation and Permeabilization Kit (with BD GolgiPlug™ protein transport inhibitor containing brefeldin A): Brefeldin A solution, 1 mL. The Brefeldin A solution is supplied at 1000 × working concentration. Prior to addition to the cell cultures, a 1:1000 dilution in the culture media should be made. Fixation/Permeabilization solution (125 mL): supplied at working concentration and contains 4% paraformaldehyde. Contact with eyes, skin, and mucous membranes should be avoided. BD Perm/Wash ™ buffer (100 mL): The Permeabilization Buffer is supplied as a 10 × concentrate. Prior to use, this should be diluted tenfold in distilled water.
4. Fluorescence labeled antibody (Becton, Dickinson and Company).
   (a) Isotype control.
      - FITC Mouse IgG3 κ Isotype Control (50 tests, 20 μL per test).
      - PerCP-Cy™5.5 Mouse IgG1 κ Isotype Control (50 tests, 20 μL per test).
      - APC Mouse IgG1 κ Isotype Control (0.1 mg, 0.2 mg/mL, 500 μL, 0.125 μg per test).
      - PE Mouse IgG1 Kappa Isotype Control (50 tests, 20 μL per test).

   (b) CD and cytokine antibody.
      - FITC Mouse Anti-Human CD3ε (50 tests, 20 μL per test).
      - PerCP-Cy™5.5 Mouse Anti-Human CD4 (50 tests, 20 μL per test).

- APC Mouse Anti-Human IL-4 (50 tests, 5 μL per test).
- PE Mouse Anti-Human IFN-γ (50 tests, 20 μL per test).

5. Complete RPMI 1640: RPMI 1640 containing 100 μg/mL each of penicillin and streptomycin and 5–10 μg/mL gentamicin and 10% FBS.
6. Wash buffer: Sterilized phosphate-buffered saline (PBS).
7. Staining buffer: PBS with 1% heat-inactivated FCS and 0.09% (w/v) sodium azide (*see* **Note 3**), adjust PH to 7.4–7.6, and filter. Store at 4 °C.
8. rSm-p80: 12 μg/mL for 5×10^6 cells/mL.

2.5 Systems Vaccinology

2.5.1 Future Perspectives: Application of a Broad, Holistic Approach to Predicting Vaccine Efficacy

1. RNA isolation kit (GenElute™ mammalian total RNA miniprep kit, MilliporeSigma, St. Louis, MO).
2. Qubit® 3.0 Fluorometer and RNA HS assay kit (Thermo Fisher Scientific).
3. TruSeq® Stranded mRNA LT kit (Illumina).
4. HiSeq SBS Kits (Illumina).
5. HiSeq 2500 sequencer (Illumina).
6. epMotion robot (Eppendorf).
7. SuperScript® III reverse transcriptase (Thermo Fisher Scientific).
8. AMPure XP purification beads (Beckman Coulter).
9. NaOH.
10. 10 mM Tris–HCl, pH 8.5, 0.1% (v/v) Tween-20.

3 Methods

3.1 Establish Animal Model

The use of an animal model is inevitably required for an efficacy study of the vaccine. Due to the ease of accessibility, availability, and susceptibility to the infection, murine model is popularly used in the protection study.

The murine model has also been used to elucidate the interaction of schistosomes with the host immune system in the course of the disease, particularly Th1 and Th2 responses. A nonhuman primate model, the baboon, is susceptible to both experimental and natural schistosome infection. The similarities of clinical symptoms and immune responses throughout the infection between nonhuman primates and humans make the nonhuman primate model more invaluable and practical in examining vaccine efficacy.

3.1.1 Mouse Model

1. Blood collection: Restrain mice and puncture submandibular vein with 5 mm lancet. The pierced point should be located on an intersect between a line from the corner of the mouth and a line from the lateral canthus of the eye. Collect 50–100 μL blood in an Eppendorf tube for serum isolation. After bleeding, release the pressure on the scruff and compress a sterile cotton ball to the puncture site for hemostasis. Blood collection is performed before every time point (pre-bleed, primary immunization, every immunization boost, challenge, and euthanasia).
2. Vaccine preparation: Prepare vaccine according to manufacturer's instruction. Put vaccine vials on tube rotator to fully mix. Pull the plunger back to fill less than 100 μL with the vaccine and gently tap the syringe to remove any air bubbles.
3. Vaccination: Hold the mouse by its scruff to expose quadriceps of two legs, clean each quadricep with alcohol pad, and allow to air dry. Inject less than 50 μL vaccine into each quadricep (*see* **Note 4**). Apply light pressure on injection sites using sterile gauze (*see* **Note 5**). The times and intervals of immunization depend on vaccine design.
4. Cercariae collection: Forcefully spray the shells of the infected snails with water to remove rotifers and other attachments on the shells. Put the snails into a 100 mL beaker and expose the snails to a strong light source for 15 min; then, add purified water at a density of one snail per 2 mL to the beaker for an additional 30 min exposure to the light. With feather forceps, remove the snails from the beaker and return them to their container. Gently swirl the cercarial suspension, and with a large-orifice pipet tip fastened to a pipette, withdraw 20 μL of the liquid. Place the liquid drop onto a slide with an additional drop of iodine solution for killing and staining the cercariae. Repeat this step two more times for three total drops on the slide. Count the shedding cercariae under dissection microscope. Calculate the average number of cercariae in the three 20 μL aliquots and the volume of cercariae suspension needed for 150 cercariae to infect one mouse. Use cercariae to infect mouse within 5 h after harvest (*see* **Notes 6** and 7).
5. Challenge.

   Method 1: Percutaneously expose mice to *S. mansoni* cercariae via tail.

   (a) Pipet the calculated volume of cercariae suspension containing 150 cercariae into a 12 × 75 mm glass tube, and add purified water to 1 cm from the top of tube. Place the tube in a test tube rack.

   (b) Place a mouse in a plastic restraining tube (*see* Fig. 1), with its tail extending from the bottom of the tube to the glass

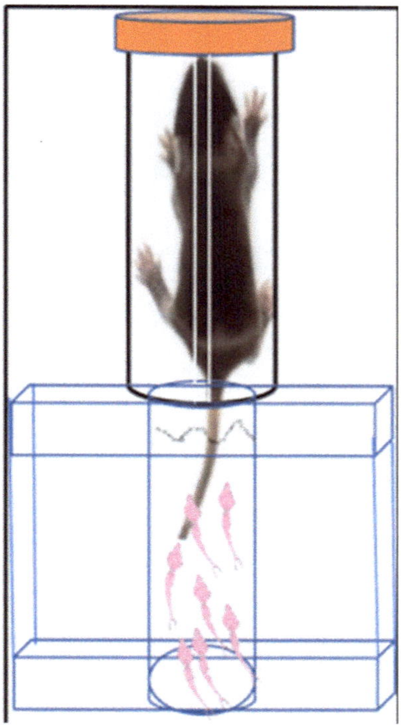

Fig. 1 Percutaneously expose mice to *Schistosoma mansoni* cercariae via tail. Place mouse into a restraining device for tail exposure to *Schistosoma mansoni* cercariae. The top rack holding the restraining tube can be made using Corning™ Foam Premium Quality 50 mL Centrifuge Tube Rack with pierced hole which allows the tail through to infected water. The bottom rack holding the glass tube can be replaced by the Corning™ Foam Premium Quality Centrifuge Tube Rack for both 15 mL and 50 mL centrifuge tubes

tube containing cercariae suspension. Attach adhesive tapes to the base of the tail to keep it in place (*see* **Note 8**).

Expose mouse to cercariae for 45 min, then release the mouse from the restraining tube, and put it back to its cage without wiping the tail. Most cercariae will penetrate within 30 min of exposure, but 45 min allows enough time for maximal numbers of cercariae penetrating through skin (*see* **Note 9**).

(c) Empty the liquid in the glass tube into a Petri dish and gently swirl with Pasteur pipet. Pipet 3 20 μL drops onto a slide and add a drop of iodine solution to each. With a dissecting microscope, count the remaining intact cercariae and the bodies without the tails to estimate the penetration ratio of cercariae after the 45 min exposure period.

(d) Monitor the mice for about 30 min. If no pain, distress, or discomfort is observed, return mice to their housing room.

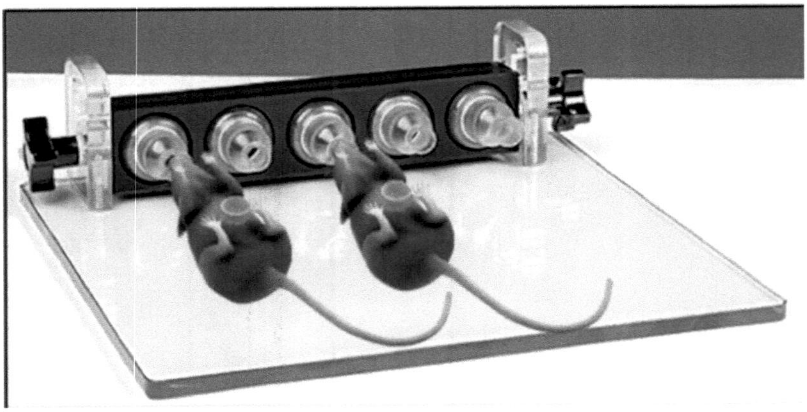

Fig. 2 Percutaneously expose mice to *Schistosoma mansoni* cercariae via abdomen

Method 2: Percutaneously expose mice to *S. mansoni* cercariae via abdomen.

(a) Lay down the mouse with its belly facing upward and insert its nose into the face mask with connection to isoflurane anesthesia system. Adjust the isoflurane volume to induce the mouse into deep sleep (*see* Fig. 2).

(b) Carefully shave the abdomen with a razor and wipe the abdomen with a gauze moistened with purified water.

(c) Place a stainless steel ring on the abdomen and drop cercariae suspension containing 150 cercariae into the ring.

(d) Expose mouse to cercariae for 45 min. Keep the mouse warm throughout the procedure with a warming lamp or heated pad. Remove the cercariae suspension from the ring with a Pasteur pipet and then release the mouse from the multichannel anesthesia system stand and return it to its cage without wiping the penetration area. Most cercariae will penetrate within 30 min of exposure, but 45 min allows enough time for maximal numbers of cercariae penetrating through skin (*see* **Note 9**).

(e) Monitor mice for about 30 min. If no pain, distress, or discomfort is observed, return mice to their housing room.

6. Euthanasia: 6–7 weeks post challenge, collect adult worms or eggs after euthanasia. Place mouse into the euthanasia chamber and introduce 100% carbon dioxide with a flow rate of 10–30% of the chamber volume/min till respiration ceases. Take animal out from the chamber, perform cervical dislocation, and place it on a glass pan.

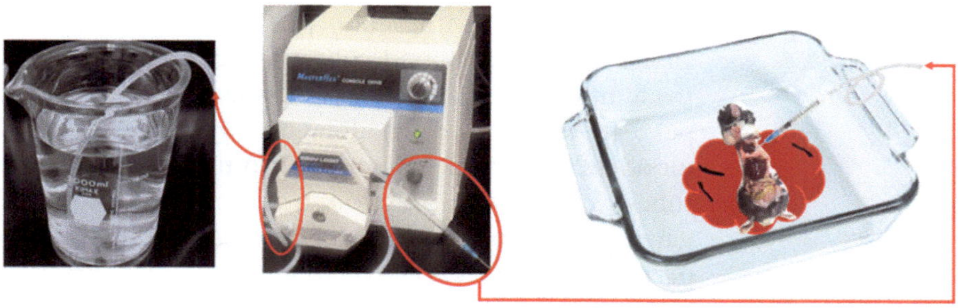

Fig. 3 Diagram of perfusion procedure for collecting adult *Schistosoma mansoni* worm from the hepatic portal system. A 26-G needle fitted on 1 cc syringe with an inserted silicon tubing is fixed in the descending aorta. Adult worms are perfused out from the severed portal vein

7. Perfusion [6, 7] (*see* **Note 10**) (*see* Fig. 3).
   (a) Clip the skin and open the cavity from the abdomen and thorax for exposing the left half of the thoracic cavity. With the dissection scissors, sever the ribs near the spinal column and carefully cut a small slit in the hepatic portal vein, and then insert 26-gauge needle attached to the silicon tubing into the descending aorta.
   (b) Press the driver to move perfusion fluid through the needle to the vessel. With a gentle massaging of liver and intestine, the worms exit from the hepatic portal vein to the glass pan.
   (c) Stop perfusion when no more worms flow out and the color of liver completely changes to a pale color. Carefully check the mesenteric vessels and pick up the trapped worms using fine-tip forceps. Transfer the perfusate in a container (Petri dish, graduated cylinder, or Pilsner glass with visible bottom).
8. Tissue collection: Remove the intestine and liver after perfusion. Completely rinse liver with 1.2% (w/v) NaCl and determine the weight. Open the intestine lumen, wash out the content with 1.2% (w/v) NaCl, and determine the weight. All the cleaned tissues can be temporarily stored at 4 °C for egg burden determination and other studies.

3.1.2 Nonhuman Primate Model

1. Blood collection: Apply chemical restraint to sedate the baboon. Gently squeeze the baboon cage followed by intramuscular injection of ketamine and acepromazine (or just ketamine depending upon the procedure to be performed). Touch the muscle tone and observe the response to stimuli to determine proper sedation. Scale the weight and transfer the baboon to procedure stage. Evaluate the physical condition by measuring the blood pulse, breathing rates, and body temperature. Pose the baboon in an appropriate position, clean the

venipuncture site with alcohol pad, and allow it to naturally dry. Insert the winged blood collection set safety lock needle into the femoral vein. Based upon a calculation of 6–10 mL of blood per 1 kg of body weight, draw no more than 10% of the total blood volume to vacutainer heparin tubes or serum separation tubes. After blood collection, compress sterile cotton ball and gauze to the puncture site for hemostasis. Blood collection is performed before every time point (pre-bleed, primary immunization, every immunization boost, challenge, and euthanasia).

2. Vaccine preparation: This step is similar to Subheading 3.1.1, **step 2**. The volume of vaccine is less than 500 μL for each baboon.
3. Vaccination: Lay the baboon on his or her side, shave the thigh muscle area, and wipe it with an alcohol pad and wait until dry. Intramuscularly inject vaccine into the thigh muscle and gently massage the muscle to help the absorption of vaccine (*see* **Note 4**). Apply light pressure on injection sites using sterile gauze. Evaluate the physical condition again. Hold the baboon till he or she can straighten his or her head before sending him or her back to the cage. The times and intervals of immunization depend on the vaccine design.
4. Cercariae collection: The procedure is the same as Subheading 3.1.1, **step 4**. The difference is that 1000 cercariae per each baboon is used for challenge in baboon model.
5. Challenge:
   (a) Following chemical restraint as used in blood collection, place the baboon into dorsal recumbency.
   (b) Shave area around the armpit to be exposed and clean it with gauze moisturized with purified water.
   (c) Hold the arm upright to form a "bowl" in the axillary region.
   (d) Drop the cercariae suspension containing 1000 cercariae onto the prepared armpit region and immobilize animal for about 45 min (*see* **Note 11**).
   (e) Following the 45 min exposure period, remove the solution in the armpit area into a beaker filled with bleach and discard appropriately (*see* **Note 9**).
   (f) Assess the baboon's physical condition at the middle and the end of the exposure period. Send the baboon back to the cage without wiping the axillary region when all the physical indices are stable after challenge.
   (g) Monitor the infected baboon for diagnostic symptoms post challenge (*see* Fig. 4).

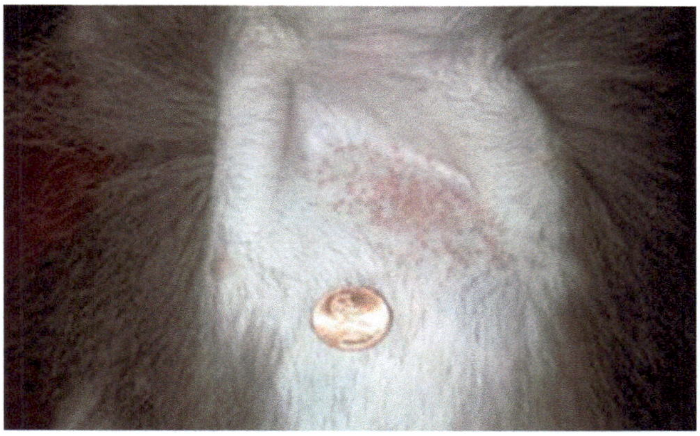

Fig. 4 Dermatitis occurred post percutaneous exposure to *Schistosoma mansoni* cercariae in baboon model

6. Euthanasia: 8 weeks post challenge, collect adult worms and eggs after euthanasia. Withhold food from baboon for 24 h before intramuscularly injecting ketamine (15 mg/kg) and acepromazine (0.5 mg/kg) to immobilize animals. Shave abdomen, chest, flexor surface of calves, and upper arms. Once the baboons reach a surgical depth of anesthesia, a catheter of appropriate size is placed into the cephalic vein and 100 U/kg of heparin is administered for anticoagulation. Move the baboon to appropriate space for surgery, such as a surgical suite. A cut is made to access the right femoral vessels. Once the femoral vessel is isolated, a second catheter of appropriate size is placed into the isolated femoral artery while a second dose of heparin is administered at 100 U/kg through the existing catheter into the cephalic vein. Gentle vacuum aspiration of blood through the catheter into femoral vein is done. Following blood exsanguination, SomnaSol or Euthasol is administered through the initial catheter (*see* **Note 12**).
7. Perfusion (*see* **Note 10**).
   (a) After euthanasia, remove the left rib cage with bone snips and scalpel to expose the descending aorta and ligate it. Note and photograph gross pathology of liver. Make a small longitudinal incision below the ligation and insert the syringe fitted into silicone tubing.
   (b) Ligate the abdominal aorta and inferior vena cava together immediately above the bifurcation to the legs. Dissect carefully as it is easy to tear the vena cava. Place two surgical sutures under inferior vena cava just above the liver and carefully ligate vena, leaving enough space between the ligatures to cut vein for later isolation and perfusion of liver.
   (c) Place two surgical sutures under the hepatic portal vein from the lesser omentum and common bile duct which are

just proximal to the liver. Ligate the proximal part of hepatic portal vein to the liver and make the small incision in the hepatic portal vein which allows the exit glass cannula to be inserted. The exit glass cannula should be positioned toward the intestinal part and is securely fixed using the other string.

(d) Press the pump to load perfusion buffer; the blood will flow through the transparent tubing attached to the exit cannula and drain into the collection receptacle, the Pilsner glass. During the perfusion, gently massage intestines, continually inspect intestines, and manually dislodge worms from mesenteric vessels. At the beginning of perfusion, press pump gently and slowly to check for obstruction or leaks.

(e) Stop perfusion when no more worms can be seen passing through the glass cannula and transparent tubing. After the worms stop coming out from the exit cannula for at least 5 min, remove the cannula from hepatic portal vein and drain the remaining fluid into a Pilsner glass.

(f) Dissect the liver to free it from the diaphragm and other moorings. Cut or loosen the ligatures, proximal to the livers, at the inferior vena and the hepatic portal vein. Insert the aortic cannula into the inferior vena cava stump and perfuse. Collect the perfusate in a pan and transfer it to the Pilsner glass.

8. Tissue collection: Ligate the small intestine at the cecum and the site a little above the cecum. From the initial site of the stomach to the cecum, gently separate the lesser omentum and pancreas from the small intestine. Bisect the small intestine between the ligatures. Ligate the large intestine at the rectum and remove the greater omentum. Free the large intestine from the mesentery and bisect the large intestine below the rectal ligature. Dissect the small and large intestines out of abdominal cavity. The liver of baboon has been taken out in the step of liver perfusion.

9. Manual examination of intestines: Note and photograph gross pathology of intestines. Carefully inspect the entire intestines and search for any worms, which appear as a black thread (for females) in mesenteric venules. Recover those worms by making a transverse cut across the venule and then squeeze them out using the back of the scalpel blade. Place those worms in the Petri dish containing perfusion buffer. After that, empty the intestines' contents. Open their lumens with scissors and flush with 1.2% NaCl. The cleaned intestines and livers can be also stored at 4 °C for other studies.

3.2 Parasitological Measures

Assessment of the vaccine efficacy using parasitological measures can aid in addressing the potential anti-fecundity, blocking of diseases transmission, and reduced pathology. Such measurements as a decreased hatching rate can indicate the anti-fecundity as well as the potential role of vaccine in the blocking of disease transmission; and lesser worm burden and egg burden in tissue and feces can help to explicate the reduced pathology. The efficacy of vaccine can also be assessed through the detection of CAA and CCA in host urine or sera. The importance of using such tests is due to the fact that the quantification of CAA and CCA antigen levels has been shown to be more sensitive, specific, and reliable method to determine worm burden and egg production of the parasite.

3.2.1 Worm Burden Determination

1. After settlement of worms in the Pilsner glass, decant the top fluid into a 150 mm Petri dish with a grid for worm inspection and selection and then dump the fluid into a liquid waste tank.
2. Wash the bottom fluid containing the settled worms in the Pilsner glass with perfusion buffer.
3. Following settlement of worms again, repeat **step 1**.
4. Repeat Subheading 3.2.1, **steps 2** and **3** one more time.
5. Use perfusion buffer and fine-tip forceps to flush and move the worms clinging onto the glass wall and bottom to the Petri dish.
6. Count and identify unpaired females and males and immature and paired worms under a dissecting microscope and record the numbers. A male worm does not contain pigment and its length varies from 6 to 13 mm; female worm contains dark pigment and its length ranges from 10 to 20 mm; immature female contains little to no pigment and its length is less than half of that of a mature female worm.
7. The worm protection (P) is calculated by comparing control (C) and experimental (I) groups with a standard formula: $P\% = \frac{C-I}{C} \times 100\%$ [8, 9].

3.2.2 Egg Burden Determination in Tissues

1. Visually divide the cleaned liver or intestine of mouse into three sections. Take one section and measure the weight. Cut the tissue into small pieces using a razor blade and place them into a 50 mL conical tube. For the cleaned intestine and livers of baboon, cut about 3–5 g tissue from small intestine, large intestine, and livers into small pieces, and place each into a 50 mL conical tube.
2. Add 10 mL 4% (w/v) KOH to each individual tube. Digest tissues at 37 °C for 16–24 h in a shaker at a speed of 150–200 rpm (*see* **Note 13**).

3. Centrifuge the tubes at 300–400 × *g* for 5 min and carefully dump the supernatant into a waste tank.
4. Resuspend the precipitation with 10 mL 1.2%(w/v) NaCl for washing the pellet and centrifuge at 300–400 × *g* for 5 min.
5. Dump the supernatant and repeat washing one more time.
6. Finally dilute the pellet in 1 mL 1.2% (w/v) NaCl. Vortex to mix the suspension thoroughly, and start enumerating eggs under a dissecting microscope or a compound microscope with a 10× objective.
7. Pipet and spread 20 μL suspension on microscope slide and cover it with coverslip [6]. Systematically scan the coverslip and record the number using counter at the same time. For each sample, scan at least three coverslips.
8. Calculate egg number in per gram based on the weight of digested sample. The egg burden (P) is calculated by comparing control (C) and experimental (I) groups with a standard formula: $P\% = \frac{C-I}{C} \times 100\%$ [8, 9].

3.2.3 Egg Hatching in Tissues

Method 1

1. Weigh 3–5 g tissue from liver, large intestine, and small intestine of baboon, or one-thirds portion of liver and intestine from mouse.
2. Mince tissue individually with scalpel blades in a Petri dish and then move into 50 mL conical tube containing 50 mL PBS with 10 mg collagenase B, 10 μg penicillin, and 20 μg streptomycin. Incubate the mixture in a shaker at 37 °C overnight at 150–200 rpm.
3. Next day: Wash the digested tissue. Centrifuge the tube at 800–1000 × *g* for 5 min, remove supernatant, and resuspend the pellet in 25 mL fresh cold PBS.
4. Repeat the wash process of digested tissues (detailed in **step 3**).
5. Serially pass the resuspend through 250 μm and 150 μm sieves. Collect the filtrate and centrifuge at 800–1000 × *g* for 5 min, and discard supernatant.
6. Resuspend the pellet in 10 mL PBS and apply to the top of Percoll column (8 mL Percoll +32 mL of 0.25 M sucrose in 50 mL tube). Centrifuge the tube at 800–1000 × *g* for 10 min. Remove liver cells and fatty tissues from the top layer of Percoll with pasture pipette.
7. Mix the egg pellet in 25 mL PBS containing 1 mM EDTA, 1 mM EGTA in a 50 mL tube, and centrifuge it at 300 × *g* for 5 min.
8. Discard supernatant, dissolve the pellet with 10 mL of above buffer, spin at 300 × *g* for 3 min, and repeat once again.

9. Dissolve the egg pellet in 5 mL PBS and apply on to second Percoll column (2.5 mL Percoll plus 7.5 mL Sucrose in 15 mL tube), and centrifuge at 2000 × *g* or 10 min. Remove the supernatant and wash egg pellet with 10 mL PBS, spin at 300 × *g* for 3 min, and repeat the wash three times.
10. Dissolve final pellet in 4 mL of purified water, and divide the volume into 4 wells of 24-well plate.
11. Keep under bright light for 30–60 min.
12. Observe egg hatching under microscope and enumerate hatched miracidium and unhatched mature eggs.
13. Egg hatching rate is calculated as the percent of hatched eggs versus mature eggs. Egg hatching reduction is evaluated by means of the formula as previously published: $P\% = (1 - E/C) \times 100\%$ where (P) represents the protection level, (E) represents the hatching rate of the control group, and (C) represents the hatching rate of the experimental group [8, 10, 11].

Method 2 [12, 13]

1. Weigh the tissue, cut it into small pieces, and resuspend in 50–100 mL 1.2% (w/v) NaCl.
2. Place the mixture into blender and blend for 30 s at low speed. Successively pass the homogenate through the tiers of stacked sieves (the pore sizes of 425 μm, 180 μm, 106 μm to 45 μm in sequence from top to bottom) (*see* Fig. 5).
3. Blend the remaining tissue from the surface of the widest sieve for 30 s at medium speed followed by passing through sieves again.
4. Blend the tissue trapped on the top of sieve an additional time for 30 s at the highest speed setting followed by passing the sieves again while rinsing the top sieve use a plastic spout bottle filled with 1.2% NaCl. After repeating the rinse and gently agitating the stacked sieve during the entire procedure, the eggs will be retained on the lowest sieve (pore size of 45 μm).
5. Remove the top sieves and collect the fluid containing the eggs in the lowest sieve into 50 mL conical tube. Put the lowest sieve upside down in a glass Petri dish, rinse the screen with 1.2% (w/v) NaCl, and pour the filtrate into the same 50 mL tube.
6. Pass the mixture through the gauze into a new 50 mL conical tube. Centrifuge the tube at 800–1000 × *g* for 5 min at 4 °C.
7. Place 40 μm nylon cell strainer in a glass Petri dish half full of cold 1.2% (w/v) NaCl and transfer the egg mixture into the strainer using the Pasteur pipette to concentrate the eggs.

Fig. 5 Stainless steel sieves stacked in sequence (sizes from top to bottom 425 μm, 180 μm 106 μm, and 45 μm)

8. Discard the waste content of Petri dish and gently pipet the clean eggs from strainer to 15 mL tube and centrifuge at 2000 × *g* for 5 min at 4 °C.
9. Wash eggs two times following the procedure: Carefully discard the supernatant, gently resuspend the pellet in 10 mL 1.2% (w/v) NaCl. Centrifuge at 2000 × *g* for 5 min at 4 °C.
10. Resuspend the egg with 4 mL purified water and aliquot the eggs into 4 wells of 24-well plates.
11. Keep under bright light for 30–60 min.
12. Observe egg hatching under microscope and enumerate hatched miracidium and unhatched mature eggs.
13. Egg hatching rate and reduction will be calculated as in Method 1.

3.2.4 Egg Burden Determination in Feces

1. Weigh 3–5 g fecal sample and fix it in 10 mL 10% formalin.
2. Vortex the mixture and filter the mixture through two layers of gauze into a fresh 50 mL conical tube. Centrifuge the tube at 1500–2000 × *g* for 3–5 min and discard the supernatant.

3. Add 12 mL distilled water and 3 mL ethyl acetate and mix them with the pellet. Centrifuge again at 1500–2000 × *g* for 3–5 min. Gently dispose the supernatant and wipe the inside wall of tube with Kimwipe.
4. Finally dissolve the pellet in 1 mL 1.2% (w/v) NaCl.
5. Observe and count the eggs under a dissecting microscope or a compound microscope with a 10× objective. Pipet and spread 20 μL suspension on microscope slide and cover it with coverslip [6].
6. Scan the slide and calculate egg burden (*P*) as same as described in egg determination in tissue (*see* Subheading 3.2.2, **step 8**).

3.2.5 Egg Hatching in Feces

1. Weigh 5 g fecal sample and dissolve it in 50 mL conical tube with 25 mL 1.2% (w/v) NaCl.
2. Vortex the mixture and filter it through two layers of gauze into a fresh 50 mL conical tube. Repeat the filter step an additional time.
3. Pass the filtrate through the tiers of stacked sieves (the pore sizes of 425 μm, 180 μm, 106 μm to 45 μm in sequence from top to bottom).
4. Rinse the top sieve use a plastic spout bottle filled with 1.2% (w/v) NaCl. After repeating, rinse and gently agitate the stacked sieve during the entire procedure; the eggs will be retained on the lowest sieve (pore size of 45 μm).
5. Remove the top sieves and collect the fluid containing the eggs in the lowest sieve into 50 mL conical tube. Place the bottom sieve upside down into a glass Petri dish and flush the mesh with 1.2% NaCl. Collect the fluid in the glass Petri dish.
6. Pass all the collected fluid through the gauzes and other clean tiers of sieves again. Repeatedly rinse the top and collect the fluid on the bottom sieve into 50 mL conical tube. Place the bottom sieve upside down into a glass Petri dish and flush the mesh with 1.2% NaCl. Collect the fluid in the glass Petri dish into the same conical tube.
7. Centrifuge the tube at 400–500 × *g* for 5 min. Carefully discard the supernatant and resuspend the pellet in 4 mL purified water.
8. Aliquot the egg fluid into 4 wells of 24-well plate and set the plate under bright light for 30–60 min.
9. Observe egg hatching under microscope and enumerate hatched miracidium and unhatched mature eggs.
10. Egg hatching rate and reduction will be calculated as the same as egg hatching in tissues (*see* Subheading 3.2.3, Method 1, **step 13**).

Table 1
Example for standard series spiked urine/serum

Experiment tube number	Concentration (pg/mL)	Urine/serum (μL)	CAA
1	10,000	51.75	5.75 μL of 100,000 pg/mL
2	1000	67.5	7.5 μL of 10,000 pg/mL
3	316	50	23 μL of 1000 pg/mL
4	100	50	23 μL of 316 pg/mL
5	32	50	23 μL of 100 pg/mL
6	10	50	23 μL of 32 pg/mL
7	3.2	50	23 μL of 10 pg/mL (discard 23 μL)
8	0	50	–

3.2.6 Circulating Anodic Antigen Assay (CAA)

1. Prepare standard curve (CAA standards spiked in urine/serum for 20 μL assay). Procedure is shown as Table 1.

   Prepare eight microcentrifuge tubes with label of 1–8. Add different volume of negative urine or serum sample to corresponding tube (volume and related tube number shown on the column 1 and 3). The serial dilutions are done in the similar manner. For example, for tube No. 1, with addition of 5.75 μL concentrated CAA antigen (100,000 pg/mL), the final concentration of tube No. 1 is 10,000 pg/mL. From tube No. 1, take 7.5 μL to tube No. 2; the final concentration of tube 2 is 1000 pg/mL. The final dilution of each tube is demonstrated in column 2.

2. TCA treatment [trichloroacetic acid (TCA) treatment of urine/serum samples for 20 μL assay].
   (a) Pipette 50 μL urine/serum/standards into an Eppendorf tube.
   (b) Add 50 μL 4% TCA and vortex.
   (c) Incubate for 5 min at room temperature.
   (d) Centrifuge the tubes at 152,000 × *g* for 5 min.
   (e) The supernatant should be well separated from the pellet.
   (f) Transfer 20 μL supernatant into the CAA UCP assay.

3. Up-converting phosphor lateral flow (UCP-LF) CAA [CAA in 20 μL assay (dry UCP)].
   (a) Prepare the run buffer: Add 10 mL milliQ water to the lyophilized HSLF buffer tube. Vortex until everything is dissolved properly.

(b) Open the aluminum bag and take out the amount of CAA UCP wells you need. Place them in the holder. Re-close the aluminum bag properly.

(c) Pipette 100 μL HSLF run buffer in each well.

(d) Add 20 μL of supernatant from standard or samples to the corresponding wells.

(e) Seal the plate with white covers and incubate on a shaking incubator at 37 °C for 1 h at 150 × *g*.

(f) During incubation, mark the reverse side of the CAA UCP strips. Write the sample identification number on the upper backside of the strip; make sure not to touch the middle of the front part of the strip. The strips are composed of three parts: top part is sample pad of the strip (1 cm glass fiber); middle part is nitrocellulose (2.5 cm); bottom part is absorbent pad (2-cm-thick paper).

(g) After the incubation, remove the cover carefully to avoid splashing. Insert the strip with the sample pad down into the corresponding sample well (*see* **Note 14**).

(h) Incubate for at least 3 h or overnight (best results are obtained with completely dry strips; if possible, consider overnight incubation) before scanning the strip.

(i) The LF strip will be analyzed by using a modified Packard Fluorocount meter. Test line signals will be normalized to control line signals of each individual strip; the results are expressed as ratio signals (Up-converting Phosphor Technology, UPT value). The test line signal will be normalized to the control line signal for each individual sample. Standards of known concentrations made in each assay will be used to create a standard curve and to determine the cutoff point, which approximately is 30 pg/mL. The value above cutoff point will be considered positive.

3.2.7 Point-of-Care Circulating Cathodic Antigen Assay (POC-CCA)

POC-CAA cassettes are provided by Rapid Medical Diagnostic Company. The procedure follows the instructions (*see* **Note 15**).

1. Transfer 1 drop of urine sample to the circular well of the cassette and allow the sample to absorb entirely into the specimen pad.
2. Shake the buffer bottle and mix well. Transfer 1 drop of buffer to the circular well.
3. Incubate for exactly 25 min and read results. Any results read outside 25 min should be considered invalid and must be repeated.
4. After buffer is absorbed into test area, the blue control line should turn pink as control line. If the control line stays blue, it

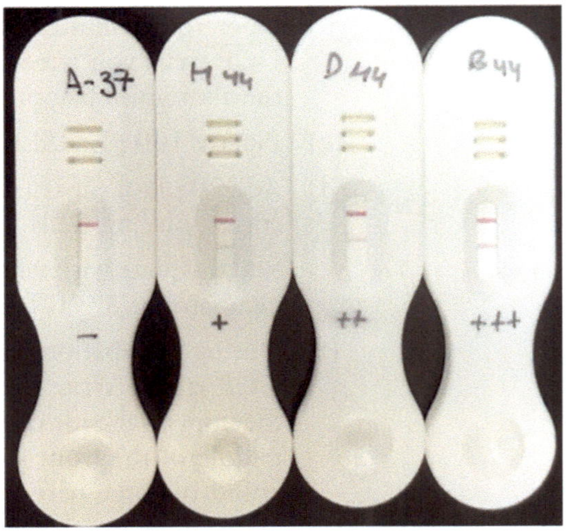

Fig. 6 Rapid test of POC-CCA from baboon urine samples. The picture shows different reactions from negative to strong

means the test is invalid. Any result line appeared in test T area should be considered positive.

5. Test result can be scored according to the strength of result band color. No line showed in test T area should be considered negative. An extremely light band is scored as trace; a visible and still weak band is scored as weak (+); a dark result line in T area but is lighter than control line is scored as medium (++); if the result line is dark as control line or darker than control line, it is scored as strong (+++) [14] (*see* Fig. 6).

3.3 Humoral Immune Responses

Data from immunological assays and tests are important to develop an effective vaccine, specifically to target immune responses relating to protection. The various methods listed below are all used to determine the specific immune response that the vaccine initiates in our in vivo models.

3.3.1 ELISA for Immunoglobulin Endpoint Titers

Conduct the experiment at room temperature, but make sure to keep your samples on ice until time of use.

1. Binding: Dilute antibody 1:50 in ECB for 12 μg/mL antibody working solution by mixing 240 μL of antibody with 11.760 mL ECB for each plate. Coat microtiter plates with 100 μL/well of antibody (1.2 μg/well) working solution in ECB. Seal the plate and incubate for approximately 4 h at room temperature.
2. Blocking: Aspirate wells and wash three times with 300 μL/well EWB. After the third wash, invert plates and blot on towel to remove residual buffer. Block plates with 250 μL/well EBB. Seal the plates and incubate overnight at 2–8 °C overnight.

3. Addition of primary antibody.
   (a) Prepare your antibody samples using fresh cluster tubes. Start with a 1:50 dilution and using a dilution factor of 5, create an 8-point titration curve. The resulting dilutions will be as follows: 1:50, 1:250, 1:1250, 1:6250, 1:31,250, 1:156,250, 1:781,250, 1:3,906,250. To do this, dispense 297 μL DEBB to tubes A1:A10, 495 μL DEBB to tube A11, 240 μL DEBB to tubes B1:H11, and 450 μL DEBB to tubes B11:H11. Then, dispense 3 μL of each test sample to tubes A1:A10 and 5 μL of positive control to tube A11 bringing the final dilution to 1:100. Dilute samples serially 1:5 down columns 1–10 by mixing the first well 5 times (without introducing bubbles) before transferring 60 μL to the next row. Dilute samples serially 1:10 down column 11 by mixing the first well five times (without introducing bubbles) before transferring 50 μL to the next row.
   (b) Remove the plates from the 2–8 °C incubation. Aspirate wells and wash three times with approximately 300 μL/well EWB. After the third wash, invert plates and blot on towel to remove residual buffer. Using a multichannel pipet, dispense 100 μL dilutions into each 96-well plate. Dispense 100 μL DEBB into column 12 which will serve as a plate blanks (no primary serum or conjugate) or ELISA negative controls. Seal the plates and incubate for 2 h at room temperature.
4. Addition of secondary antibody conjugate: In the last 10 min of primary incubation, dilute goat anti-Mouse IgG secondary conjugate 1: 1000 in DEBB. Aspirate wells and wash three times with approximately 300 μL/well EWB. After the third wash, invert plates and blot on towel to remove residual buffer. Add 100 μL secondary HRP-conjugated antibody solution to each well except A12:D12. Add 100 μL DEBB to each well in the range A12:D12. Seal plates and incubate for 1 h at room temperature.
5. Color development: Aspirate wells and wash three times with approximately 300 μL/well EWB. After the third wash, invert plates and blot on towel to remove residual buffer. Add 100 μL/well substrate solution to entire well and incubate for 10 min at room temperature. After 10 min, stop the reaction by adding 100 μL/well 1 N H_2SO_4 to each well in rapid succession.
6. Reading the data: Open SoftMax Pro® file, BioTek Microtiter Plate Reader or equivalent. Within 20 min of stopping the reaction, take readings at 450 nm. Results will be expressed as

endpoint titers calculated from a curve of optical density versus antibody dilution to a cutoff of 2 standard deviations above background control values.

3.3.2 ELISPOT (Using PBMCs as Illustrated Example)

1. Mixing and preparing: Prepare 1 day before start of experiment. Prepare wash buffer by taking homemade, filter sterilized, or autoclaved PBS and mixing it with 0.05%(v/v) Tween 20.
2. Day 1 of experiment: Using 25 μL of 70%(v/v) ethanol, pre-wet the PVDF membranes and allow them to incubate for 1 min at room temperature. Aspirate the membrane to remove the excess ethanol and immediately rinse each well twice with sterile and pyrogen-free PBS (PBS-I). After each wash, invert plates and blot on towel to remove residual buffer. Add 50 μL of diluted coating antibodies into each well. Seal the plate and allow it to incubate overnight at 4 °C.
3. Day 2 of experiment: Remove plates from the refrigerator and decant the solution from the wells. Wash each plate three times with 200 μL PBS-I/well. Add 200 μL Blocking Buffer to each well, seal the plates, and allow them to incubate for 1 h at 37 °C. Dilute cells in culture media containing a stimulus such as polyclonal stimulus or antigen. Decant the solution from the incubated plates, and without washing the wells, fill each well with 100 μL of cells. For polyclonal stimulated culture media, ensure that there are no more than 3×10^5 cells/well. For antigen-specific responses, prepare triplicates of 3×10^6 cells/well. Seal the ELISPOT plate and place it in an incubator set to 37 °C, 5% CO_2, and 100% humidity. The incubation period typically lasts between 5 and 24 h but may take as long as 48 h. Activation conditions vary depending on cell type, cytokine of interest, and kinetics of cytokine released.
4. Day 3 of experiment: Remove the plates from the incubator and firmly shake the plates to remove a bulk of the cells. Wash the wells twice with 200 μL of room temperature PBS. Then wash each well five times with 250 μL wash buffer/well. Decant the wash buffer and add 100 μL of diluted biotinylated detection antibodies to each well. Seal the plate and incubate either at 37 °C for 1 h or 4 °C overnight.
5. Day 4 of experiment: Decant solution from wells and remove the underdrain from the bottom of the plates. Wash both sides of the PVDF membrane five times with 250 μL wash buffer/well. Place 50 μL of diluted GABA into each well. Seal the plate once more and incubate at 37 °C and 100% humidity for 1 h. Decant the solution from the wells and thoroughly wash the PVDF membrane five times with 250 μL wash buffer/well. Add 30 μL of freshly prepared Activator I/II solution into

each well ensuring that the solution is uniformly distributed over the well. Then cover the plate with a lid and incubate at room temperature in the dark. Monitor spot development by light microscopy from 25 to 30 min. When clear spots have developed, stop the reaction by rinsing the wells with demineralized water. Air-dry the plate at room temperature and count the spots using a dissecting microscope or an ImmunoSpot image analyzer.

3.4 Cellular Immune Responses

3.4.1 Cytokine Expression ProfileqRT-PCR

1. Day 1: Pool cells according to the groups that have been established and add 20 mL of complete media to the pooled cells. Centrifuge the cells at 300 × *g* for 10 min at room temperature. Decant the supernatant and resuspend the cells in 20 mL of complete media. Centrifuge once more at the aforementioned rates, decant the supernatant, and resuspend the pellet in 5 mL of complete media. From the suspension, take a 10 μL aliquot to do a cell count. Seed cells at 1×10^6 cells/well in a 6-well plate. Stimulate the cells with 12 μg/mL rSm-p80(antigen). Incubate the cells at 37 °C in 5% CO_2 for 24 h.
2. Day 2: Harvest the cells by centrifuging at 500 × *g* for 10 min. Extract total RNA using GenElute™ mammalian total RNA miniprep kit according to the manufacturer's instructions. Store RNA at −80 °C until the sample is needed.
3. Day 3: Use Thermo Scientific™ Maxima™ First Strand cDNA Synthesis Kit to carry out first-strand cDNA synthesis according to the manufacturer's instructions. Store cDNA at −20 °C until needed.
4. Day 4: Carry out real-time PCR to determine expression levels of Thl, Th2, and Th17 with specific primers and SYBR® Premix Ex Taq™ using the StepOne™ plus real-time PCR machine. Use DataAssist™ to calculate relative cytokine mRNA expression by comparing variations in the message levels of the control group with the experimental group after normalization with GAPDH.

3.4.2 FACS (Using Baboon PBMCs as Illustrated Example)

1. Preparing target cells of interest: Prepare a single cell suspension consisting of baboon PBMC in RPMI 1640 media with 10% FBS and seed in between 1 and 10×10^6 cells/well. Establish a negative control by adding media as a control. Establish a positive, isotype, and compensation control. This consists of 100 ng/mL of PMA and 1 μg/mL of ionomycin. 12 μg/mL of Sm-p80 is then added into test samples. The positive control is then incubated for between 4 and 6 h. To inhibit cytokine secretion, each well is incubated with the addition of 1:1000 dilutions of Brefeldin A in the culture media, but for the positive control, Brefeldin A is added at

the same time as PMA and ionomycin. For the test sample, Brefeldin A is added at the last 6–12 h of the incubation. The test sample incubates with Sm-p80 at 37 °C, 5% CO_2 for 6 h, 12 h, 24 h, and 48 h.

2. Harvest Cells: Label 1.5 mL Eppendorf tube. After completing the incubation process, harvest the cells from the wells to their corresponding Eppendorf tube. Then, centrifuge the samples at 400 × *g* for 5 min and wash once in ice-cold phosphate-buffered saline (PBS), suspend in a small amount of staining buffer, and distribute (50 μL per tube).
3. Stain cell-surface antigen: Some antibodies which recognize cell surface markers may not bind to fixed or denatured antigen. To account for this, perform immunostaining for the surface determinants with unfixed cells prior to staining for intracellular cytokines. For better delineation between monocytes and lymphocytes during flow cytometry analysis, add fluorochrome-conjugated MAbs against the surface determinants to each tube in the following combinations: negative, positive, and test sample (CD3-FITC and CD4-PerCP-Cy™5.5) and isotype control (appropriate FITC- and PerCP-Cy5.5™-conjugated isotype controls). Incubate the cells with MAbs for 30 min at 4 °C. After that, wash the cells twice in staining buffer of 500 μL each time.
4. Fixation and permeabilization: After the last wash, add 250 μL of fixation/ permeabilization solution for 20 min at 4 °C. Then, wash 2 times using BD Perm/Wash™ buffer (500 μL each tube), centrifuge for 5 min, and aspirate supernatant.
5. Intracellular cytokine staining: Add BD fluorochrome-labeled anti-cytokine antibodies to the appropriate tube. Mix via vortex. Incubate in the dark at 4 °C for 30 min. For all samples, add IFN-γ-phycoerythrin (PE) and IL-4-APC.
6. Wash cells two times with BD Perm/Wash™ buffer (500 μL each tube), centrifuge for 5 min, and aspirate supernatant.
7. Resuspend the cell pellet in 0.5 mL of flow cytometry staining buffer (PBS with 0.1% (w/v) bovine serum albumin or PBS with 1% heat-inactivated FCS, 0.09% sodium azide pH 7.4–7.6).
8. Analyze samples on a flow cytometer.

3.5 Systems Vaccinology

Vaccines are known to induce adaptive immunity by triggering specific innate immune response, in which these responses ultimately decide the fate of the type of immune protection induced (cell-mediated or humoral) [15]. Therefore, identifying early key gene signatures induced by vaccination serve as key predictors for vaccine immunogenicity and efficacy.

3.5.1 Future Perspectives: Application of a Broad, Holistic Approach to Predicting Vaccine Efficacy

1. Total RNA Extraction.
   (a) Isolate total RNA from the cells using the RNA isolation kit according to the manufacturer's instructions.
   (b) Determine total RNA concentrations with Qubit® 3.0 Fluorometer and RNA HS assay kit according to manufacturer's instructions.
   (c) Confirm quality of RNA using RNA ScreenTapes on Agilent 2200 TapStation.
2. Library Preparation and RNA Sequencing.
   (a) Use 1 μg of total RNA for the cDNA library construction using TruSeq® Stranded mRNA LT kit and epMotion robot.
   (b) Incubate total RNA with magnetic oligo (dT) beads and denatured with heat.
   (c) Prime the fragmented mRNA with random hexamers and transcribe into first-strand cDNA using SuperScript® III reverse transcriptase and random primers, followed by the second-strand cDNA synthesis using DNA polymerase I.
   (d) 3′ adenylate the resulting blunt-ended dscDNAs to prevent the fragments from ligating to one another and to provide a complementary overhang for ligating adapters.
   (e) Dual-index the libraries using unique indexes for each sample.
   (f) Enrich the dual-indexed libraries via PCR to selectively amplify DNA fragments that contain the adapter molecules on both ends to amplify the amount of DNA in the library. Then purify the libraries using AMPure XP purification beads.
   (g) Library construction produces single-indexed libraries with a median insert size of ~300 bp, which can then be validated by gel electrophoresis on an Agilent 2200 TapeStation instrument using D1000 Screen Tapes.
   (h) Quantify all libraries in triplicate using SynergyH1 fluorescent plate reader (BioTek, Winooski, VT).
   (i) Prior to cluster generation for sequencing, normalize all samples to an equimolar concentration.
   (j) Dilute the validated and indexed stock libraries in 10 mM Tris–HCl, pH 8.5 with 0.1% Tween 20 to 10 nM concentrations, pool, and quantify using Qubit® 3.0 Fluorometer.
   (k) Dilute the resulting libraries to 4 nM before denaturation with 0.1 N NaOH.

(l) Load the pooled denatured cDNA libraries on a cBot for cluster generation followed by 2 × 125 bp read length paired-end sequencing using HiSeq SBS Kits with V4 chemistry on a HiSeq 2500 sequencer.

3. Bioinformatics.

(a) Demultiplex sequencing run data containing base call information using bcl2fastq software (Illumina). Analyze the samples automatically using the Illumina workflow apps according to the designed sample sheet.

(b) Decompress the resulting fastq.gz files using 7-zip software to generate fastq files for read 1 and read 2 for each sample.

(c) Using QC statistics with FastQC software tool (Babraham Bioinformatics), preform raw-read quality assessment.

(d) For raw-read alignment and differential gene expression analysis fastq files, import into QSeq® Version 15.0 software (DNASTAR, Madison, WI, USA) and align to *Mus musculus* (if murine samples) or the *Homo sapiens* (if baboon samples) reference genome (assembly GRCm38.p4, GRCh38.p12, respectively) with reads assigned per kilobase of target per million mapped reads (RPKM) normalization.

(e) Use student's two-tailed unpaired t-test with Benjamin Hochberg FDR correction to compare the means of gene expression values for five individual replicates for a given gene.

(f) Analyze the RNA-Seq QSeq data by creating replicate group sets according to the different animal time points and treatment groups.

(g) Identify the genes as differentially expressed with $\geq$ two-fold change if they had the false discovery rate method for multiple test correction p-value of ≤ 0.05 through DNA star software.

(h) For each observation, export the resulting differentially expressed genes (DEGs) with annotations, Log2 fold-change, and p-value for input for into pathway mapping into Ingenuity Pathway Analysis Software (IPA®, QIAGEN Redwood City) tool core analysis using standard settings with duplicates resolved.

(i) Consider the network molecules associated with an immune system response in the Knowledge Base for analysis and identification of key molecular signatures as well as molecules and pathways with greatest activation z-score difference. These molecules are selected for further analysis.

(j) Use the right-tailored Fisher's exact test to calculate the p-value to reflect the likelihood of an overlap between a set of molecules and a given process is due to random chance.

(k) Use the activation z-score statistic to predict the directional effect of one molecule on another or on a process to predict the activation state (inhibition/activation).

(l) Determine immune-related DEGs by gene lists from the Immunology Database and Analysis Portal and the Immunogenetic Related Information Source, which are connected into networks using IPA. Cross-analyze DEGs of interest between the made comparisons in Subheading 3.4.1, **step 3** to potentially identify markers specific to the vaccine immunogenicity or efficacy.

4 Notes

1. All animal procedures shall be conducted in accordance with the Institutional Animal Care and Use Committee (IACUC) Guidelines and need approval by the according Animal Ethics Committee. Baboons used are housed in Association for Assessment and Accreditation of Laboratory Animal Care (AAALAC)-accredited facilities. Institutional, local, and national guidelines should always be followed for animal work.
2. Volumes for both buffers can be scaled as long as the concentration remains the same. Buffers may be stored for approximately 3 months; however, they are not recommended for use if turbidity or precipitation is evident. Finally, prepare the Diluted ELISA Blocking Buffer (DEBB), and dilute EBB 1:10 in EWB for dilution of samples and secondary antibody conjugate. DEBB must be made fresh each day. Incubation periods can occur at room temperature (20–25 °C). Bring all buffers and SureBlue TMB to room temperature. Use a fresh bottle of TMB for each assay.
3. The following reagents must be prepared before the start of the protocol. For sodium azide, use appropriate personal protective equipment (PPE) such as goggles, gloves, and laboratory coat. Cells are frozen promptly after animal euthanasia. Collect the spleen from individual mice, both spleen and lymph node from individual baboon for isolating splenocytes and lymph node cells. This can be done by crushing the spleen or lymph node in a cell strainer, repeatedly washing, and centrifuging in RPMI 1640 media (supplemented with 100 μg/mL penicillin G, 100 μg/mL streptomycin, and 10 μg/mL gentamycin). Post wash, the cells must be resuspended in freezing media (RPMI-1640 supplemented with 30–40% fetal bovine serum

and 10% DMSO) and stored in a −80 °C freezer. For peripheral blood mononuclear cells (PBMCs) of baboon, twofold volume of diluted blood is gently loaded on the top of onefold volume of Histopague-1077 (Sigma) and centrifuged at 400–500 × *g* room temperature for 30 min. Aspirate opaque interface containing the mononuclear cells, wash two times with RPMI 1640 media, and store in freezing media.

4. Insert the needle to the muscle and slightly pull the plunger back. If there is no blood, it means the needle is in a muscle. If blood moves into the syringe, it means the needle is in a blood vessel. Change the syringe and injection site to restart.
5. Massaging the injection site after injection can help the absorption of vaccine into the muscle. It is normal to see slight bleeding at the injection site. After immunization, return the mice from the procedure room to the housing room where the animals can completely recover form pain, distress, and discomfort and actively move around in the cage.
6. Keep the room temperature at 25–26 °C during the snail shedding procedure. The infected snails are kept in the dark 24 h before shedding them for harvesting high yields of cercariae.
7. Schistosomes are a biohazard. Personal protective equipment (PPE) is required, like gloves, goggles, and lab coats. Workers should wear PPE at all times when handling cercariae suspensions, infected snails, or any material related to infected snails.
8. Mouse urine can kill cercariae and contaminate the suspension. A piece of absorbent tissue or tissue paper can be placed in the bottom of the restraining tube in avoid of mouse urine dripping to cercariae suspension.
9. For the challenge through abdomen, it is impossible to remove all the non-penetrating cercariae with the Pasteur pipet. Therefore, it is less reliable to determine the number of non-penetrating cercariae for abdominal exposure.
10. A foot pedal can be used to facilitate the perfusion process to free both hands to perform the procedure. It is very difficult to perfuse out every adult worm from hepatic portal system. Additional steps are necessary for acquiring maximal worm recovery since some worms may still be stuck in the mesenteries or swept into the liver. Worms trapped in the liver can be recovered by compressing freed liver between two glass plates. Examining the worms under a dissecting microscope is another way to account for the un-perfused worms.

11. During the exposure period, if extended chemical restraint is necessary as determined by the muscle tone and response to stimuli provided by the procedure, then additional sedative will be administered at half the initial dose (ketamine only).
12. If necessary, perform chest compressions to help the euthanasia drug reach the brain, and then open chest to ensure death.
13. For tissue digestion, shaker or incubator should be used without CO_2 because absorbed CO_2 can neutralize KOH.
14. Place the marked strips in a clean 96-well microtiter plate (not the plates and/or strips from the aluminum bags) in the correct order such that after incubation one only needs to transfer the strips into the corresponding sample-UCP wells.
15. Ensure all reagents are equilibrated to room temperature (20–25 °C) before commencing the assay. Remove the test cassette and collection device from their pouches just prior to use.

Acknowledgments

This work was supported in part by grants from the Bill and Melinda Gates Foundation (OPP1097535) and from the NIAID/NIH SBIR (R43/R44 AI103983). The snails were supplied through a NIH-NIAID contract (HHSN272201000005l) to Biomedical Research Institute. The authors state that there are no conflicts of interest to declare. Authors thank the assistance of Dr. Govert van Dam of Leiden Medical University Center in helping us develop the CAA assay in our laboratory.

References

1. Colley DG et al (2014) Human schistosomiasis. Lancet 383(9936):2253–2264
2. Pennisi E (2018) Hybridization may give some parasites a leg up. Science 361(6405):832–833
3. Molehin AJ et al (2016) Development of a schistosomiasis vaccine. Expert Rev Vaccines 15(5):619–627
4. Mo AX et al (2014) Schistosomiasis elimination strategies and potential role of a vaccine in achieving global health goals. Am J Trop Med Hyg 90(1):54–60
5. Siddiqui AA, Siddiqui SZ (2017) Sm-p80-based Schistosomiasis vaccine: preparation for human clinical trials. Trends Parasitol 33(3):194–201
6. Tucker MS, Karunaratne LB, Lewis FA et al (2013) Schistosomiasis. Curr Protoc Immunol 103:Unit 19.1. https://doi.org/10.1002/0471142735.im1901s103
7. https://www.afbr-bri.org/wp-content/uploads/2014/02/Perfusion-of-adult-worms-from-mice-Protocol.pdf
8. Zhang W, Molehin AJ, Rojo JU et al (2018) Sm-p80-based schistosomiasis vaccine: double-blind preclinical trial in baboons demonstrates comprehensive prophylactic and parasite transmission-blocking efficacy. Ann N Y Acad Sci 1425(1):38–51. https://doi.org/10.1111/nyas.13942
9. Zhang W, Ahmad G, Molehin AJ et al (2018) *Schistosoma mansoni* antigen Sm-p80: prophylactic efficacy using TLR4 agonist vaccine adjuvant glucopyranosyl lipid A-alum in murine and non-human primate models. J Investig

Med 66(8):1124–1132. https://doi.org/10.1136/jim-2018-000786. Epub 2018 Jul 10
10. Cao Y, Qiao H, Shi Y et al (2016) Evaluation of lethal giant larvae as a schistosomiasis vaccine candidate. Biomed Res Int 2016:4680812
11. Le L, Molehin AJ, Nash S et al (2018) Schistosoma egg-induced liver pathology resolution by Sm-p80-based schistosomiasis vaccine in baboons. Pathology 50(4):442–449. https://doi.org/10.1016/j.pathol.2018.01.004. Epub 2018 May 5. MID: 29739616
12. https://www.afbr-bri.org/wp-content/uploads/2014/02/Isolation-of-Schistosoma-spp.-eggs-from-tissues-1.pdf
13. Fu CL and Hsieh MH (2012) Schistosoma haematobium egg isolation. http://www.bio-protocol.org/e279
14. Casacuberta M, Kinunghi S, Vennervald BJ et al (2016) Evaluation and optimization of the circulating Cathodic antigen (POC-CCA) cassette test for detecting *Schistosoma mansoni* infection by using image analysis in school children in Mwanza region, Tanzania. Parasite Epidemiol Control 1(2):105–115
15. Zepp F (2016) Principles of vaccination. Methods Mol Biol 1403:57–84

Correction to: Measuring Histone Modifications in the Human Parasite *Schistosoma mansoni*

Ronaldo de Carvalho Augusto, David Roquis, Marion AL Picard, Cristian Chaparro, Celine Cosseau, and Christoph Grunau

Correction to:
Chapter 9 in: David J. Timson (ed.), *Schistosoma mansoni: Methods and Protocols*, Methods in Molecular Biology, vol. 2151,
https://doi.org/10.1007/978-1-0716-0635-3_9

The affiliation of Dr. Ronaldo de Carvalho Augusto has been changed to LBMC, Laboratoire de Biologie et Modélisation de la Cellule Univ Lyon, ENS de Lyon, Université Claude Bernard Lyon 1, CNRS, UMR 5239, INSERM, U1210, Lyon, France.

The updated online version of this chapter can be found at:
https://doi.org/10.1007/978-1-0716-0635-3_9

David J. Timson (ed.), *Schistosoma mansoni: Methods and Protocols*, Methods in Molecular Biology, vol. 2151,
https://doi.org/10.1007/978-1-0716-0635-3_20,

INDEX

David J. Timson (ed.), *Schistosoma mansoni: Methods and Protocols*, Methods in Molecular Biology, vol. 2151, https://doi.org/10.1007/978-1-0716-0635-3,

S

T

U

V

W

X

Z

MIX
Papier aus verantwortungsvollen Quellen
Paper from responsible sources
FSC® C105338

If you have any concerns about our products,
you can contact us on
ProductSafety@springernature.com

In case Publisher is established outside the EU,
the EU authorized representative is:
Springer Nature Customer Service Center GmbH
Europaplatz 3, 69115 Heidelberg, Germany

Printed by Libri Plureos GmbH
in Hamburg, Germany